LIVING WITH PLANTS

LIVING WITH PLANTS

A Gardener's Guide to Practical Botany

By Donna N. Schumann

Introduction by Richard W. Pippen
Chairman of the Biology Department,
Western Michigan University

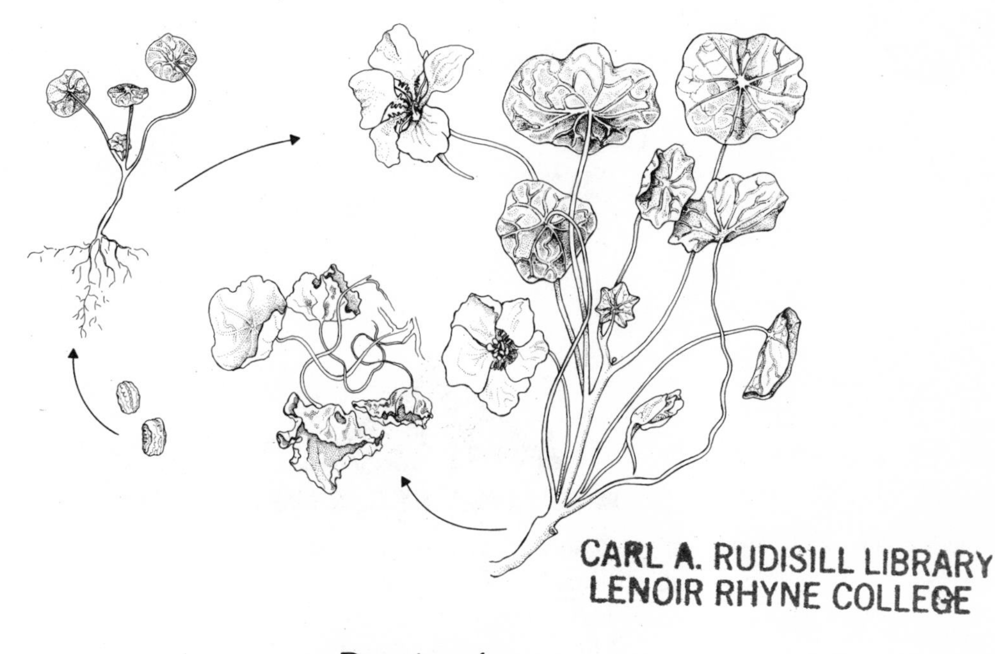

Drawings by
Kathryn Simpson
Wendy Smith-Griswold
Nancy Wright

Published by Mad River Press
Route 2, Box 151-B
Eureka, California 95501
Printed and typeset by Eureka Printing Company, Inc.
Eureka, California
ISBN 0-916422-20-8

Dedicated to my mother
whose love of plants was so contagious
and to my late father
whose faith in me was so encouraging.

PREFACE

Growing plants is an art, a craft, a science, and a joy, and here is a book to shed some light on the fundamentals of growing and using plants in a home setting. The text contains well-illustrated information extracted from the fields of botany, horticulture, pest control, plant physiology, soils, gardening, and landscaping so that the reader learns not only how to handle plants but why. Just as a general biology book selects from the various fields of botany and zoology, so also does this book select from the fields associated with growing and using plants, providing a background of information which may encourage either a student or general reader to begin a hobby or avocation or may even create an interest that will lead to a career.

Since the book is intended for non-professional readers, there is no chemistry, physics, or mathematics used in any technical way, and no previous background in botany, agriculture, or horticulture is assumed. The emphasis is on understanding the how and why of plants in commonly understood language. Basic botanical information is in the first chapter, and other necessary scientific terms or information are incorporated, with clear explanations, into other chapters when appropriate or in the separate glossary.

Plants are IN these days, and courses in both universities and community colleges that offer this kind of material are enrolling many students. This book will provide a suitable text for such a course. Even instructors who do not cover all of the material in the book will find a great deal of useful information for even a limited course in, for example, Indoor Plants or Outdoor Gardening. Knowing and growing plants can provide pleasant intellectual stimulation and a great deal of satisfaction. It will also lead to a better understanding of the ecological problems with which the world is struggling today. It is therefore quite appropriate that we try to inform as many people as possible about the values and pleasures to be found in plants. Plants add a dimension to our lives that is quite beyond the factory-made decorative doodads, reproduced prints, and antique bottles with which we so often surround ourselves. Plants share the world of life with us and exhibit complexities and characters to which we can respond and contribute, so that the more we know about them the more satisfying our association with them will be. *Living with Plants* will provide the reader with the knowledge of what plants are, their structure, how they behave, what their needs are and why, and how to utilize plants in one's surroundings for aesthetic effects and enjoyment.

There are many books on all kinds of gardening and plant growing, and a selected bibliography is included in the appendix. A good garden reference book containing encyclopedic information about garden and plant care, cultural conditions, and lists of plants is usually the mainstay of any serious gardener's library. Gardening books are however descriptive only and do not provide the botanical background for understanding the instructions and information given. In fact, many gardening do's and dont's still fall into the category of "old wives' tales" and often have little basis in botanical fact. Understanding plants will lead to success in growing and using them and to a more positive attitude about plants in general. When you read this book we hope you too will agree that "plants are neat".

I am indebted to many people who

assisted and encouraged me in producing this book. Mary Wilcox critically read the entire first draft and assured me the effort was worthwhile. Eric Anderson and Connie Beaubien produced some of the instructive photographs. Julie Medlin supplied the darkroom and enviable photographic expertise for producing suitable prints. Dr. Richard Pippen, department chairman, who reveals his enthusiasm for botany so skillfully in the first chapter, was totally supportive of my literary efforts at all times. Dr. Gail Schumann provided considerable assistance with biological information in several chapters. Kathryn Simpson interpreted our ideas with uncanny judgment to produce the exceptionally fine drawings, a project to which Wendy Smith-Griswold and Nancy Wright contributed. Dr. James F. Waters coordinated an editing staff par excellence and firmly but gently guided me to produce a readable manuscript. My fellow members of the Michigan Botanical Club, the Kalamazoo Garden Club, and the Landscape Critics Council also deserve recognition for allowing me to participate in and learn from their many activities over the years. Finally, to my husband, who not only encouraged my writing but tolerated the many inconveniences it so often caused, and to my faculty colleagues who assisted in so many ways, I say a sincere and appreciative thank you to all.

March, 1980

Donna N. Schumann
Western Michigan University
Kalamazoo, Michigan

TABLE OF CONTENTS

Chapter 2. Soils and Plants .... page 50

Chapter 3. Seeds and Seed Plants 82

Chapter 4. Vegetative Propagation

List of Figures

List of Tables

Figure Credits

Kathryn Simpson—1-13, 1-15, 1-16, 1-17, 1-20, 1-23, 1-27, 1-28, 1-29, 2-4, 2-11, 2-14, 3-1, 3-5, 3-6, 3-10, 4-3, 4-4, 4-5, 4-6, 4-8, 4-18, 4-19, 4-20, 4-22, 4-23, 5-4, 5-5, 5-8, 5-9, 5-13, 6-2, 6-3, 6-5, 6-7, 6-9, 6-10, 6-11, 6-12, 6-14, 6-15, 6-16, 6-17, 6-18, 6-19, 6-20 6-21, 6-22, 7-13, 8-6, 8-10, 8-11, 10-1, 10-4, 10-6, 10-7, 10-8, 10-9, 10-13.

Wendy Smith-Griswold—1-17, 1-18, 1-21, 1-22, 6-13, 7-4, 8-7, 8-9, 9-10, 9-12, 9-13, 10-3, 10-5, 10-10, 10-11, 10-12, 10-14, 10-18, 11-11.

Nancy Wright—1-1, 1-2, 1-4, 1-5, 1-6, 1-9, 1-12, 1-14, 1-17, 1-19, 1-26, 2-7, 2-8, 2-12, 2-15, 11-6.

Photography by **Donna N. Schumann** and her associates. Dover Publications gave permission to use chapter initials. Illustrations previously published in other sources are credited in the legend to the particular figure.

1. INTRODUCTION TO A PLANT

To study the practical aspects of plants successfully we must understand something of the structure and organization of the different types of plants that exist around and with us.

What do we mean by a plant? For a simple question it is embarrassing that there is no neat and simple answer. Some would say that plants are living organisms that differ from animals in that they are immobile and cannot move under their own power—but some plants can. Others would say that plants are living organisms that contain chlorophyll (a green pigment) and can make their own food, but there are some organisms traditionally considered plants that do not have this ability. Most people would agree that plants definitely include those living organisms that (1) possess chlorophyll and have the ability to manufacture their own food, (2) have a rigid cell wall usually containing the chemical compound cellulose, and (3) have the ability to grow continually. The major problem is that the more we learn about all living organisms the less able we are to put them into neat little categories, and the more people try, the greater the number of opinions that result, and thus there seems to be little agreement as to what really is a plant. For our purposes here, we are going to consider that plants definitely include: flowering plants, gymnosperms, ferns, clubmosses, horsetails, whisk ferns, mosses and liverworts, and most algae. In our discussions we'll include as "plant allies" the fungi, bacteria, and blue-reen algae.

There are approximately 500,000 different kinds of plant-like organisms sharing our world. The most familiar of these organisms are the flowering plants, angiosperms, which are the most noticeable and abundant plants on Earth today. On the basis of major physical differences, each of the 250,000 presently recognized species of angiosperms maybe assigned to one of two major subdivisions, the **monocotyledons** (monocots) including lilies, grasses, and orchids, or the **dicotyledons** (dicots) including buttercups, roses, and daisies. When you hear the word "plant," most probably the image you conjure up in your mind is of these flowering plants.

We will say something about the other organisms considered to be plants or plant-like later, but for now let's concentrate on the flowering plants. Despite the tremendous diversity that might be expected with over 250,000 different kinds, flowering plants do show a remarkable amount of similarity, at least in basic organization. Let us examine the **vegetative** (non-reproductive) structure of a typical flowering plant. (See Fig. 1-1.)

STRUCTURE AND ORGANIZATION

The Plant Plan

The body of most flowering plants consists of two systems: root and shoot. The root system provides the basic functions of

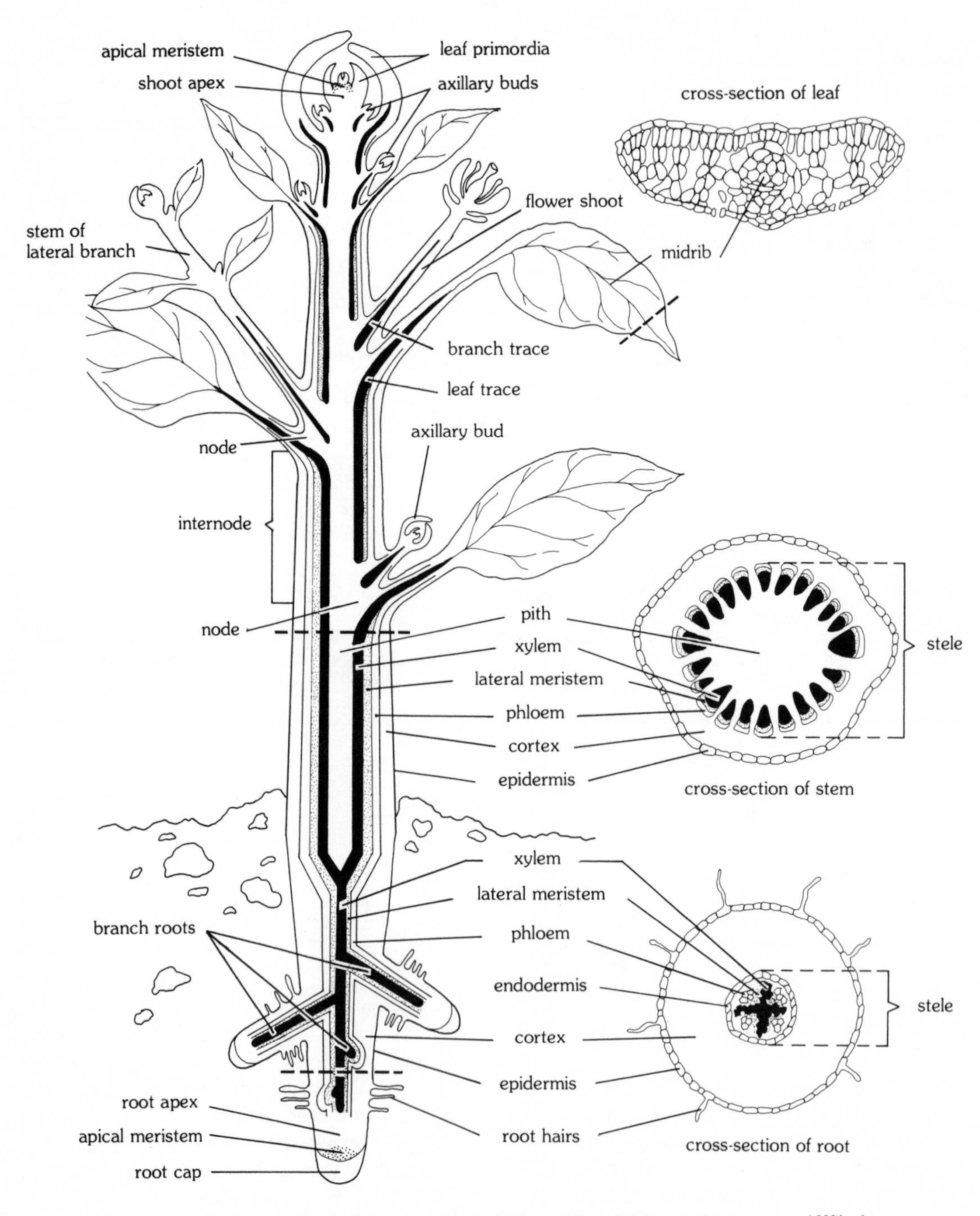

Fig. 1-1. Structure of typical flowering plant. (Adapted from Weier *et al.*, courtesy of Wiley).

(1) **anchorage**—holding the plant in the substrate (most often the ground), (2) **absorption**—taking up water and minerals from the environment into the plant, and (3) **storage** of extra food. The shoot system consists of two organs: leaves and stems. The former are essentially the food factories of the plant, the sites of photosynthesis where inorganic nutrients are converted into organic compounds—sugars, starches, and so forth. The stems form the main axis of the shoot and provide support for leaves and a pathway for translocation of water and food to and from the leaves.

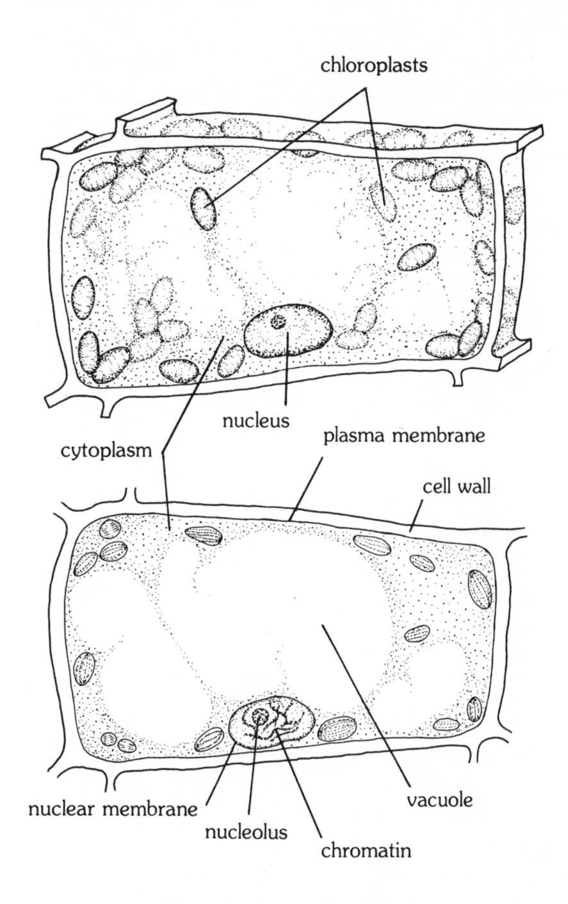

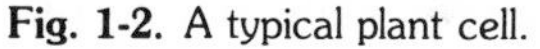
Fig. 1-2. A typical plant cell.

The Plant Cell

Before we can discuss roots and shoots in more detail we must first consider the basic unit of organization of plants—the **cell**—and consider how cells work together to form the organs of the plant body. Hooke, an Englishman, applied in 1665 the word "cell" to the basic units of plants, but it was Schleiden and Schwann, both German scientists, who announced to the world in 1838 that all plants and animals were composed of basically similar units called cells; and then Virchow, another German scientist, added in 1858 that all cells arise from pre-existing cells. The **Cell Theory** established a basis for further understanding of the microscopic organization and function of plants.

The protoplasm or living substance of flowering-plant cells is typically composed of two major units: nucleus and cytoplasm. The **nucleus**, a body surrounded and bound by a membrane and located within the cytoplasm, contains mostly deoxyribonnucleic acid (DNA) and protein. This unit controls all of the chemical activities of the cell and provides the inheritable information that is passed on from one generation of cells to the next. The **cytoplasm** is all of the rest of the protoplasm surrounding the nucleus. The cytoplasm is also bounded by a membrane, the cell or **plasma membrane**, outside of which is a more rigid **cell wall** (called the primary cell wall) that contains a number of chemical compounds, the most important of which is **cellulose.** Between adjacent cells is the **middle lamella** composed of cell cement (calcium pectate) that holds cells together.

Inside the cell, the cytoplasm is usually restricted to a relatively narrow region pressed against the cell wall by a large fluid-filled sac, the **vacuole**, which occupies most of the central portion of the cell. The vacuole contains mostly water with materials such as sugars, pigments, and salts dissolved or suspended in it. When the concentration of dissolved substances (called **solutes**) in the water of the vacuole is great, then water migrates into the vacuole from regions outside the cell. The additional water in the vacuole exerts a pressure against the cell wall, called **turgor pressure**, and this pressure against the rigid cell wall gives the cell its characteristic shape. All of the living cells in the plant exerting such pressure against each other give shape to the whole plant. When there is not enough water available in and around the plant to maintain this pressure, then the whole plant begins to droop or wilt. When a plant wilts because of a lack of water, it does so because each living cell cannot maintain its turgor pressure. Cell walls and large central vacuoles are features found in plant cells that are normally lacking in animal cells.

Within the cytoplasm there are a number of membrane-bound units called **organelles** in which specific chemical processes occur. One of the most characteristic organelles in many plant cells is the **plastid**. Plastids are ovoid bodies, easily visible under the light microscope. They usually contain pigments, and there are several types which may be distinguished by their color. Therefore, **chloroplasts** are green plastids, the color due to green pigments, **chlorophylls**. Chloroplasts are the sites of the food-manufacturing process, **photosynthesis**. Plastids which are a color other than green are called **chromoplasts**. These are usually red, orange, or yellow and provide the colors of many flowers and fruits such as marigolds, zinnias, and tomatoes. Plastids which are clear or colorless are called **leucoplasts** and function primarily as food-storage bodies.

There are several other important cell organelles, such as: mitochondria (sites of cellular respiration, the source of chemical energy, see Fig. 1-17), dictyosome (packaging and distribution centers), ribosomes (sites of synthesis of protein molecules), and the endoplasmic reticulum (a highly complex series of membranes), but they are beyond the scope of this book. You can learn more about these in any general biology or botany textbook.

Tissues and Systems

Let us now consider how individual cells work together to form the whole plant. You may recall from other sources that: a **tissue** is a group of cells that are similar in appearance and function; an **organ** is a group of tissues working together with a common function; an **organ** system is a group of organs functioning together; and an **organism** is composed of a number of coordinated organ systems.

The complex organ systems that we associate with animals, such as reproductive, circulatory, and digestive systems, are not found in plants. Instead, most plants, as mentioned on page 1, have only two organ systems, roots and shoots; and these, in turn, are formed from three basic organs—roots, stems, and leaves (the latter two making up the **shoot**). These organs are formed from four basic **tissue systems**: the **meristematic**, **dermal**, **ground** (or **fundamental**), and **vascular**. These form continuous systems throughout the various organs.

The cells of the **meristematic tissues** are capable of undergoing cell division continually to produce new cells. In fact, these are essentially the only tissues in the plant with such capabilities, an ability that gives plants the characteristic of open or continuous growth. The meristematic tissues give rise to the other tissue systems (dermal, ground, and vascular) and may be distinguished by their locations within the plant.

The **apical meristems** are located at the apices (tips) of roots and shoots. Thus, growth, in terms of formation and development of new cells and the resulting increase in height or length of the plant, occurs only at the tips of plants. The apical meristem of the shoot produces the tissues which develop into stems and leaves, and once the cells are formed the new organs "grow" by taking water into their vacuoles thereby enlarging and elongating. New tissues and growth resulting from the apical meristems are called **primary tissues** and **primary growth**.

In the **axil** of each leaf (the angle where the leaf joins the stem) is a structure known as an **axillary** or **lateral bud**. This is actually a potential shoot apical meristem and when stimulated to grow may produce a **lateral branch** which is a shoot composed of stem and leaves. Quite often these lateral buds may be stimulated to grow by pinching or cutting off the shoot apex. This treatment will cause the plant to have a more branched or bushy appearance.

The **lateral meristems** are formed in the roots and stems of many plants and account for the increase in thickness or diameter of these organs. Tissues and growth produced by the lateral meristems are called **secondary tissues** and **secondary growth**. The **vascular cambium** is a lateral meristem that produces secondary vascular tissue (which we will discuss shortly). Wood, in trees and shrubs, is formed from secondary vascular tissue produced by vascular cambium.

Another lateral meristem is the **phellogen**, often called **cork cambium**, located just under the surface of stems and roots. It accounts for the formation of **periderm** (part of the dermal system) which forms an outer protective covering in the stems and roots of woody plants. Most of the bark of trees and shrubs is formed by cork-cambium growth.

The **dermal system** contains the tissues that form the outer protective covering of the plant. There are two types of dermal tissues: epidermis and periderm. The **epidermis** is a primary tissue (formed by the apical meristem) usually forming a single layer of cells encompassing the whole plant. The epidermis of plant parts exposed to air secretes a waxy outer coating called the **cuticle**. The cuticle forms a non-cellular layer that protects the plant from water loss and invasion by bacteria and fungi. The substance making up the cuticle of the epidermis is **cutin**, a type of wax. Certain plants produce very excellent, hard waxes that have been economically important as waxes for floors, walls, cars, and so forth. Carnauba wax is obtained from the cuticle of the carnauba palm (*Copernicia cerifera*) of northeastern Brazil.

To allow water, carbon dioxide, oxygen, and other gases to get in and out of the shoot, pores, called **stomata**, are found in the epidermis. These pores are surrounded by special epidermal cells, called **guard cells**, that are able to expand and contract and thereby control the opening and closing of the pores, like the automatic vents in a greenhouse or windows in a house. The stomata are not covered by cuticle. See figs. 1-3 and 1-4.

In woody plants where the vascular cambium is active and secondary growth abundant, the epidermis, which cannot stretch or grow, is usually torn apart as the plant increases in girth. To replace the protective outer layer, the cork cambium or phellogen produces **periderm**, which is another type of dermal tissue. The periderm forms the bulk of the bark of woody plants and, like epidermis, protects the plant from desiccation and invasion by bacteria and

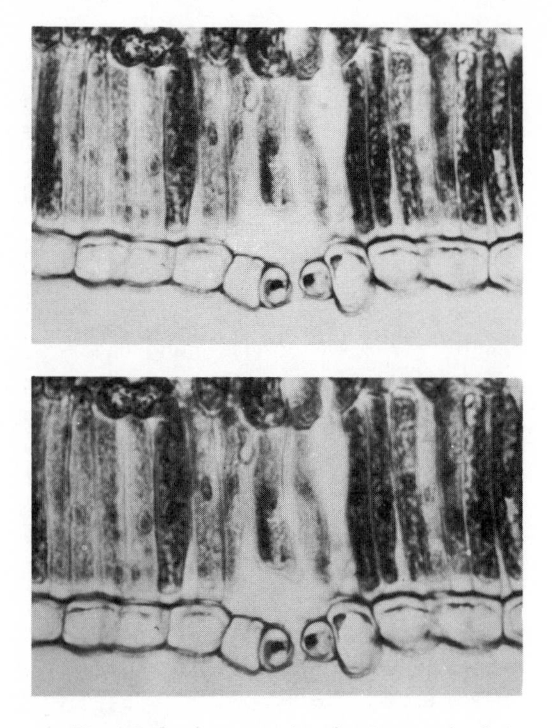

Fig. 1-3. Leaf cross-section showing guard cells in lower epidermis.

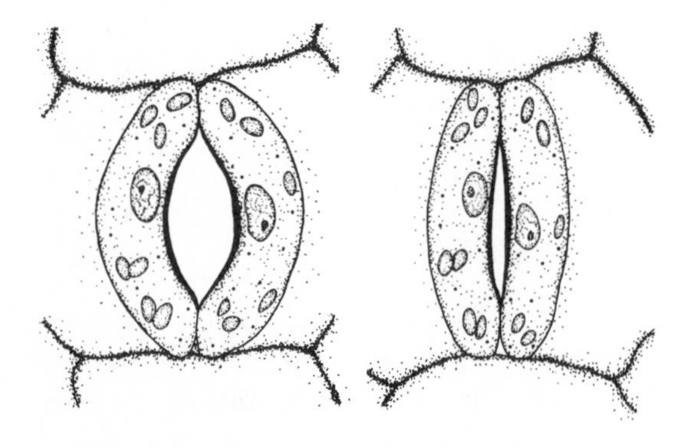

Fig. 1-4. Surface of epidermis to show stomate action. (left) When water is abundant, turgor pressure within the sausage-shaped guard cells forces them into a curve and opens the pore. (right) When guard cells dry out, the thick elastic strip of cell wall next to pore flattens out, closing pore and minimizing water loss from deeper tissues.

fungi. The outermost layers of cells of the periderm contain in their cell walls a waterproofing substance, called **suberin**. **Lenticels** are areas of loosely aggregated cells in the periderm that allow gases like carbon dioxide and oxygen to move in and out of the stem.

Some plants, particularly the cork oak (*Quercus suber*) from the Mediterranean region, produce abundant layers of periderm that can be carefully removed without damaging the tree and processed to provide the commercial cork that we use as bottle stoppers, shoe soles, bulletin boards, and other items.

Three tissues compose the **ground** or **fundamental tissue** system: parenchyma, collenchyma, and sclerenchyma. **Parenchyma** is probably the most widespread, versatile, and flexible tissue in plants, usually forming a continuous tissue throughout large portions of roots, stems, and leaves. Parenchyma cells are living, have thin walls (that is, only a primary wall), possess a nucleus, cytoplasm, vacuole, and most of the membrane-bound organelles found in the cytoplasm, and are more or less spherical or angular (but nearly round) in shape. They are capable of carrying on most of the chemical processes associated with living cells, including photosynthesis (to manufacture food), respiration (to obtain chemical energy), storage, and even regeneration (to heal wounds).

Collenchyma is more or less similar in appearance to parenchyma except that the cells are more elongated in shape and have irregular thickenings of pectins (a carbohydrate compound) in the cell walls. The thick walls provide strength and support; thus collenchyma cells function mainly as a supporting tissue. Since the cells are living—possessing a nucleus, cytoplasm, and cytoplasmic organelles—they also carry on some chemical processes, including respiration and even regeneration. Collenchyma is usually found in leaves, around the veins, and just under the epidermis of stems where it may form a continuous layer around the stem or occur in distinct strands.

Sclerenchyma is a major structural or supporting tissue in plant organs. These cells have thick **secondary walls** formed inside the primary cellulose cell wall and are usually not alive at functional maturity. The secondary wall contains a compound, **lignin**, which is very hard. This substance gives the sclerenchyma tissue its tough, strong characteristics that allow it to function in support. There are two kinds of sclerenchyma: sclerids and fibers. **Sclerids** or **stone cells** have various shapes from spherical to angular and occur in leaves, stems, and fruits. Sclerids are abundant around the core of some fruits, such as the pear, protecting the seeds. **Fibers** are long, thin, spindle-shaped cells that occur in bundles forming support columns in stems, roots, and leaves. Fiber bundles are often quite long (up to 16 cm.) and durable, and some natural plant fibers are used to make twine and rope. Manila hemp is made from the sclerenchyma fibers of the Manila hemp plant (*Musa textilis*). The fibers in marijuana (*Cannabis sativa*) have also been used to make rope. In fact, during World War II, marijuana was grown legally and commercially in the Midwest as a fiber source.

The **vascular system** comprises the tissues that carry food (products of photosynthesis) and water throughout the plant. The water-conducting tissue is called **xylem** and the food-conducting tissue **phloem**. Xylem consists of two kinds of water-conducting cells: tracheids and vessel elements. **Tracheids** are long, spindle-shaped cells with tapered ends that are attached to one another vertically to form columns, several cells thick. These cells are present in all plants with vascular tissue. **Vessel elements** occur only in the more advanced plant groups such as flowering plants and a few gymnosperms and ferns. They are usually arranged much like tracheids and are often adjacent to them. They can be distinguished from tracheids in that the end walls of the vessel elements are squared off and perforated with one or more pores so that they essentially form continuous vertical "water pipes" (**vessels**) without cell walls between the stacked cells. Lack of end walls makes vessels more effective than tracheids in conducting water. At maturity both tracheids and vessel elements have thick secondary walls containing lignin (much as sclerenchyma tissue does) and the cells usually lack cytoplasm and nuclei. Not only do they form a continuous system to transport water up from the roots through the stems and into leaves, but with the thick secondary walls they also provide structural support for the plant. (See Fig. 1-5.)

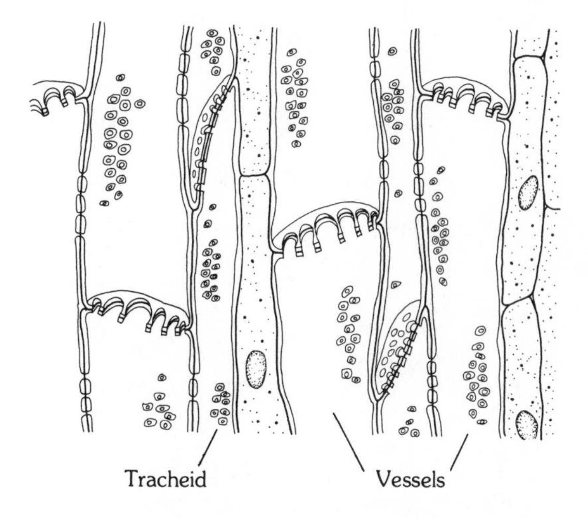

Fig. 1-5. Xylem, cut longitudinally.

Phloem is composed of two kinds of cells that function in the translocation of food: sieve-tube elements and companion cells. **Sieve-tube elements** are long cells with squared-off end walls, attached end to end, forming **sieve tubes**. The end walls, called **sieve plates**, are perforated with several pores that are continuous with the pores of the adjacent cell. The food flows from cell to cell, through the pores in the sieve plates. The sieve-tube elements are cell walls and cytoplasm but contain no nuclei. Adjacent to each sieve-tube element is a **companion cell** which is a smaller cell with an unusually large nucleus. It is thought that the nucleus of the companion cell controls the activities of the sieve-tube element. (See Fig. 1-6.)

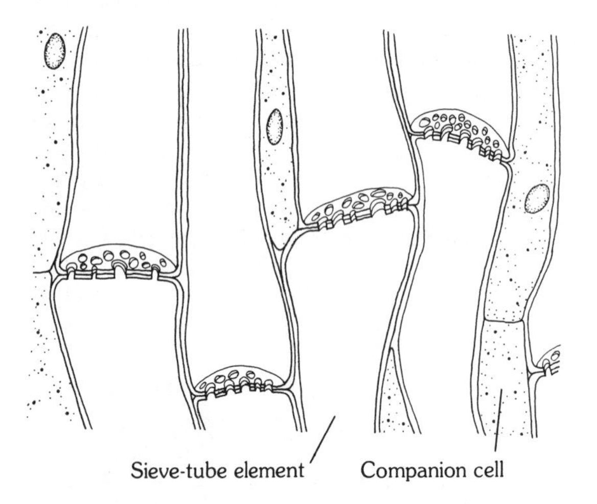

Fig. 1-6. Phloem, longitudinal section.

Plant Organs

Now that we've briefly considered the four tissue systems let us see how they work together to form a plant. The plant in Fig. 1-1 consists of a continuous main axis and lateral branches (roots and stems) with flattened lateral appendages on the stem (leaves). All are bounded externally by an **epidermis**. The stem and root have a **cortex** region inside the epidermis which is composed mostly of parenchyma and occasionally collenchyma and sclerenchyma, and a central cylinder of vascular tissue, the **stele**. The leaf, being flattened, has an upper and a lower epidermis with the region between called the **mesophyll**. We will now examine each of these organs in greater detail.

Roots have three major functions: anchorage, absorption, and storage. As anchorage structures, the roots penetrate the substrate (usually soil), branch, and rebranch, often forming extensive systems that hold the plant firmly in place. Basically two types of root systems may be recognized: tap-root and fibrous root. The **tap-root system**, such as that found in the dandelion (*Taraxacum officinale*) or carrot (*Daucus carota*), has one main, larger root with many smaller lateral branches. This has a dual function, of storage as well as anchorage and support. The **fibrous-root system**, such as that in the tomato (*Lycopersicon esculentum*) or marigold (*Tagetes*), has many roots branching and rebranching in all directions without any single main root. (See Fig. 1-7.)

We can understand the functioning of roots better if we examine their internal structure as shown in Fig. 1-8. The roots of both dicotyledonous and monocotyledonous plants have essentially the same basic organization, although a few significant differences will become evident later. As an adaptation for absorption, the epidermis of the roots forms only a very thin cuticle so that the root may absorb water and minerals directly from spaces between soil particles. In many plants the epidermal cells near the tips of the roots develop lateral outgrowths, called **root hairs**, that are able to penetrate smaller spaces and greatly increase the surface area for absorption (Fig. 1-9). These are very delicate structures that must be carefully protected from drying out when transplanting or digging up a plant, or the effectiveness of the root for absorption will be severely reduced. Some plants that do not produce root hairs have developed a mutually beneficial arrange-

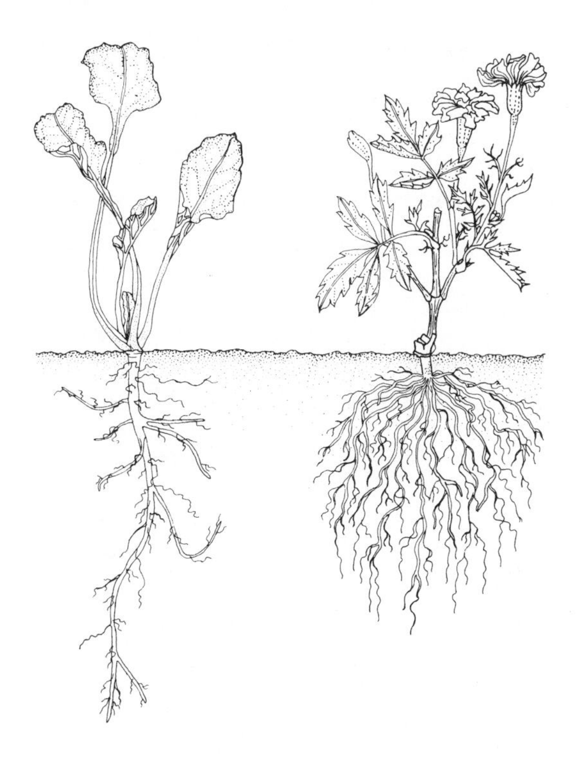

Fig. 1-7. Plant on left has tap root; plant on right, fibrous roots.

ment with fungi that develop a mantle around the root and aid in the absorption of some materials, particularly minerals. This association of root and fungus is called a **mycorrhiza**.

Inside the epidermis the cortex forms a wide cylinder, mostly of parenchyma tissue, in which much food is usually stored. At the boundary between the cortex and the stele is a very interesting and important layer of cells called the **endodermis**. The function of the endodermis is to control the movement of materials between the cortex and the stele. Substances may move through roots (from the epidermis to the vascular tissue) in two ways: either through the middle lamellae and cell walls or through the plasma membranes and the protoplasts of the cells. In the former route there are normally no impediments to the movement of substances, but in the latter the plasma membranes may *select* which substances will move through the cells. Within the cell walls of the endodermis (especially in the absorbing regions of the roots) is a bandlike zone of suberin that extends completely around the cells and through the middle

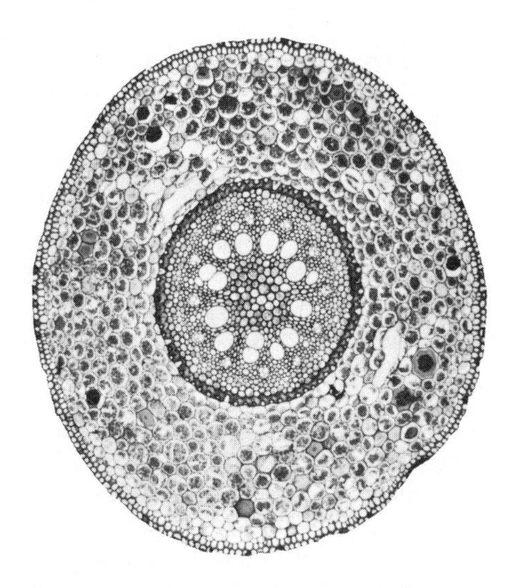

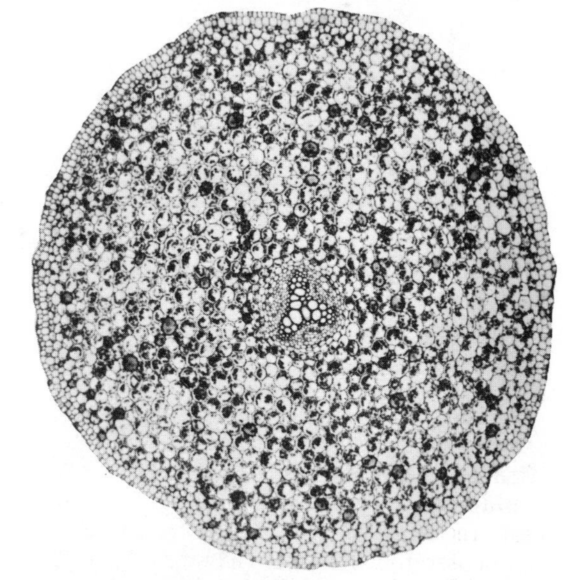

Fig. 1-8. Roots in cross section—monocot on left, dicot on right.

lamellae to adjacent endodermal cells. This band of suberin is called the **Casparian strip**. Its presence forms a barrier that prevents substances from moving through the cell walls and middle lamellae of the endodermis and forces the substances to pass through the plasma membranes and protoplasts of the endodermis cells. Thus the endodermis layer controls the movement of substances between cortex and stele much as a bouncer at a bar controls the entrance and exit of people. (See Fig. 1-9.)

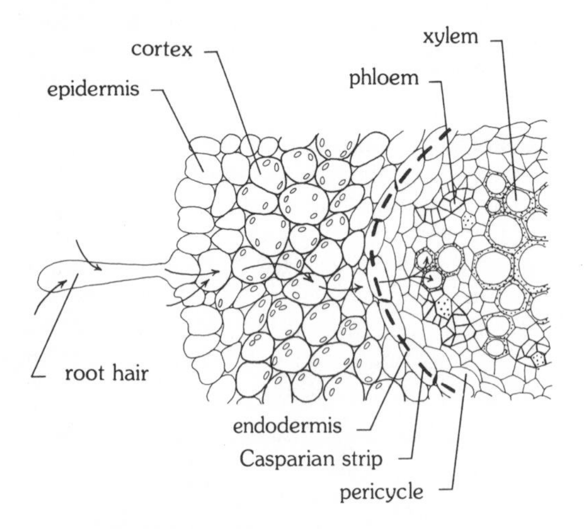

Fig. 1-9. (above) Movement of water and nutrients through cells of root. (below) Root hairs increase surface area of root and thus rate of water uptake.

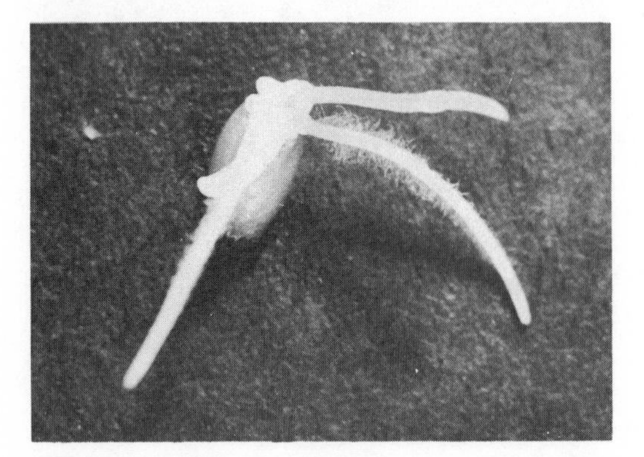

The stele is a cylinder inside the endodermis containing the vascular tissues. In the center of the stele the xylem consists of a core with radiating arms. Alternating with the bands of xylem are zones of phloem. Between the xylem and phloem *in the dicotyledonous roots only* is a single layer of lateral meristem—the **vascular cambium**. These cells are capable of dividing to form new xylem cells (to the outside of the existing xylem) and new phloem (to the inside of the existing phloem). In this way these roots are able to grow larger in diameter by secondary growth. The vascular cambium is *not present in monocotyledonous* plants. Outside of the xylem and phloem but inside the endodermis in both dicots and monocots is another layer of cells, the **pericycle**. This layer also has meristematic properties, forming part of the vascular cambium, phellogen, and lateral roots. When a lateral branch of a root develops, it originates from the pericycle and pushes its way out through the cortex and epidermis. In this way the vascular system of the lateral root is connected directly to the vascular system of the main root.

Many roots, particularly tap roots, have become specially modified as storage organs. They usually become very large through a special type of secondary growth in which the cambium produces large amounts of storage parenchyma (surrounding the secondary xylem and phloem) in which quantities of sugars and starches are stored. Examples, of course, are carrot, turnip, beet, and sweet potato.

Stems and leaves form an integrated system (the shoot). The location on the stem where a leaf occurs is a **node**; the region of the stem between two nodes is the **internode**. The stem essentially is the structure that connects the leaves, which supply food, to the roots, which supply water and minerals. That may not appear to do justice to the stem, since in many plants, particularly woody ones, the stem makes up the most obvious and bulkiest part of the plant. Functionally the stem forms the vertical and lateral axes of the shoot, supports the leaves, and conducts water and food to and from them. Shoots are exceedingly diverse and various in size and shape. They may be slender and upright as in many wildflowers, garden flowers, and weeds; weak, clinging or creeping as in a vine; horizontal and underground as in the lily-of-the-valley (a **rhizome**); thick and fleshy as in a cactus; or tall and sturdy as in a tree. Some shoots may be especially modified as storage organs such as a **tuber** (white potato), **corm** (*Gladiolus*), or **bulb** (lily).

Plants may be considered **herbaceous** if the growth of the shoot is essentially primary, that is, with little or no secondary growth, or **woody** if the shoots (and roots) exhibit extensive and continual secondary growth. In most geographical regions where there are alternating growing seasons and dormant seasons, such as alternating warm and cold periods or wet and dry periods, plants may live for only one season (**annuals**) or for several seasons (**perennials**). Most annuals are herbaceous, but there are both herbaceous and woody perennials. In most herbaceous perennials of temperate regions the aerial shoots usually die back at the end of the growing season and the plants survive the adverse season by dormant underground structures such as tubers, corms, rhizomes, or bulbs. (There are some exceptions, of course, such as many desert plants that don't die back but merely go dormant during adverse conditions.) At the beginning of the next growing season new aerial shoots will be produced by primary growth.

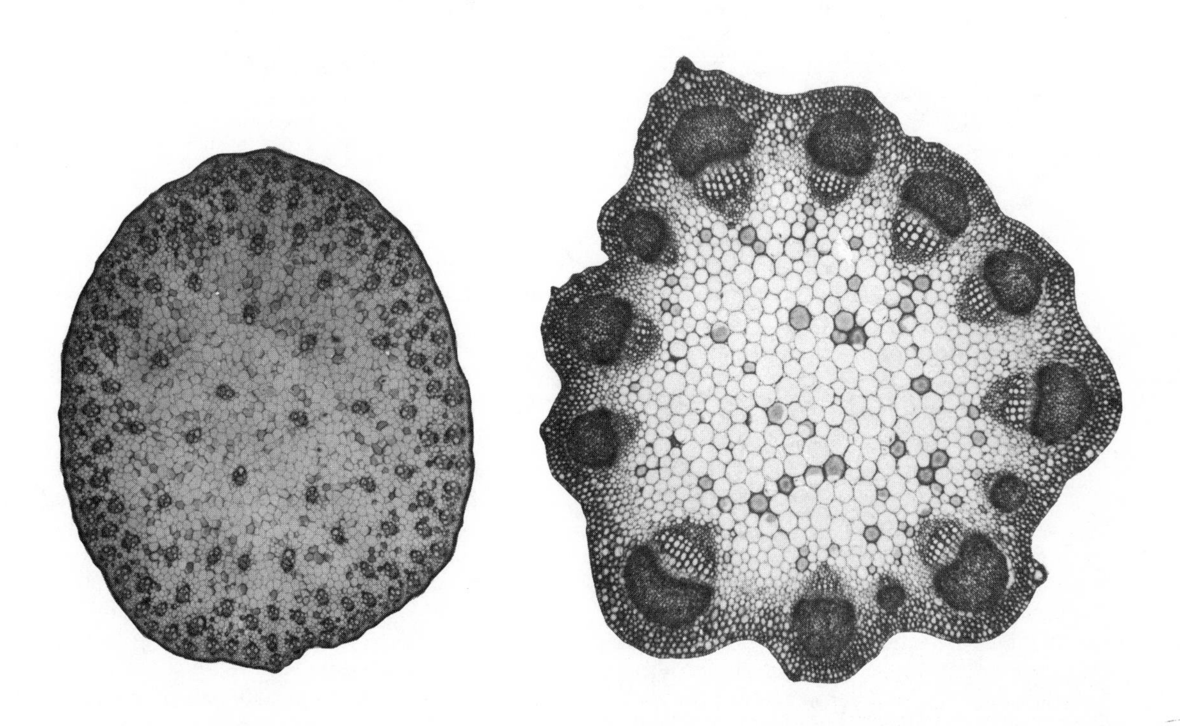

Fig. 1-10. Herbaceous stems in cross section—monocot on left, dicot on right.

In **woody perennials** the aerial shoots remain above ground and every year add new layers of secondary vascular tissue and periderm. They survive the adverse growing periods by protecting their apical meristems in dormant buds and "closing up shop" or going dormant until environmental conditions are favorable for growth. During these adverse periods some plants (**deciduous**) may drop their leaves while others (**evergreen**) may retain their leaves, but in the latter the leaves are usually especially modified to resist the adverse conditions.

Let us examine the stem more closely by taking a look at its internal organization. We will first consider a cross section of an herbaceous dicotyledonous stem as shown in Fig. 1-10. As in the root, the epidermis and cortex are external to the cylinder of vascular tissue, the **stele**. Unlike the stele in the root, which you recall is an uninterrupted central cylinder containing only vascular tissues, the stele in the stem is composed of discrete vascular bundles separated by bands of parenchyma (pith rays). Inside the ring of vascular bundles is a central zone mostly of parenchyma tissue, called the **pith**. Each vascular bundle consists of a zone of xylem cells closest to the pith, outside of which is a single layer of lateral-meristem cells, the vascular cambium. Beyond that is a zone of phloem cells. In some plants a zone of sclerenchyma tissue, the **bundle cap**, occurs to the outside of the phloem. When present, the vascular cambium becomes active as the plant gets older and produces secondary xylem cells to the inside and secondary phloem cells to the outside as in the root. Usually the cambium will form more secondary xylem cells than secondary phloem, so that the xylem is more abundant. As the cambium layer becomes active, it invades the pith rays between the vascular bundles and produces secondary xylem and phloem there also. Eventually a complete, uninterrupted cylinder of secondary xylem and phloem will be formed.

In most herbaceous plants only one such cylinder will form, because at the end of the growing season the stem normally dies back. But if we examine the cross section of a woody dicot stem (Fig. 1-11) we

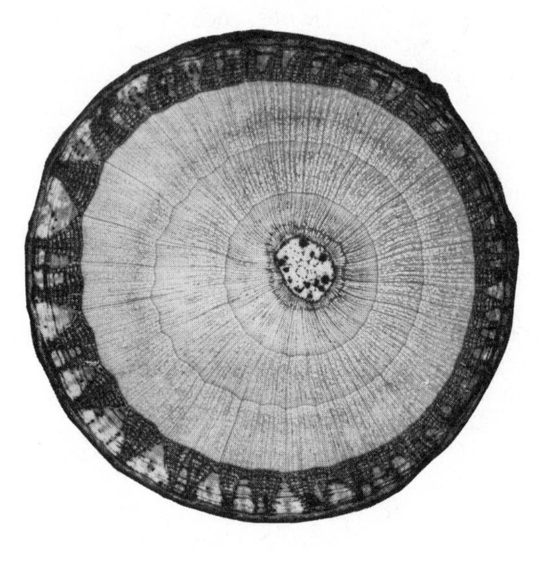

Fig. 1-11. 3-yr.-old woody dicot stem in cross section.

notice that the situation is somewhat different, because of the permanence of the stems and the continual activity of the vascular cambium. In this cross section of a three-year-old stem, you will note that the pith is still present in the center, but now the stele consists of primary xylem, three complete, concentric layers of secondary xylem, the vascular cambium, three concentric layers of secondary phloem, and the original primary phloem. (The delicate phloem layers are crushed and do not show up well.) At the beginning of each growing season the vascular cambium resumes the production of new secondary vascular tissue (xylem toward the inside and phloem toward the outside of the cambium). Each year's growth of secondary xylem forms a distinct concentric cylinder surrounding that of the previous year, and this series of

annual rings of secondary xylem may be used to determine the age of the plant, at least in temperate climatic zones. Since the secondary xylem tissue is tough, long-lasting (because of the lignin in the secondary cell walls), and profusely produced, it accumulates to form the bulk of the aerial stems (and older roots) and is known as **wood**. The term **bark** is used to describe all of the tissues outside the vascular cambium. As the secondary xylem is annually produced and expands, causing the stem to increase in thickness, the primary tissues to the outside (primary phloem, cortex, and epidermis) are not able to expand or stretch to keep up with the internal growth, and gradually they split and are sloughed off. To replace these tissues, and to continue to protect the stem from desiccation, the phellogen forms periderm. As the stem continues to expand in thickness, not only is some of the periderm continually sloughed off and constantly replaced but so are the older secondary phloem cells. In older stems of woody dicots, then, secondary phloem and periderm contribute to the bark (Fig. 1-12).

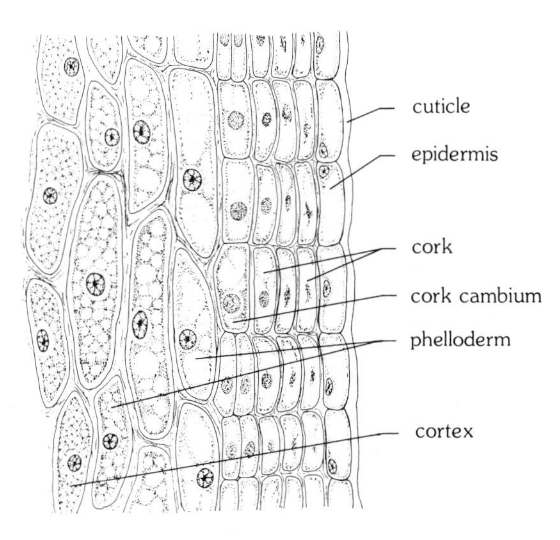

Fig. 1-12. Bark tissues.

In monocotyledons we find quite a different situation when we examine the cross section of the stem (Fig. 1-10). Inside the epidermal layer the vascular bundles appear either scattered throughout the ground-tissue parenchyma or in a few (usually two) concentric rings near the outer edges of the stem. It is thus difficult to distinguish regions such as cortex or pith clearly since the vascular bundles are not arranged in a single ring or cylindrical fashion. A closer look at the vascular bundles shows absence of a vascular cambium and thus a lack of secondary growth. A few plants are exceptions, however. A certain small group of monocots closely related to lilies that includes the century plants (*Agave*), Joshua trees (*Yucca*), mother-in-law's tongue (*Sansevieria*), *Cordyline*, *Dracaena*, and *Aloe* produce secondary vascular tissue from a cambium. This cambium is unusual, though, because it develops from parenchyma tissue outside of the primary vascular bundles and it divides to produce discrete secondary vascular bundles (much like the primary ones in appearance) and secondary parenchyma. Thus there is no solid core or series of concentric cylinders of secondary xylem to form wood as we observed in the dicots. Palm trees are monocots that develop thickened, tough stems; but these are produced by a special type of thickening growth of parenchyma cells and not by a vascular cambium. Internally the palm stem is similar to that of a corn stem with discrete vascular bundles scattered throughout the ground tissue. At the periphery of the stem, however, periderm, much like that in the woody dicots, is produced as an outer protection.

Leaves have already been defined as the lateral appendages of stems, parts of the shoot system. Leaves vary a great deal both in appearance (or structure) and in function. While the major function of most leaves is photosynthesis, the food-manufacturing process within plants, some leaves are highly modified for other func-

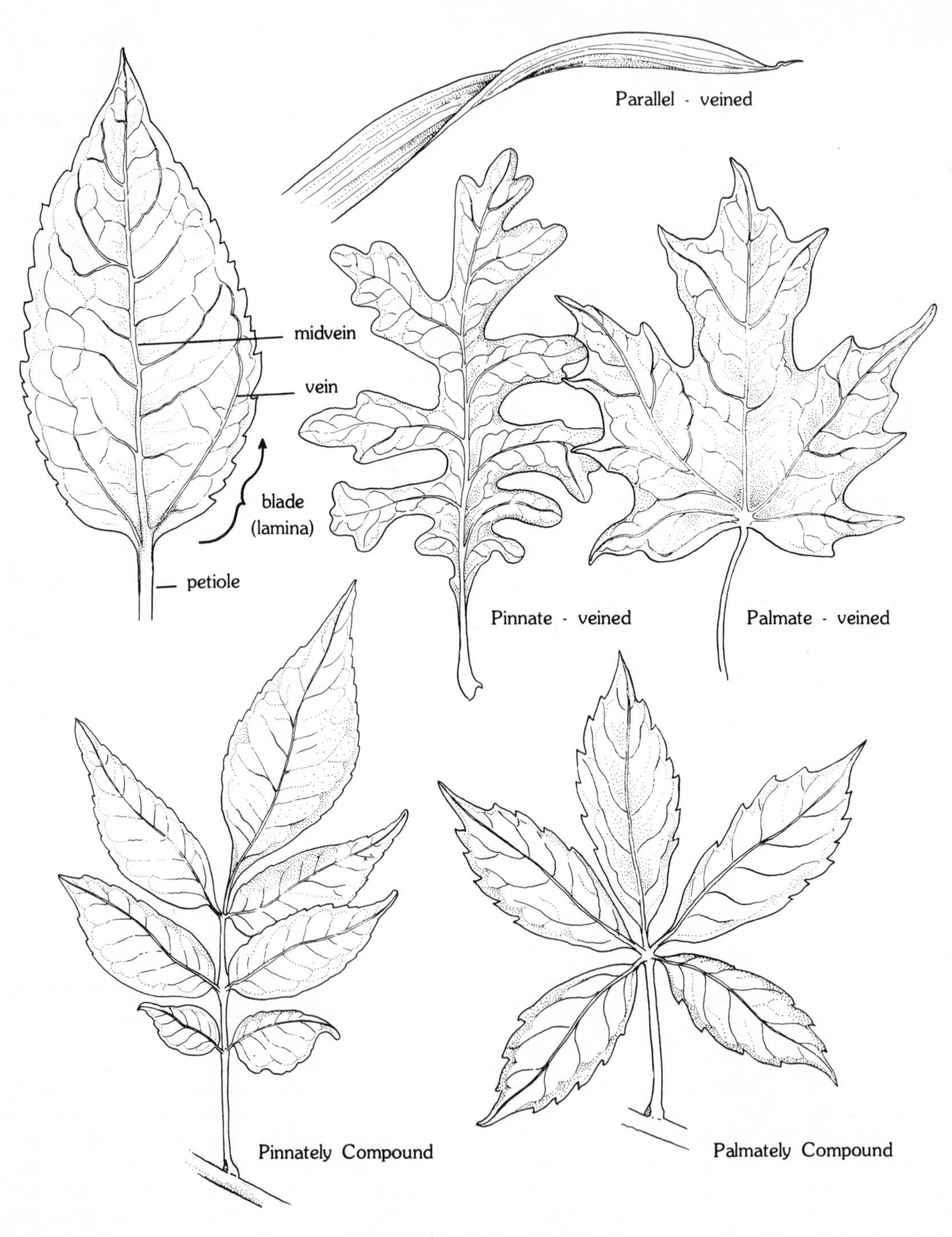

Fig. 1-13. Types of leaves.

tions such as protection or reproduction. Externally, although leaves do vary, we can usually distinguish two distinct parts: the **petiole**, a stem-like segment that attaches the leaf to the stem, and the **blade** (or lamina) which is usually a flattened, thin segment attached to the other end of the petiole. In some plants the petiole may completely encircle the stem forming a **sheath**; in other plants the petiole may be lacking and the blade attached directly to the stem (the leaf is said to be **sessile**). The blades exhibit a great deal of variation (Fig. 1-13) and may be **simple**, if not subdivided, or **compound**, if the blade is divided into several smaller leaf-like segments. Blades of both simple and compound leaves may have **teeth** or **lobes** breaking up the margins. The blades have an extensive, interconnected system of **veins** (which contain bundles of vascular tissue) that are arranged in a characteristic fashion in different plants. Venation patterns may be **pinnate**, like a feather with one main vein (**midrib**) and several lateral veins (as in oaks, *Quercus*); or **palmate**, with three, five, or more major veins originating from one point at the base of the blade and extending outward like the fingers from the palm of the hand (as in maples, *Acer*); or **parallel**, in which there are several or many main veins, all aligned parallel to each other (as in grasses like corn, *Zea*, or tulips, *Tulipa*). We may generalize that most monocots have simple leaves with parallel venation patterns. However, there are numerous exceptions, and venation alone may not always distinguish monocots from dicots.

In the axil of the leaf, usually just above the junction of the petiole and the stem, is an axillary or lateral bud. As mentioned earlier, these buds contain shoot apical meristems which may be stimulated to develop into lateral branches of the shoot. Leaves of many dicotyledonous plants have at the base of the petiole a pair of appendages called **stipules**. These are usually much smaller than the leaf itself and may resemble small leaves (and carry on photosynthesis) or they may resemble thorns. Their function appears to be to protect the young leaf during its early stages of development.

Let us now examine the internal organization of a typical leaf as shown in Fig. 1-14. Between the upper and lower epidermis

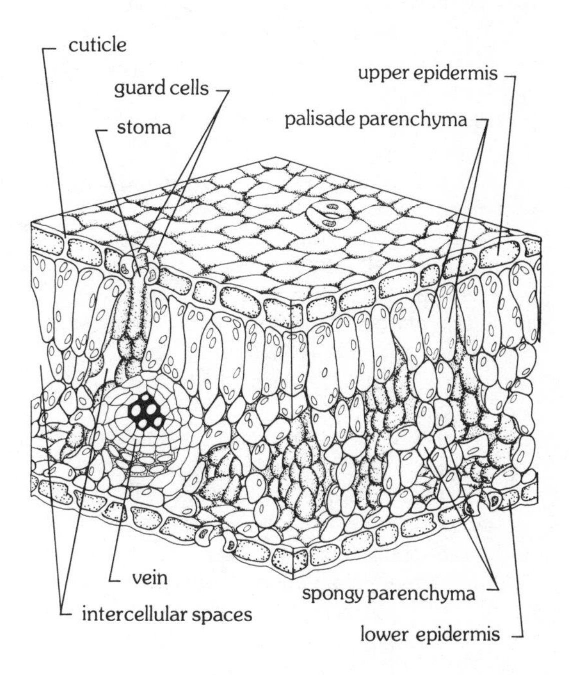

Fig. 1-14. Internal structure of leaf.

is a region called the **mesophyll**. Normally two distinct zones can be distinguished in the mesophyll: the **palisade parenchyma**, a densely packed layer of columnar cells immediately below the upper epidermis, and the **spongy parenchyma**, an area of more loosely arranged parenchyma cells with abundant intercellular air spaces opening to the outside through the stomata. Throughout the mesophyll is a network of veins containing vascular bundles. Cells of both the palisade layer and the spongy mesophyll are filled with chloroplasts; thus these are the regions within the leaf where photosynthesis occurs.

Leaves are very vulnerable to environmental conditions, and thus much of the variation in leaf structure is associated with the adaptation of leaves to survive in particular ecological situations. Leaves exposed to full sunlight will often be smaller and thicker than leaves of the same species (or even plant) in full shade. The smaller surface and thicker body of the sun-leaves reduces the amount of water that can be lost by evaporation from the leaf. The shade-leaves have thinner bodies and greater surfaces, maximizing the exposure of chloroplast-containing mesophyll cells to available light and air, since, comparatively speaking, water loss is not quite the serious problem in the shade that it is in full sunlight. Leaves on plants that grow in deserts will often be narrow and very thick, with few intercellular air spaces in the mesophyll, guard cells and stomata sunken below the level of the regular epidermis, and thick cuticles. All of these adaptations function to reduce the amount of water that will be lost by evaporation from the leaves and thus allow the plants to exist better in areas of low soil-water availability. In some desert plants such as cacti, the foliage leaves may be completely lacking or modified into thorns, and photosynthesis occurs in chloroplast-containing cells in the stem. Leaves on plants growing submerged in water, on the other hand, have few, large, widely-spaced mesophyll cells with an extensive system of intercellular air spaces. These air spaces function to hold gases such as oxygen and carbon dioxide that are used in some of the vital chemical processes within the cells. These gases are harder for the plant to get from the water.

PLANT FOOD AND FUEL

Now that we have examined the basic structure and organization of flowering plants, from cells to organs, let us return to the cell to consider some processes that occur therein, particularly those that involve energy. Some chemical reactions within cells require no energy, while others release it when they occur, but most require energy. Energy is necessary for cells to stay alive, carry on their vital processes, and make new cells. An energy source, or fuel, is necessary so that its stored energy may be transformed into energy to do work. Common non-cellular examples of energy sources for such transformations are gasoline which makes the automobile engine operate, fuel oil in the furnace to warm us, and electricity which lights a lamp. In cells, **glucose**, a type of sugar and an organic compound, is the fuel. Organic chemicals are compounds containing carbon, hydrogen, and oxygen as a basic skeleton to which other elements may be attached. Sugars, starch, cellulose, proteins, fats are examples of organic chemicals in living cells.

The energy needed by cells is **chemical energy** which is transported by **adenosine triphosphate (ATP)**. When a chemical reaction requires energy, the ATP **molecule** (a chemical unit) supplies it by transferring its energized part to the reacting substances, thus activating the substances and allowing them to react with other substances. An example of such an energy exchange is starch formation. When glucose is abundant within the cell and not needed for fuel, many glucose molecules may be joined together to form **starch**, a more compact and efficient fuel-storage unit. ATP energy is necessary to fasten the glucose molecules together. This is accomplished by the transfer of the energized part of the ATP to the glucose, activating the glucose and allowing it to react with another glucose unit and become joined to it. By repeating this process a number of times, about 1000 glucose units are bound together to form a starch unit (molecule). (See Fig. 1-15.)

All chemical reactions within cells,

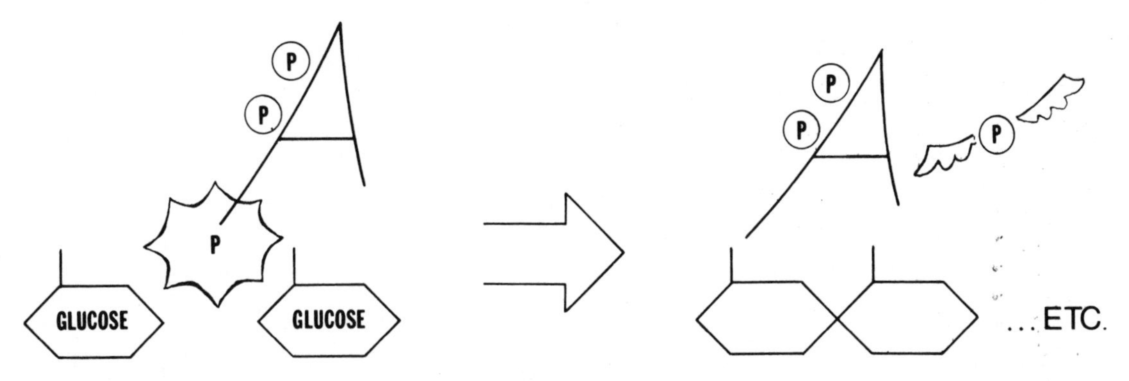

Fig. 1-15. An adenosine triphosphate molecule (ATP) consists of a complex unit of adenosine with three units of phosphate attached to it, two of them by "high-energy" bonds. When the bond of the third phosphate is broken, its "high energy" is released to do work such as bonding new molecules together. The remaining molecule with its two phosphates is called adenosine diphosphate (ADP) and may regain a third phosphate by other energy-acquiring reactions (see Fig. 1-18).

In this illustration the energy of the broken third phosphate is used to bond together two units of glucose as the first step in producing a starch molecule. Many more glucose units must be added, requiring more ATP energy, before the starch molecule is completed.

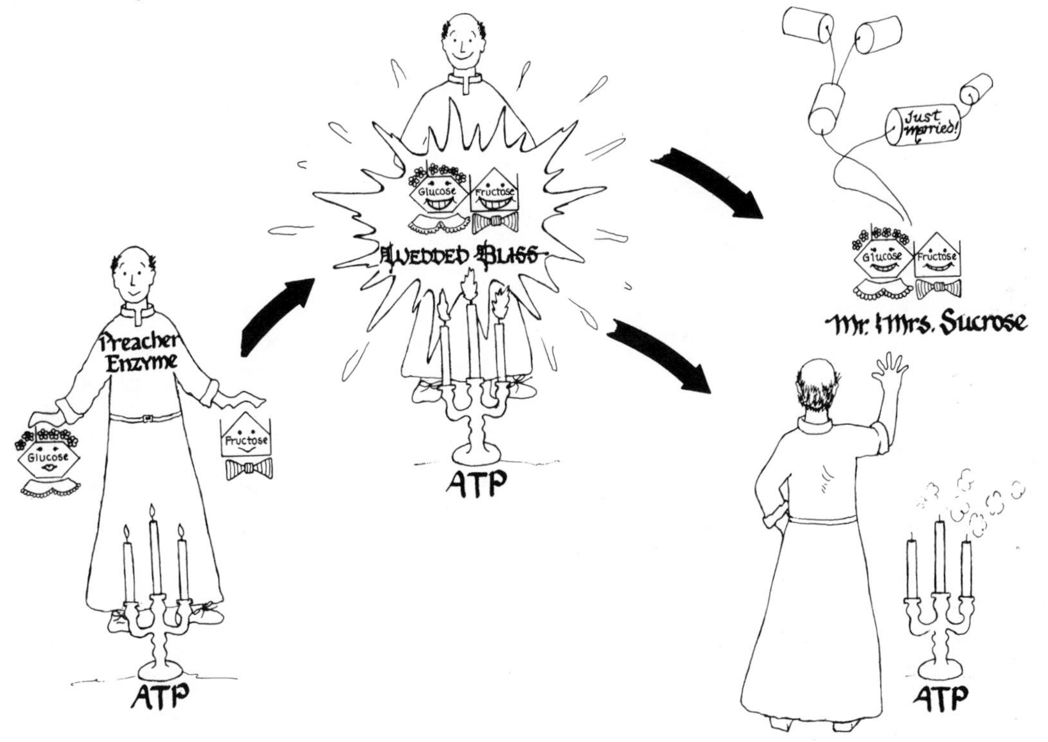

Fig. 1-16. Enzymes are special molecules that *regulate* the chemical work of a cell. Here a new molecule of sucrose will be produced from units of glucose and fructose utilizing the energy of ATP. The reaction is catalyzed (accelerated) by the activities of particular enzymes.

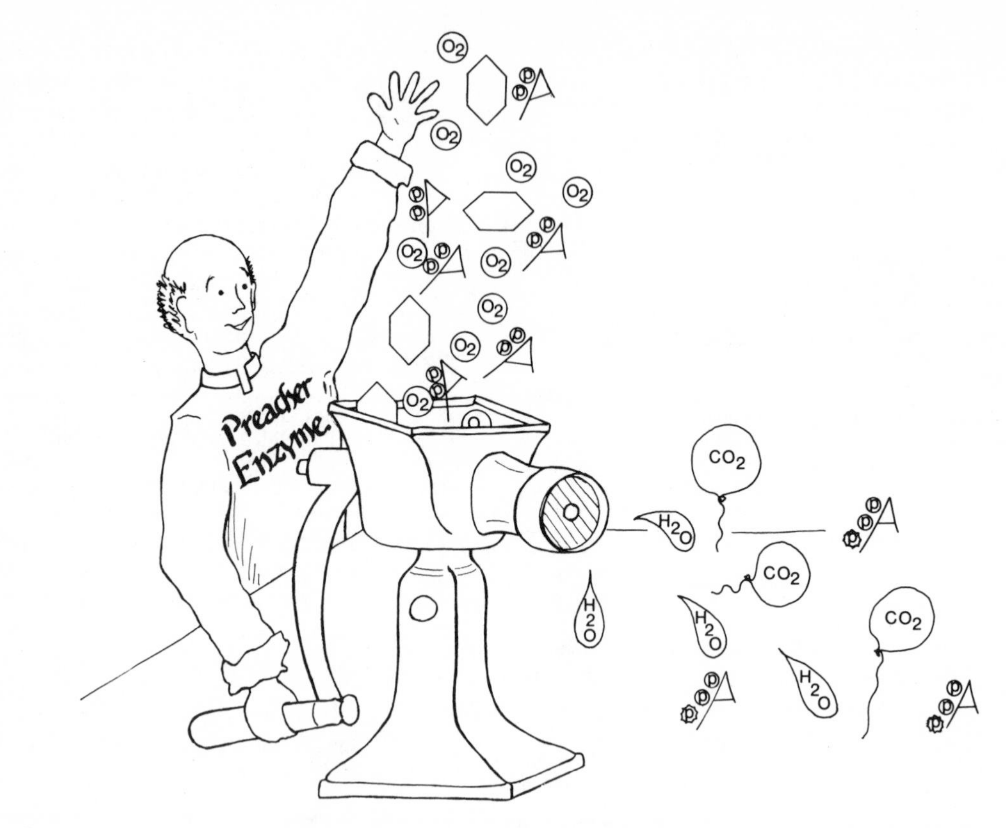

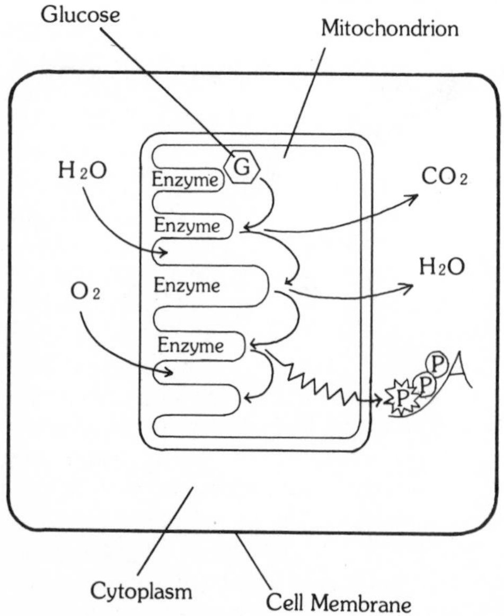

Fig. 1-17. Two diagrammatic approaches to cellular respiration. Cellular respiration is a series of enzyme-regulating chemical reactions that release energy stored in molecules such as glucose and transfer it as active energy in molecules of ATP. Glucose, oxygen, ADP, and special enzymes are involved in a complex series of reactions that result in many new units of energy in the form of ATP molecules. Carbon dioxide and water are released as by-products.

whether energy-using or -releasing, are carefully regulated by special chemicals called **enzymes**. In a reaction the enzyme controls the rate of the reaction by becoming temporarily involved with the reactants, but once the reaction is completed, the enzyme disengages, becomes free, and is no longer involved. The involvement is somewhat like that of a minister in a wedding joining the bride and groom together and then removing himself from the process. (See Fig. 1-16.)

Cellular Respiration

When chemical energy is needed in a cell, it is obtained by breaking up glucose molecules into simpler forms in a series of precise, enzyme-regulated reactions. At many steps in these reactions, stored energy is released and transferred to ATP molecules. In the degradation of glucose, carbon dioxide and water are released. (See Fig. 1-17). For every one molecule of glucose broken up, six molecules of oxygen are used, and six molecules of carbon dioxide, six molecules of water, and 38 new molecules of ATP are produced. Chemically stated this reaction is $C_6H_{12}O_6 + 6O_2 \rightarrow 6CO_2 + 6H_2O + 38$ ATP. Not all of the energy released is captured by ATP; about 33% is released as heat.

In some cells, especially under certain stressful situations when oxygen is not available for the degradation of glucose, an alternate pathway is used so that the cell may still obtain energy. Instead of being converted into water for which the oxygen is necessary and carbon dioxide, the glucose molecule is degraded into ethyl alcohol and carbon dioxide. This process results in the release of only a little energy, 2 new molecules of ATP, which is still enough to keep the cell alive.

These processes described above are called **cellular respiration**. The process that uses oxygen is called **aerobic respiration**. The process occurring in the absence of oxygen is called **anaerobic respiration** or **fermentation**. One or the other of these processes, depending upon the availability of oxygen, occurs in *all* living cells *all* of the time.

Anaerobic respiration, as we have seen, is not nearly as effective as aerobic respiration in making energy available, but it is better than nothing. Many cells are capable of utilizing the anaerobic pathway at least temporarily. Root cells, for instance, will respire anaerobically when the soil is waterlogged after a flood, or when a potted plant with inadequate drainage is overwatered. However, if these conditions last too long, the alcohol concentration will become high enough to kill the cells. Other organisms, such as yeasts (a type of fungus) and bacteria, are better able to utilize this process for longer periods of time. The ability of yeasts to respire in the absence of oxygen and produce ethyl alcohol and carbon dioxide forms the biological basis for the entire alcoholic-beverage and bread-baking industries. The former capitalizes on the alcohol, the latter on the carbon dioxide which causes the dough to rise.

Photosynthesis

If glucose is the food or fuel that cells (and thus plants) use to obtain chemical energy for their various life processes, then a logical question is—where do plant cells get glucose? The answer is surprisingly simple, but the process amazingly complex—they make it! Green plants, of all the organisms on earth, have the unique ability to capture radiant energy from light and use it to make glucose from ordinary carbon dioxide and water. They key to this process is chlorophyll, a group of green pigments usually contained in chloroplasts within certain cells.Chlorophyll is capable of absorbing radiant energy from light, using this energy to split water molecules into hydrogen and oxygen, thereby releasing chemical energy which is stored in ATP molecules. (See Fig. 1-18). The chemical energy in the ATP molecules is then used along with

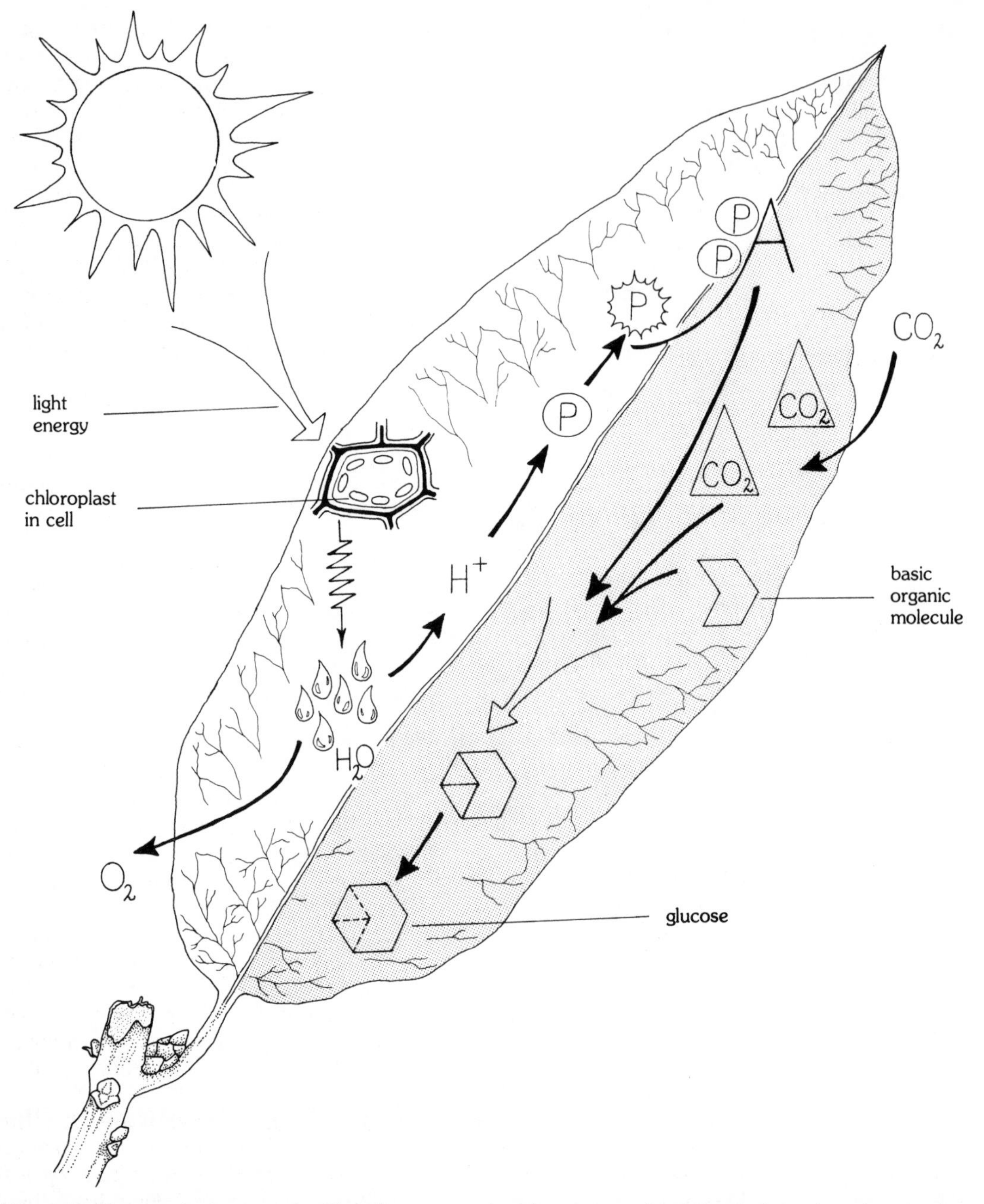

Fig. 1-18. Photosynthesis occurs in chloroplasts as a series of enzyme-regulated chemical reactions that transform radiant energy (sunlight) into chemical energy which is stored in glucose molecules. The process begins when chlorophyll absorbs the radiant energy and utilizes it to split water molecules, which releases chemical energy to molecules of ATP. ATP energy then provides the energy to incorporate carbon dioxide into glucose molecules.

enzymes to incorporate carbon dioxide into glucose.

This glucose may then become involved in a number of pathways of chemical transformations, each regulated by specific enzymes. The glucose may be used in cellular respiration to provide more chemical energy, or it may be converted into other substances such as sucrose (table sugar), starch or cellulose, or amino acids and protein, or fats, or other organic compounds that make up the protoplasm of the cell. Thus it should be evident that photosynthesis initially accounts for the formation of all the organic compounds that make up the cell and thus the whole organism and is really the source of all food and the substance of animals as well as plants. In addition, the oxygen released in photosynthesis (when water is split into hydrogen and oxygen) accounts for the oxygen present in the atmosphere. Indeed, we should all follow the advice of that old bumper sticker that says "Have You Thanked A Green Plant Today?".

PLANTS AND WATER

Diffusion and Osmosis in Plants

Earlier in this chapter, in our discussion of cells, we stated that vacuoles are sacs containing mostly water with dissolved substances. The reason water enters or leaves a cell vacuole is based upon the physical law of **diffusion** which states that molecules of a substance will move *from* a place where molecules of that substance are very concentrated *to* a place where they are less concentrated. This physical law applies to all substances, not just water. In the case of water moving into a cell vacuole, an extra feature is added—the water must move through a membrane that has pores of such size and configuration that not all kinds of molecules can get through. Therefore the membrane is **selectively permeable**. The diffusion of water through a selectively permeable membrane is called **osmosis**.

When a lot of a substance such as sugar is dissolved in the water of a cell vacuole, then the concentration of water in the vacuole will be low, and water will tend to move into the cell vacuole from the surrounding water-abundant areas at a greater rate than the water molecules will move out of the vacuole. In this case the cell and vacuole membranes function as selectively permeable membranes because they do allow the passage of water molecules into and out of the vacuole but they do not allow the sugar molecules to leave the vacuole. (See Fig. 1-19.)

In the hairs of the root epidermal cells, the concentration of water molecules in the vacuoles is usually less than outside in the soil; thus water will move into the root cells. From the epidermis the water then moves through the cortex into the stele and into the xylem. The water moves upward through the pipe-like xylem into the stems and leaves and is drawn off by osmosis into cells along the way that need more water.

It is interesting to note that of the water entering the plant through the roots, only about one percent is retained by the plant. The rest of the water evaporates from the surface of the shoots, mostly through the stomata. This process, the evaporation of water into the air through the stomata of the plant, is called **transpiration**. Thus the plant functions pretty effectively as a faucet pouring water from the tank in the soil into the sink in the atmosphere. It has been estimated that an acre (209 ft. x 209 ft., or 61 m. x 61 m.) of corn plants during one growing season would transpire enough water to cover that acre with about 11 in. (26 cm.) of water.

How Does Water Rise In Plants?

This brings us to an interesting question—how does water rise, against the pull of gravity, up the xylem from the roots to the uppermost leaf, especially in a tree such as a coastal redwood in California that is almost 300 ft. (over 90 m.) tall? Physiolo-

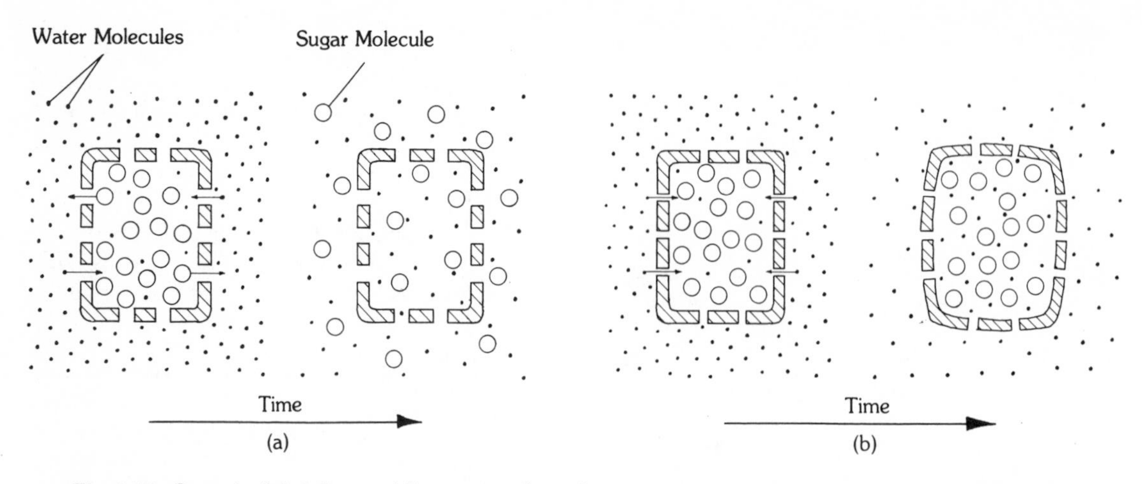

Fig. 1-19. Osmosis. (left 2 diagrams) If pores in cell membrane were large enough, water molecules (dots) and sugar molecules (circles) could diffuse freely through them. (right diagrams) In the actual situation, water molecules diffuse freely through the minute pores in the cell membrane, but sugar molecules are too large to pass. Where sugar is more abundant, water is less dense. Thus water enters the cell, increasing its internal pressure, since the sugar solution in the cell can never be diluted to match the concentration outside.

gists tell us that the water may actually move up the xylem at a rate of 30 in. (75 cm.) per minute. That speed would never be possible by diffusion or osmosis.

Several explanations have been suggested, such as the force of normal atmospheric pressure, capillary action, and root pressure (an energy-expending process in the roots that would force water up) but none seems to explain the process as well as the **cohesion-tension theory**. This theory is based upon certain physical laws and characteristics of water; (1) water molecules in small closed tubes have a strong attraction to each other (cohesion)—they tend to stick together. (2) Water molecules also have a tendency to cling to other molecules such as the sides of the tubes or cellulose (adhesion). (3) Generally the cohesive power of water is greater than its adhesive power. According to the cohesion-tension theory, then, the water molecules in the tiny tubes formed by the tracheids and vessel elements of the xylem have a strong tendency to cohere very tightly and also to adhere to the cellulose in the walls of the xylem cells, thus forming columns of water extending from the roots up into the leaves. The water columns are held up by the adhering of the water molecules at the tops of the columns to the cellulose in the cell walls of leaf cells. When molecules of water transpire into the air through the stomata of the leaves, water molecules in the cell walls of leaf cells move into the air spaces inside the leaves near the stomata to replace those molecules that passed through the stomata, and other molecules of water move into the walls from adjacent cells. This process is repeated until water molecules are pulled out of the xylem water column. This creates a tension or negative force (more precisely, a water deficit) that is actually strong enough to pull water molecules up the water column itself all the way from the root and even from the soil outside of the root. Thus when transpiration occurs a continuous stream of water is pulled from the soil into the roots, up in the xylem, and into the leaves. Experiments have demonstrated that cohesive and adhesive forces of water within the plant system are strong enough to make this soil-to-sky plant-pump work. This method is efficient in that it requires little or no energy expenditure by the plant. Minerals and other substances present in the xylem are passively carried up from the roots into stems and leaves by this transpirational pull of the water column in the xylem. (Xylem sap contains, in addition to water, about 0.1 percent solids comprising mostly minerals and occasionally some sugars and amino acids.)

In dry periods, when no further external source of water is available to the plant, transpiration still continues, and water present within the vacuoles of the cells is drawn out, reducing turgor pressure within the cells and causing the plant to wilt. If this drought is temporary, then when external water is again available, the cells' turgor will be restored. If the drought is of long duration, then the plant may lose too much water from the protoplasts, and permanent wilt will cause the plant to die.

How Is Food Translocated In Plants?

Phloem sap contains 10-30% sucrose plus lesser amounts of other sugars, amino acids, and miscellaneous solutes. Sucrose is thus the most abundant material carried as food in the phloem. The translocation (movement) of food from the leaves to other parts of the plant through the phloem is quite a different process from the translocation of water. No truly satisfactory explanation for this process has yet been offered, but the most accepted explanation to date is **mass flow**. Experiments have shown that the sugar solutions flow through the phloem under pressure and therefore the sugar particles move more rapidly than is possible by diffusion. Other experiments have shown that in leaves the concentration of sugars (especially sucrose) is much greater in the sieve-tube elements near the cells where photosynthesis is occurring than in the photosynthesizing cells themselves. This suggests that the photosynthesizing cells actually use energy (ATP) to transfer sugars into the phloem. This loading of sugars into these sieve-tube elements creates a solution of high sugar concentration, which causes a high rate of water absorption into these cells by osmosis. If this process of phloem loading continues, then the constant osmosis of water into the sieve-tube elements will create a pressure (turgor or osmotic pressure) within the sieve-tube elements and will cause the water to flow through them (through the pores in the sieve plates) toward regions of the plant where the turgor pressure within the phloem is lower. The sugars and any other substances in the sieve tubes are passively carried along by the water. In regions of the plant (such as roots, shoot tips, or fruits) where no photosynthesis is occurring and sugars are needed for growth or storage, sugars move from the phloem to other cells by diffusion. This removal of sugars lowers the concentration of dissolved substances in the sieve-tube solution, causing some of the water to osmose out into adjacent cells, thus reducing the turgor pressure in the sieve tubes. These processes create a continual flow of sugar solution in phloem from the sites of sugar manufacture to all other regions of the plant where sugar is needed for growth. The flow will continue as long as the sugar is pumped into the phloem at the manufacturing site (normally the leaves).

While this simple explanation basically accounts for the general translocation of phloem sap, there are still many mysteries and differences of opinion regarding how food is translocated within the plant.

PLANTS AND SOIL MINERALS

Minerals found in the soil and absorbed into plants through the roots are essential to plants. While we have shown that the real foods of plants are organic substances (essentially sugars) made by them and used as fuel to release energy, the "plant food" we buy in a store is not a bottle of sugar solution but of plant fertilizer, a mixture of inorganic elements which are not fuels. (These inorganic plant foods are the same regardless of the source, whether a garden compost or manufactured fertilizer, and they are the raw materials that, along with carbon dioxide, are needed for proper plant growth.)

The elements essential for plant growth are carbon (C), hydrogen (H), oxygen (O), phosphorus (P), potassium (K), nitrogen (N), sulfur (S), calcium (Ca), magnesium (Mg), iron (Fe), copper (Cu), boron (B), chlorine (Cl), manganese (Mn), zinc (Zn), and molybdenum (Mo). (The following mnemonic device is often used to recall most of the essential plant elements: C HOPK'NS CaFe, Mg. Cl Mn. Translated, this means: "C. Hopk'ns Cafe, Mighty Good! Closed Mondays.") A few additional elements have been found in plants but are not known to be essential. If the elements are present in the soil, they will be absorbed whether needed or not.

The roles of these nutrients vary considerably, from providing the building blocks or components of the various organic compounds that make up the plant to regulating certain processes. Still others play necessary roles that are not well understood. It is interesting to note that the plant is unable to absorb and use most of those elements in their pure or elemental form, such as carbon, potassium, or nitrogen, but must obtain them in the salt form such as sodium carbonate, ammonium nitrate, potassium sulfate, and so forth. You will find further discussion on the specific actions of these salts in Chapter 2.

REGULATION OF GROWTH AND DEVELOPMENT

Hormones are substances that are produced in one part or organ of the plant and are then usually transported to another location (perhaps only a few cells away, perhaps farther), where they have a powerful effect. They control most of the major processes in plants. Hormones are active in very small amounts, and although there is still a great deal we don't know, they are thought to act by controlling the rate of production of particular enzymes. We shall discuss five groups of plant hormones: gibberellins, auxins, cytokinins, ethylene, and abscisic acid.

Gibberellins. This is a group of about 30 to 40 naturally occurring growth regulators of which the best known is **gibberellic acid** (GA), first discovered in a fungus that infected rice seedlings and caused them to grow taller than non-infected ones. Gibberellins are normally produced in shoot apices, roots, and seeds. They are apparently involved in a number of plant processes, but most notably they seem to play a role in:

(1) Regulation of cell division and elongation of stem tissue.

(2) Promotion of flowering in certain plants, particularly those that flower during long days (see Chap. 6, Plants and and Light) and those that require two seasons of growth before flowering will occur.

(3) Activation of starch-digesting enzymes in germinating seeds.

Auxins. These are the most-studied and probably the best-understood plant hormones. **Indole-acetic acid** (IAA) is apparently the only auxin naturally produced in plants. It is manufactured in the shoot apex of young leaves of most plants as well as by embryos in seeds. Under normal conditions the IAA produced in the shoot apex moves more or less uniformly down the stem and regulates many growth processes including:

(1) Cell enlargement. Regulation of the enlargement of individual growing cells is probably the most significant effect of IAA.

(2) Suppression of growth of lateral buds (**apical dominance**). IAA produced by the main shoot apex suppresses

growth of lateral buds for a certain distance below it. This effect insures that the main shoot will grow most. (See Chap. 10 for practical application.)

3) Directional growth of the plant toward a light source (**phototropism**). If light strikes a plant on only one side, the IAA flow on the light side of the stem is greatly reduced, and apparently the IAA moves to the shaded side of the stem. There, in greater abundance, it stimulates the cells on the shaded side to grow more rapidly than those cells on the sunny side, causing the plant to grow toward the light. If the plant is turned around, then the IAA migrates to the newly shaded side, and cells on that side will grow more rapidly, causing the other side of the plant to grow toward the light (see Fig. 6-7).

(4) Response to gravity (**geotropism**). If a normally upright-growing plant falls over into a horizontal position, then by gravitational forces the IAA accumulates on the lower side of the stem and stimulates the cells on that side of the stem to grow faster than the other side, causing the stem to begin to grow upright.

(5) Suppression of the mechanisms that cause leaves and fruits to fall off the plant. Auxin inhibits the formation of the special zone of cells (**abscission layer**) responsible. This quality of auxin has been exploited by orchardists. By spraying the plants with an auxin solution they can keep fruits on the trees for a longer time so that they may ripen more.

(6) Development of roots from stem or leaf tissue. Such roots are called **adventitious roots**. Their formation is stimulated by auxin. This effect is useful in vegetative propagation of stem and leaf cuttings (see Chapter 4).

(7) Fruit formation. While this appears to be a rather complicated process, auxins do seem to play a major role. This may be demonstrated by cutting off the tip of the pistil of a flower (the stigma, see p. 28) before it has been pollinated and applying IAA to the cut-off end. The auxin will stimulate the pistil to develop into a fruit in which there will be no seeds.

A number of synthetic compounds produce effects similar to those of IAA. Two of these synthetic auxins are dichloro- and trichlorophenoxyacetic acid (2,4-D and 2,4,5-T). Because they stimulate rapid cell elongation, application will cause some plants to grow too rapidly and erratically and the plants will actually die. These auxins are often applied to lawns and roadsides to kill broad-leaved weeds but not the grass, because their effect is apparently directly proportional to the width of the leaves and narrow-leaved grasses show little adverse effect. They do, however, have deadly effects on garden plants as well as weeds and should be used cautiously.

Cytokinins. These hormones are of most importance in stimulating cell division in all parts of the plant. They are also very important as co-workers with auxin and gibberellins to help regulate the processes discussed under each of those groups of hormones. Thus cytokinins acting with auxin regulate cell division and elongation, abcission of leaves and fruits, fruit development, and so forth.

Ethylene. This hormone is fascinating because it seems, like cytokinins, to interact very closely with IAA in certain processes. The most noticeable effect of ethylene is causing fruit to ripen. It also plays vital roles in other processes such as inducing flowering, hastening senescence (old age), and causing abscission of leaves and fruits.

Abscisic Acid. Whereas we tend to think of the regulatory aspects of most hormones as promoting or "encouraging" growth, abscisic acid seems primarily to inhibit the activities of gibberellins and some other growth processes. It regulates the preparation of many plants for dormancy, causing the terminal bud to develop thick protective bud scales and promoting the development of the abscission zone (which other hormones had inhibited) so that leaves will fall off. As the growing season progresses, abscisic acid produced in leaves increases in concentration until it is present in sufficient strength to countermand the effects of the growth-promoting hormones. Abscisic acid present in seeds tends to keep them from germinating until it is removed or deactivated by cold, soaking in water, or some other process. It also causes stomatal closing during periods of stress. It has been suggested therefore that abscisic acid is a plant "tranquilizer" since it tends to countermand regulators that promote growth.

Thus from this brief treatment of growth regulators we can see that growth in plants is not a simple matter but involves a complete interaction of many factors. I never cease to marvel at all that must occur when we plant a seed in the soil and supply water and proper temperatures. The growth that results is truly an amazing and marvelous phenomenon.

AND LIFE GOES ON

Thus far in this chapter we have examined the organization of the plant body from cell to roots and shoots, and we have considered some of the physiological processes that make plants "go". Now let's turn our attention to another major aspect of plant life—reproduction, the perpetuation of the species in time and extension in space. This involves two distinct phases: (1) formation of the reproductive unit by the parent plant and separation of the unit from the parent and (2) development of new individual from the reproductive unit.

There are essentially two kinds of reproduction in flowering plants: (1) **vegetative propagation** (often called asexual reproduction)—the development of a vegetative **propagule** (a segment of root, stem, or leaf or whole shoot) from the parent plant that grows into a new independent plant; and (2) **sexual reproduction**—the formation of special sex cells or **gametes** (**sperm**, the male gamete; **egg**, the female gamete) that fuse together (egg and sperm) to form a single-celled **zygote** (fertilized egg). The zygote develops into a multicellular **embryo** which eventually develops into a new plant. In flowering plants, sexual reproduction occurs in flowers, and embryos develop in seeds. The basic difference between the two processes is that vegetative propagation produces new plants that are exact replicas (contain the same identical inheritable traits) of the parent plant, since the propagule is a portion of the parent plant, whereas sexual reproduction involves the fusion of two separate gametes each usually from a different parent plant.

In the formation of the gametes a special kind of cell division (**meiosis**) occurs that sorts out the inheritable traits in different combinations so that when two gametes unite no resulting offspring is exactly like another or like either of its parents but combines traits or characteristics of both. This constant introduction of variability into a population of a species through sexual recombination is an important survival factor because it means that some offspring may be better adapted to the existing environment or to a changing or new environment than the parents were. Thus sexual reproduction provides greater selective advantage in a population for the survival of the species. This can best be understood when viewed over a long period of time (geologic time) wherein plants have to adapt to changes in climate, compete with other plants and animals, or develop resistance to diseases and predators. For example, on the Galapagos Islands the species of prickly pear cacti growing on the islands where there are no tortoises to eat them are low-growing and bear succulent, flattened stems and flowers close to the ground, whereas on islands where the tortoises occur, the flattened fleshy stem joints and flowers of the prickly pear are produced on tall woody stems well above the reach of the tortoises. The latter species of cactus has had to adapt by sexual recombination and through time to contend with the tortoises.

Vegetative propagation provides the advantage of rapid proliferation of identical plants since all of the leaves, stems, and roots of an individual plant have exactly the same inheritable traits. This means of reproduction may provide competitive advantage by allowing a particular type of plant to outcompete other plants in a given habitat and possibly take over, as, for example, quackgrass takes over a lawn through its extensive rhizome system (as old rhizomes die, the once-attached plants become separate individual plants).

Since vegetatively propagated plants are exact replicas of their "parent" (clones), they possess not only the desirable traits but also the undesirable ones. Over a long period of time these undesirable traits may produce a loss of vigor and a possible decline in the desirability of the strain or variety. Vegetative propagation also means that the offspring have no built-in mechanism for selective adaptation. For instance, if a vegetatively propagated strain became susceptible to a disease or predator it would lack the ability to develop resistance since its inheritable traits can only be changed through sexual recombination. Fortunately most plants reproduce sexually as well as through vegetative propagation.

In horticulture and agriculture, vegetative propagation is tremendously useful because it means that a type of plant with desirable traits (created through sexual reproduction) may be rapidly and exactly propagated. For instance, many strawberry plants of a particularly good strain or variety may be obtained from plantlets formed by runners, or an unusually colored African violet may be propagated by leaf cutting, whereas attempt to produce numerous duplicates of these plants by sexual reproduction would only increase the variation. Plants which are sexually sterile, such as seedless oranges, are dependent upon vegetative propagation for perpetuation.

FLOWERS AND THEIR ROLE IN REPRODUCTION

In flowering plants, eggs and sperms are produced on very specialized leaves in greatly modified branches of the shoot called **flowers**. Many kinds of flowers exist, but there is still a basic similarity in the organization of most flowers. Let us look at a stylized flower as shown in Fig. 1-20 that illustrates the basic flower parts.

Flowers are borne on plants in very characteristic ways. Some plants produce only a solitary flower, like the tulip, while others produce several or many flowers clustered in various designs, like lilies, snapdragons, or Queen Anne's lace. The aggregations or clusters of flowers are referred to as the **inflorescences**. Individual flowers consist of a stem portion, the **receptacle**, perched atop a stalk, the **peduncle** or **pedicel**. The term peduncle is used for the stalk of a solitary flower in plants producing only solitary flowers, or for the stalk of an inflorescence in plants producing flowers in clusters. The term pedicel is used for the stalk of an individual flower within an inflorescence. The receptacle, like all stems, consists of nodes and internodes, but the internodes are very short so that the nodes are very close together. Arising from the nodes of the receptacle are usually four series of whorls of appendages which are modified leaves. (1) The **sepals**, the lowest whorl of leaf-appendages, are usually green and provide protection for the flower bud. All of the sepals of one flower are collectively called the **calyx**. (2) The **petals**, the whorl of flower parts above the sepals, are usually larger and brightly colored. The petals often function to attract pollinators by their bright colors or particular shape and also to protect the flower in bud. All of

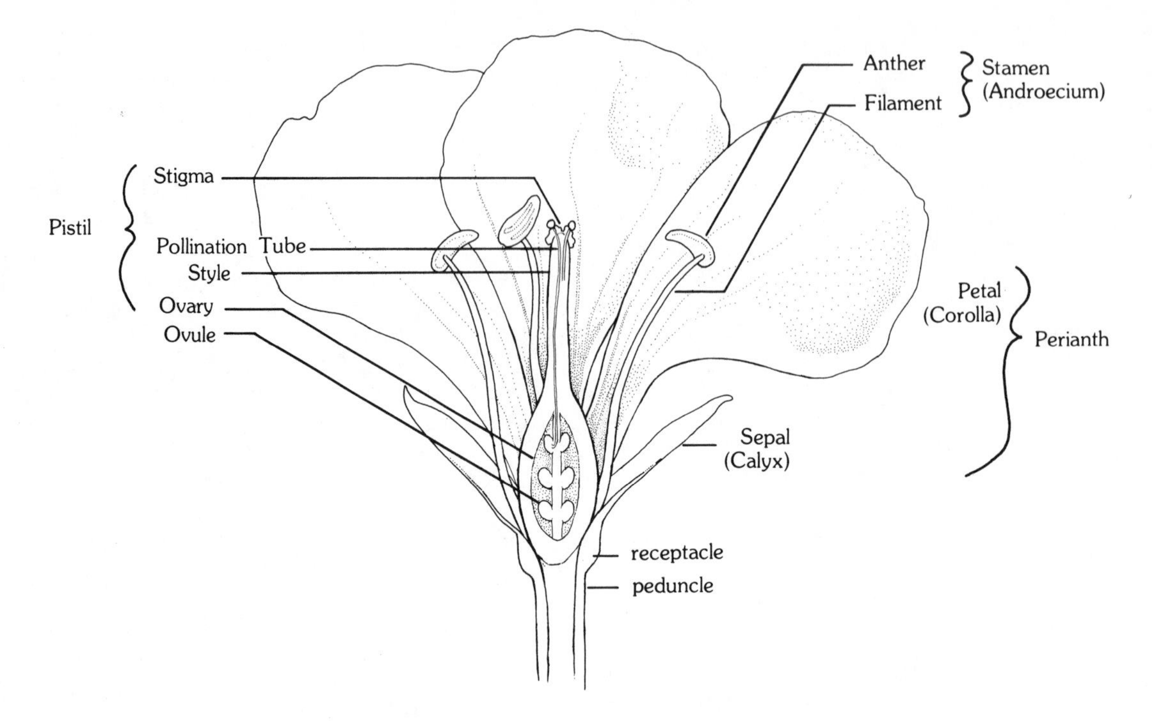

Fig. 1-20. Structure of an idealized flower.

the petals of one flower are collectively called the **corolla**. The term **perianth** is used to refer to both the calyx and corolla together. (3) **Stamens** are the flower parts just above the petals. These rarely look like leaves since they consist of only a stalk (**filament**) and pollen sacs (**anther**). Within the anthers, small, usually spherical, structures called **pollen grains** are produced. These pollen grains produce sperm cells, and thus stamens are the male sex organs of the flower. All of the stamens of a flower (which, depending upon the kind of flower, may be as many as hundreds or as few as one) are collectively called the **androecium**. (4) The **carpel** is the flower part borne at the nodes above the stamens, in the center of the flower. Some flowers have only one carpel; others may have several. When several carpels are present, they may be separate or united. Carpels are also very un-leaflike in appearance and are usually bottle-shaped. The carpel consists of an enlarged, hollow base (**ovary**), a narrow, often long, neck (**style**), and a swollen, sticky top part (**stigma**). Inside the ovary are one or more special structures called **ovules** that produce eggs (one egg per ovule). Thus carpels are the female sex organs of the flower. All of the carpels of one flower are called the **gynoecium**. Traditionally the term **pistil** has been used for the gynoecium (pistil refers to the "bottle" shape and is related to the word pestle, a tool of the same shape used by pharmacists to powder substances in a bowl called a mortar). If the gynoecium consists of one or several separate carpels, each is called a pistil; if the gynoecium consists of several united carpels, this compound structure is also called a pistil. Thus "pistil" is a confusing term.

When a flower is mature, the pollen grains are released from the anthers and carried to the stigmas of the carpels. This transfer of pollen is called **pollination**. Pollination can be accomplished by wind, water, or animals, and many elaborate devices have developed in order to accomplish it (see p. 32). Special devices have also developed to increase the chances that the pollen deposited on a stigma be from a flower on a different plant (**cross-pollination** or **out-crossing**) rather than from the same plant (**self-pollination**). Out-crossing allows the combination of inheritable traits from two different individuals and thus provides greater variation and therefore greater selective advantage to the offspring.

Once the pollen grain lands on the stigma, it begins to grow, forming a slender tube that grows down through the style and ovary until it reaches the ovule. Each pollen grain contains two sperm nuclei. The sperm nuclei are carried down the pollen tube and released within the ovule, where one unites with the egg nucleus (**fertilization**) to form the **zygote**. The second sperm nucleus moves into the ovule and unites with another nucleus, or usually two nuclei, in the ovule to form an **endosperm cell** which divides and proliferates forming a mass of cells (**endosperm**) that provides food for the zygote as the latter develops into a new potential plant, the **embryo**.

The embryo is formed by repeated cell divisions of the zygote and differentiation of the cells into distinct parts: **radicle**, **hypocotyl**, **cotyledons** (either one in monocots or two in dicots) and **epicotyl** (see Fig. 1-21). As the embryo develops within the ovule, the outer walls of the ovule become very hard and tough to protect the embryo inside. The ovule, with hardened walls and embryo inside, is called a **seed**. Most seeds are adapted to remain alive over relatively long periods of time in a dormant state until the right conditions are available for growth. However, some seeds must begin growing almost immediately after formation or they die.

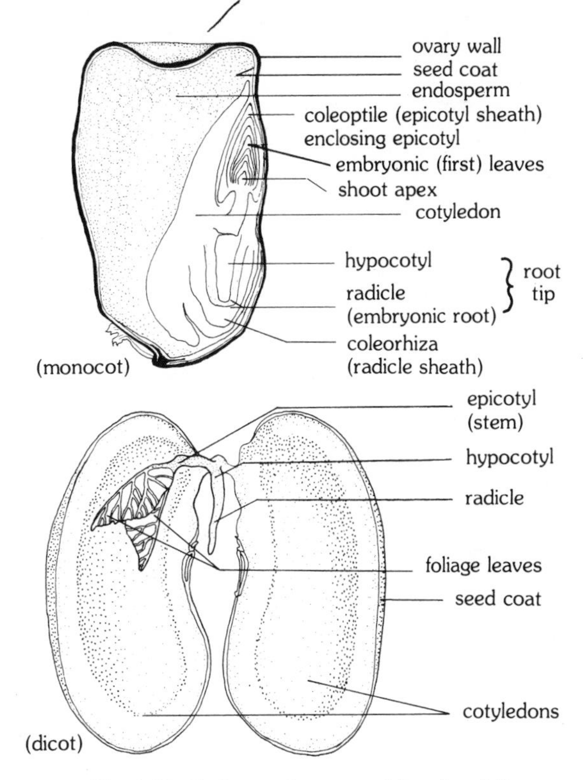

Fig. 1-21. Embryos of monocot (above) and dicot (below).

Under proper conditions of temperature, water, and other factors the embryo may grow into a new plant. The radicle, which contains the root apical meristem, develops into the root system; the hypocotyl elongates and forms the lowest portion of the stem; the cotyledons (or seed leaves) contain stored food and often function as the first leaves; the epicotyl contains the shoot apical meristem and develops into the shoot system. (Fig. 1-22).

Let us return briefly to the flower. While the zygote is developing into the embryo and the outer portion of the ovule into a seed coat, other changes occur in the flower. The sepals, petals, and stamens have played their roles and usually wither and fall off. The carpels, on the other hand,

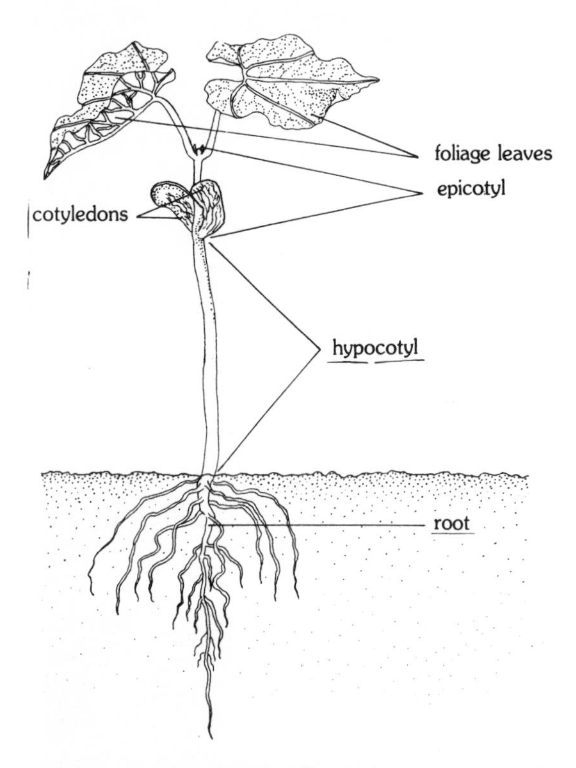

Fig. 1-22. Seedling of dicot.

since they contain the ovules, remain on the flower. The ovary changes into a structure called a **fruit**, a mature, ripened ovary usually containing seeds. Fruits may be soft and fleshy like a peach or watermelon, or dry and hard like a hickory nut or corn grain, or somewhere in between like a bean pod.

Although we all enjoy a nice juicy apple, peach, orange, or tomato, we might wonder why a plant would expend so much energy to produce such a structure. Fruits, as varied as they are, play a very important role in the reproduction and dispersal of plants. The fruit is essentially the structure that protects the seeds as they develop, and it also functions to get the seeds away from the parent plant so that they may find a suitable place to grow. For most plants, which are rooted and therefore stationary throughout their whole life, the seed stage provides the only opportunity for an individual to establish its offspring somewhere else, perhaps in a location better suited for the growth of the species because, as a plant grows, it will alter the environmental conditions around it (shade, moisture, soil nutrients, and so forth). The fruit promotes the dispersal of the seeds.

Fruits are very diverse in structure and in their mechanisms for disseminating seeds. Large, fleshy, and brightly colored fruits contain copious amounts of sugary food that animals such as birds or squirrels and other mammals may eat, carrying the seeds away to a place where they might grow. Other fruits are less fleshy but break open when mature and scatter the seeds into the wind or onto the ground (beans, milkweed, mustards). Still others are hard and strong and do not break open (nut). Some even float in water (coconut), which may take them to new fertile surroundings. Others have spines or hooks that may become caught and transported in the fur or feathers of animals (tick trefoil, *Cosmos*, cockle bur). These are but a few examples of the adaptations in structure of fruits to insure that the seeds may find suitable surroundings for new growth.

Another way of considering fruits is to examine how they are produced in a flower. If the flower has only one ovary, then the fruit is a **simple fruit**. Most of our common fruits, including those mentioned above, are simple fruits. If the flower has several separate ovaries, then each develops into a fruit, producing an **aggregate fruit**, with many separate fruits on a common receptacle. Raspberries and blackberries, aggregates of small fleshy drupelets, are good examples. In the strawberry the receptacle become the bright-red, fleshy edible structure and the small "seeds" are actually individual one-seeded fruits called **achenes** (one carpel, one ovule, fruit not opening at maturity.). In some kinds of plants many

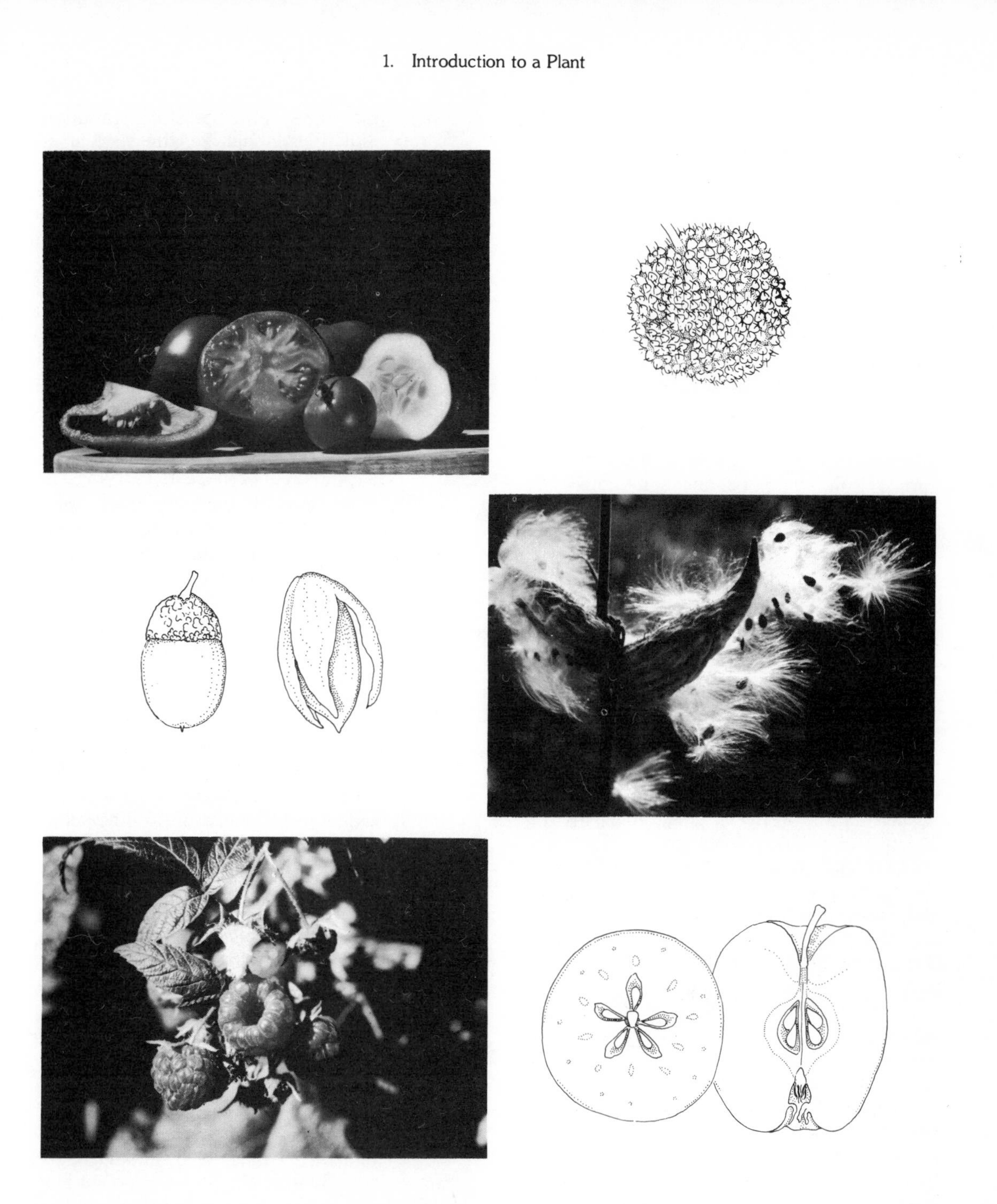

Fig. 1-23. The fleshy kinds of angiosperm fruits protect the seeds from injury and drying with firm, moist fruit tissues.

one-ovaried flowers may be clustered together into a characteristic mass inflorescence that, when "in fruit", functions as a single unit. In an ear of corn, for instance, each corn grain is actually a distinct fruit from a separate flower, but the flowers are all borne close together in a single unit. The same is true for sunflower heads, mulberry fruits, and pineapples. These are known as **multiple fruits**. The advantage of aggregate and multiple fruits is that, even though each ovary or flower produces one seed, the many seeds clustered together will be dispersed at the same time, and these types accomplish the same effect as the tomato, a simple fruit which produces many seeds. (See Fig. 1-23.)

Pollination

Let us consider a little further this business of pollination, which was defined as the transfer of pollen from an anther to a stigma (p. 28). Pollination is a necessary precursor to fertilization. At first glance this transfer would not appear to be a major problem, because, if we look at most flowers, we see that both stamens and carpels are present in the same flower and therefore it seems just a matter of the pollen grains getting across a small space to the carpels.

However, if self-pollination did occur all of the time then the advantages of sexual reproduction (recombination of inheritable characteristics between different individuals) would be defeated. Therefore most species of plants are adapted for cross-pollination. In many species internal mechanisms have developed that prevent pollen from growing on the stigma of the same flower (see discussion below of incompatibility and other methods to ensure cross-pollination). On the other hand, some plants, such as garden peas and violets, although designed for out-crossing, are also self-compatible as a back-up measure to ensure seed production. In many species of violets the "typical", familiar violet flowers are adapted for cross-pollination, but later in the season small, green, unopening flowers that are self-pollinated are produced near or just under the surface of the ground. The violet thus ensures its perpetuation both ways.

Many plants have developed special modifications or adaptations which encourage cross-pollination. For instance:

(1) The anthers in a flower may mature and release pollen as early as three or four days before the stigma of that flower is ready to receive the pollen. Since the flowering period for the whole population, composed of many plants, not all of which would bloom at exactly the same times, would probably last at least a week or more, most pollen would be transferred from one flower to another that was receptive.

(2) The flowers of some species may be **self-incompatible**. That is, internal physiological barriers inhibit the pollen from growing a pollen tube through the stigma and style of the same flower (or plant), thereby preventing self-fertilization of the eggs.

(3) Flowers may be **unisexual**, that is produce only stamens or carpels. This arrangement of course requires cross-pollination. If flowers of both sexes are produced on the same plant (for example, corn, in which the tassel is the male inflorescence and the ear is the female inflorescence), then the plant is called **monoecious** (one house). If the male and female flowers are borne on separate plants (the ultimate in cross-pollination), as in box elder or willow, then the plants are termed **dioecious** (two houses).

If a plant encourages or requires cross-pollination, then another problem develops. Since the plant cannot "get up on its roots" and change locations, a pollinating agent must be employed to carry the pollen from one flower or plant to another. Plants and flowers have become very highly adapted to attract or accommodate a particular agent.

There are three major types of pollinating agents: animals, wind, and water.

Many animals, especially insects, birds, and mammals, function as pollinating agents. Insects and birds are probably the most important, and the evolution of large, showy, or fragrant flowers has been closely correlated with their pollinating activities. Among the insects, bees are probably the best known, but butterflies, moths, beetles, and flies are also important. Of the birds we think first of hummingbirds, although orioles, honeycreepers, and sunbirds are also important pollinators. Bats are probably the most important mammals, especially in the tropics. One group of flowers is even dependent upon mice for pollination!

We must keep in mind that except for bees, most animals are not *dependent* upon flowers as their major source of food. Thus the plants must attract animal pollinators by providing a reward such as nectar or pollen or another enticement to the animal. From the animal's point of view, pollination is merely incidental to the animal's visit to the flower. Let us consider briefly certain pollinators and their flowers in more detail.

(1) **Bees** are attracted to flowers chiefly for food. Pollen is the main food (especially protein) source for the larval stages of all bees. In addition bees drink nectar (a sugary liquid produced by the flower in special glands usually at the base of the ovary or corolla, from which bees make honey). Bees generally are attracted to flowers that appear white, yellow, or blue to us. Often these flowers have special patterns which reflect untraviolet light, a color that is visible to the bee but not to us. Bees are not able to perceive red, so usually bee-pollinated flowers are not that color, although some may turn red or reddish following pollination. Bees have a well-developed sense of smell and are attracted to fragrant flowers.

Many flowers that attract bees (for example, snapdragons, violets, and mints) are irregular in shape, with petals modified to serve as landing platforms.

(2) **Butterflies** visit flowers primarily to drink the sweet nectar, and it is not uncommon to see bees and butterflies both visiting the same flowers.

(3) **Moth**-pollinated flowers exhibit another type of specialization since many moths are nocturnal. These flowers tend to be pale-colored or white (maximizing their visibility in little light) and produce very potent, pleasant fragrances, but only around dusk and later into the evening. They also have long corolla tubes containing nectar that is accessible only to the long tongues of the moths. Some types of *Nicotiana*, jasmine, mother-in-law's tongue, and orchids are examples of plants dependent upon moths for pollination.

(4) **Flies** come in many sizes and shapes, including some that look and act much like bees. These bee-flies are attracted to the same types of flowers visited by bees, but most flies are not the type that enjoy sweet-smelling flowers. They prefer a more "meaty" food and are often attracted to decaying flesh, especially for egg-laying. Thus some flowers have developed strong, unpleasant, fetid stenches and mottled, fleshy or maroon colors to mimic a food source for flies. The flower really puts on a sham because the flower does not actually supply food. Rather the fly is duped into believing that the flower is a suitable place to lay eggs and that it will supply food for the young. The carrion flowers (*Stapelia*) from South Africa are good examples of this kind of flower. In other flowers, like the Dutchman's pipe *(Aristolochia)*, the fly becomes trapped within a chamber made by the corolla and is not released until the flower is pollinated.

(5) **Beetles** crawl around on flowers chomping away at whatever parts they find to eat and in the process pick up pollen. Flowers adapted to beetle pollination are usually large with many flower parts so that some will be left after the beetles have eaten their fill. Water lilies and magnolias are examples of such flowers.

(6) **Birds** usually visit flowers to drink nectar. They are strongly attracted by red, and most bird-pollinated flowers are usually red and showy but with essentially no fragrance since birds have a poorly developed sense of smell. To provide enough nectar for the birds' large appetites, bird-pollinated flowers tend to have long corolla tubes that hold copious amounts of nectar. Bird pollination is much more common in warmer climates where more kinds of hummingbirds abound. Trumpet-creeper vine and *Hibiscus* are classical examples of hummingbird-pollinated flowers.

(7) **Bats** are also important pollinating agents, especially in tropical areas where there are many kinds of fruit-and nectar-eating bats. Flowers adapted to bat pollination are usually large and pale, with heavy musky odors. The flowers are usually borne in clusters away from the foliage on elongate upright stalks or suspended and dangling. Since bats are nocturnal, the flowers usually open and produce the fragrance from dusk into the night.

(8) Artificial pollination by **people** has been a major factor in the domestication and horticultural improvement of many agricultural and garden plants. Many new, larger, more marketable flowers and agricultural crops have been developed as a result of artificial pollination, which is really a big industry. Almost any flower is susceptible to "people-pollinators", since we are attracted to color, fragrance, size, complexity, and bizarreness.

Orchids are very highly adapted for animal pollination. There are over 20,000 species of orchids, each one highly specialized to attract a particular kind of animal—bee, butterfly, moth, fly, or bird. Within the family essentially all of these types of pollinators are utilized. People have artificially cross-pollinated different species of orchids with great success. In fact it is rapidly becoming apparent that the major factor that maintains species distinction in orchids is their extreme degree of specialization for a particular pollinator.

Some plants have adapted to wind rather than an animal as a pollinator. Such flowers often have very small sepals and petals but long stamens and long feathery stigma-styles. Large amounts of pollen must be produced which float in the breezes and lodge by chance against receptive stigmas. Many wind-pollinated plants are monoecious or dioecious. Most grasses are wind-pollinated, as are many trees, at least in temperate regions.

Most water-pollinated plants grow submerged. Flowers are usually produced on or near the surface of the water, and the pollen floats on the surface to the stigma. Water-pollinated flowers usually have poorly developed sepals and petals and are pale in color. Eel grass and water milfoil are examples of such plants.

THE WORLD OF PLANTS

Now that we have examined the structures and functions of the flowering plants, including reproduction, let's take a look at the other groups of living organisms that are plants or considered to be plant-like and see how they differ from the flowering plants. This brief overview is not intended to present a modern classification of plants but merely to distinguish the different kinds of plants. Consult a recent botany textbook for a more detailed classification.

The Seed Plants

Remember that the flowering plants are called angiosperms. This word means "vessel-seed" because the seeds are enclosed within a vessel, the fruit. This is the most sophisticated and highly advanced group of plants. Angiosperms are distinguished by producing flowers that include the sex organs and by forming the seeds within fruits.

Only one other group of plants produces seeds. These are the cone-bearing seed plants or **gymnosperms**. They differ from flowering plants in that the reproductive structures are usually produced in cones, not flowers, and the seeds are borne exposed on the surface of a receptacle or a special modified leaf or shoot, the latter called a **seed scale**. In fact, "gymnosperm" means "naked-seed". In addition, the xylem in the roots and shoots of these plants consists of tracheids only, no vessel elements. There are 700-800 species of gymnosperms, all trees and shrubs. Three important groups of gymnosperms are:

(1) **Conifers**. Trees or shrubs that have needle- or scale-like leaves. This group includes pine, spruce, fir, cedar, cypress, and redwood, to mention a few. In this group the ovules are borne on seed scales grouped to form cones. Many of these are very important commercially as lumber trees and ornamentals for lawn plantings. Most are adapted to grow in dry conditions and can withstand periods without water. Conifers form extensive forests (or did in the past) in the Northern Hemisphere. The largest and tallest plants (redwoods) and the oldest plants (bristle-cone pines) are in this group. The giant redwoods *(Sequoiadendron giganteum)* of central and eastern California are the largest known plants, reaching heights of over 300 feet (100 m.) and diameters of more than 35 feet (10+ m.). These trees have lived an unbelievably long time, as much as 4,000 years. However, these, the largest trees in volume, are not the tallest. The coastal redwoods *(Sequoia sempervirens)*, along the Pacific coast of northern California and southern Oregon, grow to the astounding height of 380 feet (114 m.) and have diameters of up to 26 feet (8 m.) and are known to be over 2,000 years old. The record for the oldest known plants is held by the bristle-cone pines *(Pinus aristata)* that grow in the southwestern United States (eastern California, Nevada, Arizona, and New Mexico). These gnarled ancients are as old as 4500 years! That means these trees began their life about 2600 years before Christ was born!

(2) **Ginkgo** or **Maidenhair Tree.** This single species of plant is in a group by itself. Unlike conifers, this tree has broad, deciduous, fan-shaped leaves. Ginkgo is a "living fossil", unchanged since ancient times. It grows natively only in one restricted region of China; however, it is a popular lawn and street ornamental tree and is widely grown commercially. The plants are dioecious, with the pollen-producing branches and ovule-bearing branches developing on separate plants. The seeds, unlike those of other gymnosperms, are not borne in cones but are produced in pairs on short branches in the axils of the leaves of the female tree. They are about the size of small plums and have an unpleasant odor, but Orientals prize the seeds as food once the fleshy outer covering is removed. Because of the unpleasant odor of the ripe seeds and the general mess created when they fall to the ground, male trees are most frequently propagated to plant as ornamentals.

(3) **Cycads.** These tropical plants are most unusual in that they are very palm-like in appearance. The group

contains about 85 species, all tropical or subtropical. Some shrubby species, often called sago palms, are used as lawn ornamentals in warm climates as in California and southeastern Gulf states.

Other Vascular Plants

Flowering plants and gymnosperms share the characteristic presence of vascular tissue with a number of other plants, and all of this group are referred to as **vascular plants**. Among them, however, only flowering plants and gymnosperms produce seeds. The non-seed-forming vascular plants include ferns, club mosses, horsetails, and whisk ferns. Instead of forming seeds, these plants produce reproductive cells, called **spores**, in spore cases, called **sporangia**. Spores are reproductive cells that, unlike gametes, may develop into a new individual plant without first fusing with another cell. The spores are released from the parent plant and carried away by wind or other agents. If they land in a suitable location, each spore developes into another individual plant that produces male and (or) female sex organs and is different in appearance from the spore-producing plant. Thus, these plants have a life cycle in which there is a distinct alternation between a spore-

Fig. 1-24. Representative plants. Left-hand page: (top, left) Diatoms and other fresh-water algae. (top, right) A fungus, the poisonous mushroom *Amanita*. (center, left) Moss. (center, right) The primitive vascular plant *Psilotum* or whisk "fern". (bottom left) *Equisetum*, the horsetail. (bottom, middle) Clubmoss. (bottom, right) Boston fern.

Right-hand page: (above, left) Cycad. (above, middle) *Ginkgo*. (above, right) Coniferous gymnosperm (spruce). (below) Flowering plant (lily).

producing phase (sporophyte) and a gamete-producing or sexual phase (gametophyte). The ferns illustrate this type of life cycle. Most commonly the characteristic plant we associate with the name, such as fern, is the spore-producing phase, and the gamete-producing phase is smaller, quite different in appearance, and usually neither seen nor recognized by most people as related to the spore-producing plant.

(1) **Ferns**. These well-known plants number about 10,000 species and occur all over the world in both tropical and temperate regions. The stem (a rhizome) is usually horizontal and underground or on the surface. The leaves, produced at nodes, are usually quite large and complex. Most ferns vary in size from a height of 2-3" (5-8 cm.) to 1-3 feet (30-100 cm.). The tree ferns are quite spectacular plants of tropical regions, particularly of the Southern Hemisphere. They are tree-like and often reach heights of 60-80 feet (18-25 m.), with individual leaves often 15 feet (about 5 m.) long and broad. The sporangia of ferns are usually produced in clusters on the undersides of leaves. Many of the smaller ferns are very shade-tolerant and make good houseplants. Ferns are discussed more fully in Chapter 7.

(2) **Club Mosses.** These plants, about 1,000 species distributed in five genera, are mostly tropical although some species occur in temperate regions. The plants are usually small and have simple scale-like leaves. The sporangia are produced on the upper surfaces of reproductive leaves **(sporophylls)** that are often clustered together in a club-shaped structure (cone) at the tip of the shoot, hence the name club moss.

(3) **Horsetails.** About 23 species, all in one genus (*Equisetum*), constitute this group of unusual-looking vascular plants. The leaves are tiny, in whorls, forming small sheaths around the widely-spaced nodes. The green stems have a very jointed appearance due to these nodes and are somewhat corduroy-like because the stem surface consists of a series of longitudinal ridges and furrows. Some species have whorls of lateral branches at each node, giving the plant a bushy appearance (hence the name horsetail), while others have essentially unbranched single stems. Some have deposits of silicon dioxide along the stem and are harsh to touch. These are called scouring rushes and in pioneer times were used to scour pots and pans. Horsetails produce the sporangia on special cone-like clusters of branches at the tip of the stem, and the plants are usually 1-3 feet (30-100 cm.) tall. They occur in both temperate and tropical regions and in both wet and dry areas.

(4) **Whisk Ferns**. These are perhaps the simplest vascular plants, having neither true roots nor leaves but consisting of a cluster of horizontal and vertical forking stems. The large sporangia are borne on short lateral branches along upper portions of the stems. There are two genera *(Psilotum* and *Tmesipteris)* and fewer than ten species, all of which are tropical epiphytic plants. *Tmesipteris* grows natively only in Australia and New Zealand, while *Psilotum* is more widespread.

The Bryophytes (Mosses and Liverworts)

There is one other group of plants, the bryophytes, which includes the mosses and liverworts, that share the characteristic of multicellular organization, including multicellular embryos, with the vascular plants discussed above, but differs from them in lacking organized vascular tissues, although many have leaf-like, stem-like, and root-like organs. The group consists of about 25,000 species of mostly low-growing plants. Water moves to various parts of the plant externally by capillarity or internally by diffusion and osmosis, or is absorbed into the plant from the atmosphere. Some forms, particularly some liverworts, have flattened, ribbon-like bodies that grow along the surface of the soil or on rocks or trunks of trees. Because of their low, mat-like growth and their ability to get adequate water from the atmosphere when it rains or from dew or fog, many mosses and liverworts are ecological pioneers, since they may grow on bare sand or rock and establish a foothold where eventually other plants may become established.

The Simple Plant-like Organisms and their Allies

There are a number of other organisms that have some characteristics of plants but differ from those plants already discussed in a number of ways, including not being organized into multicellular tissues and organs. They are essentially **unicellular** (single-celled), although some may exist as aggregates of cells organized into a simple structure called a **colony**. Generally, the sex organs are single cells, and no multicellular embryo exists. Bacteria and some algae, and fungi, fit into this category. While at one time these were all thought to be related to each other, botanists now consider them quite distantly related.

(1) **Algae**. There are about 20,000-30,000 different species of algae. They are mostly aquatic, although some forms are terrestrial. They occur in both fresh and salt water and exist as either single-celled organisms or colonies of cells aggregated together. They have pigments, including chlorophyll, and are thus photosynthetic. Algae are important ecologically as a food source for many other organisms in water, and also as a significant source of atmospheric oxygen.

The algae are divided into a number of groups (green algae, golden algae, brown algae, red algae, and blue-green algae) based largely upon the kinds of pigments they possess as well as on a variety of other characteristics. Algae and single-celled animals (protozoans) are often placed in a separate kingdom, **Protista.**

(2) **Fungi** and **Bacteria**. These two groups of organisms have been classified with plants because of the presence of cellulose in some kinds and the production of spores. They differ from all other plant-like organisms in that they are incapable of carrying on photosynthesis to produce organic compounds (one small group of bacteria is exceptional and is discussed below). In fact some botanists have now placed fungi and bacteria each into separate kingdoms on their own (**Fungi** and **Monera**, respectively), distinct from plants and animals and protists.

Organisms such as green plants capable of manufacturing their own food are said to be **autotrophic**. Organisms such as fungi and bacteria (and animals) that cannot manufacture their own food but must obtain it from some other source are said to be **heterotrophic**.

Fungi, including about 90,000 species, are heterotrophic organisms that obtain organic food mostly from plant sources but in some cases from animals. If the fungi are adapted to obtain food from a living source (usually to the detriment of the host) they are **parasites**. If the fungi are adapted to obtain their food from a dead organism, then the fungus is called a **saprotroph** (saprophyte). Parasitic fungi are major causers of plant diseases and are very important as pests. Saprotrophic fungi obtain food by digesting the dead host and, therefore, contribute to the decay of dead organic matter, both plant and animal.

In fungi the basic unit or organization is not a simple, single cell but a thread-like cylindrical filament that may contain many nuclei or may be divided into cell-like compartments each of which contains one or two nuclei. The thread-like filaments are called **hyphae** (sing. hypha), and the body of the fungus, composed of a mass of hyphae, is called a **mycelium** (see Fig. 2-11). The hyphal walls are mostly composed of chitin (another complex carbohydrate) rather than cellulose, but spores and the fruiting body that produces spores contain cellulose. Toadstools and mushrooms are fruiting bodies of certain kinds of fungi.

Bacteria, with about 2,000 identified species, are ecologically similar to fungi in that most are heterotrophic, either parasitic or saprotrophic, and therefore many are very important in decay and as causal agents of many diseases. (Some bacteria are autotrophic and contain a special chlorophyll that allows them to carry on photosynthesis to make their own organic compounds. This is quite a different process from the photosynthesis discussed earlier, since water is not split to obtain chemical energy, but other compounds such as hydrogen sulfide or organic acids are used. These bacteria live on the bottom of ponds or streams and the ocean where no oxygen is present.).

Bacteria differ from most fungi, however, in being microscopic and consisting only of single cells, and their cellular organization is simpler than that of most other organisms. Bacteria (and the blue-green algae, which are closely related) lack a true nucleus (that is, there is no organelle within the cell containing chromosomes and walled off by a membrane) as well as most of the other organelles mentioned at the beginning of the chapter. The cell wall lacks cellulose but contains other substances, including carbohydrates and special amino acids. This type of cell, found only in bacteria and blue-green algae, is called **prokaryotic**. All other organisms have cells with true nuclei and are called **eukaryotic**.

The saprotrophic bacteria and fungi should not be looked upon as being bad like the disease-causing parasitic types, because these organisms play a vital beneficial role in recycling essential elements. The decay of dead organisms returns usable materials to the atmosphere and soil. Thus plant, animals, fungi, and bacteria work together to form a basic ecological cycle:

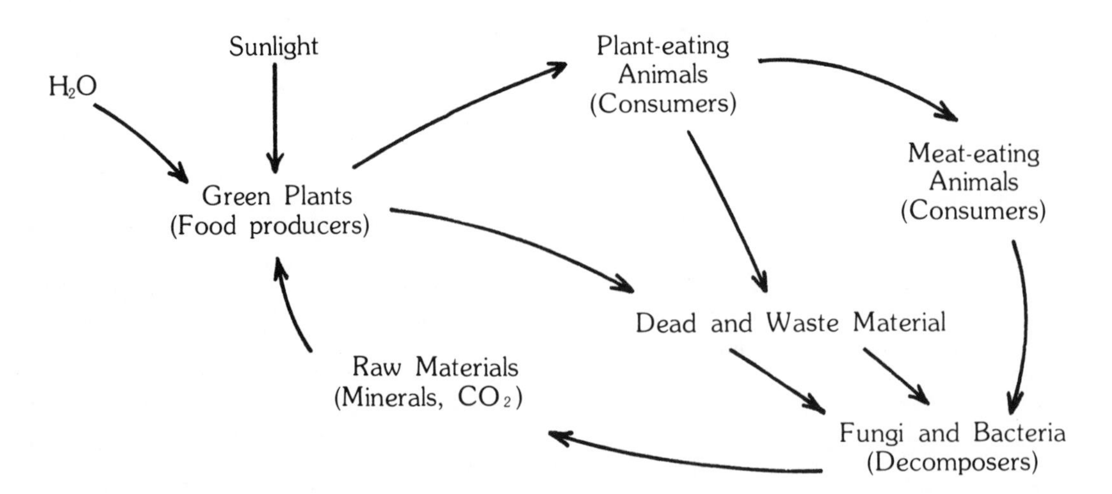

Fig. 1-25. The basic ecological cycle.

SUMMARY OF MAJOR PLANT DIVISIONS

A. Plants without roots, stems, or leaves. No embryo formed.
  B. Plants lacking chlorophyll.
    C. No organized nucleus. Single-celled organisms .... BACTERIA
    C. True nucleus present. Thread-like multicellular plant body . FUNGI
  B. Plants with chlorophyll.
    C. No organized nucleus. Pigments determine color. .... BLUE-GREEN ALGAE
    C. True nucleus present. Pigments determine color. .... ALL OTHER ALGAE
A. Plants with stems, usually with roots and leaves. Embryos formed.
  B. Plants without vascular tissue. .... BRYOPHYTES (Mosses & Liverworts)
  B. Plants with vascular tissue.
    C. Plants without seeds, dispersing by spores.
      D. Plants lacking true roots or leaves. Plant body arises from rhizomatous stems. Sporangia on short lateral stems. .. WHISK FERNS
      D. Leaves and roots present.
        E. Sporangia grouped in cone-like structures.
          F. Leaves scale-like, closely arranged in spiral around stem..... CLUB MOSSES
          F. Jointed stems, leaves in whorls at nodes. .... HORSETAILS
        E. Spores produced in clusters on undersides of leaves. FERNS
    C. Plants with seeds.
      D. Seeds not covered by fruit .... GYMNOSPERMS
        E. Seeds fleshy on short stems. Leaves fan-shaped. . *Ginkgo*
        E. Seeds in cones.
          F. Trees palm-like, tropical .... Cycads
          F. Trees or shrubs with needle leaves, range widespread .... Conifers
      D. Seeds covered in a fruit ....ANGIOSPERMS
        E. Two seed-leaves, vascular tissue in cylindrical arrangement, leaves usually net-veined, flower parts in 4's or 5's .... Dicotyledons
        E. One seed-leaf, scattered vascular bundles, leaves usually parallel-veined, flower parts mostly in 3's. .... Monocotyledons

PLANT INTERACTIONS (ECOLOGY)

We have now briefly examined those living organisms that are considered to be plants or have plant-like characteristics, and we have studied one group of these, the flowering plants, in particular. So far we have emphasized the form and function of plants—their structure and chemical activities. We have also very briefly considered the relationships between different plants. All plants, and for that matter animals, are related to each other to greater or lesser extents. The more characteristics two or more organisms share, the more closely they are related. Two daisies are much alike and hard to distinguish; they have many common characteristics and are therefore closely related. A daisy is much different from a pine tree but still shares some characteristics with it. A daisy has more characteristics in common with the pine tree than with a red squirrel. Thus the daisy and pine tree are more closely related to each other than to the red squirrel, but nevertheless the daisy, pine tree, and red squirrel do share some characters and are therefore somewhat related.

To examine this from a different point of view, it is interesting to consider that the differences between organisms are often related to evolutionary changes organisms have undergone **(adaptations)** in order to compete successfully with other organisms for "a way of life", "a place in the sun" so to speak; a way to obtain enough light, water, space, and food to survive along with other organisms. Plants and animals have to interact with each other: Plants with other plants, animals with other animals, plants with animals, and animals with plants (and fungi and bacteria, too). Plants and animals also interact with their environment—the non-living things around them like air, soil, water, and energy. I am always reminded of those famous words of John Donne— "No Man is an Iland, intire of it selfe". That really is a profound biological statement. We can broaden the idea and say no organism (plant or animal) stands alone. No plant or animal is able to function properly without interacting with other plants and animals and environmental factors. We have already seen examples of this when we talked about such things as plants and minerals, plants and water, plants and light.

The study of the interactions of plants, animals, and environmental factors is called **ecology**. A group of interacting plants and animals is called a **community** (the fish, snails, and aquatic plants in a pond; the cacti, succulents, lizards, snakes, and insects in a desert; the mosses and invertebrates on a decaying log), and a community plus the non-living environmental factors (energy, air, soil, and water) make up an **ecosystem** (a pond, forest, grassland, desert). (Ecologists use the word "community" in several senses, depending upon which level one wants to look at—from decaying log to oak-hickory forest.) We cannot conclude this chapter without a few paragraphs about the concepts and principles of ecology, particularly as they concern plants.

People have always been interested in ecology, although that may not have been what they called it. The earliest form of it was actually plant geography. When plant explorers (mostly European) began traveling to various parts of the world, they discovered that there were distinct types of plant communities. They began to draw maps to show the limits of the types and they assigned names to each, including tropical forests, temperate forests, needle-leaf evergreen forests, savannas (grasslands), deserts, and tundra. Regardless of whether it was in the northern, southern, eastern or western hemisphere, and even though the plant species might have been different, the peculiar form and appearance of each type was recognizable. They called these distinct vegetation forms **formations**.

The question logically arose why do certain regions have special formations? Initially people thought formations were where they were because of climate (available water and temperature). Early plant geographers soon found that they could predict the climate in a region upon the basis of the vegetative formation and vice versa, the vegetative formations could be predicted from climatic data. As these formations became better known through further studies, it became obvious that the formations were not always predictable from climatic data but that there were complex interactions between plants, animals, and an array of environmental factors that caused formations to be where they were.

Let us examine an ecosystem, any ecosystem, more closely to gain some idea of the complex interactions involved. An ecosystem consists of two major components: The **abiotic** factors, which are the non-living environmental components—air, soil, water, light, and so forth; and **biotic** factors, which are the plants and animals. There are four principal abiotic factors that we should briefly consider: Light, temperature, precipitation, and soil.

Light. We have already talked about light, especially in regard to its role in photosynthesis and regulation of growth. The amount of light a plant receives is important ecologically. Many plants are adapted to grow in full sunlight; they are **shade-tolerant**. Plants of this nature would inhabit open areas such as fields, prairies, roadsides, deserts, and so forth where little or no shade is present. Many weeds and grasses fit into this category, as do the mature, full-grown trees in a forest whose leaves form the canopy at the top of the forest and receive full sunlight. Basically these kinds of plants would not make good houseplants unless they could be placed near south-facing windows and receive adequate light. Other plants are adapted to grow best in low light intensities, and they actually cannot cope with full sunlight. These plants are called **shade-tolerant**. The plants that grow on the forest floor, under the shade of the big trees, are of this type. So are the plants that thrive under rock ledges or on the north sides of buildings. Many of these plants make good houseplants because they accept the lower light intensities available in most homes. There are many degrees of gradation between shade-tolerant and shade-intolerant plants.

Temperature. Temperature is a very important environmental factor. The response of plants to cold is the most obvious example. Herbaceous perennials adapt to cold by "going underground" and surviving as bulbs, corms, rhizomes, and so forth. Other herbaceous plants complete their life cycles in one season and survive the cold as seeds. Woody plants survive by going dormant and protecting their meristems with leaves modified into bud scales. In plants adapted to cold winters, the breaking of dormancy and subsequent flowering will not occur until they have been exposed to at least four to six weeks of near-freezing (5° C.) temperature. Plants which have not adapted to survive freezing temperatures naturally will not be able to grow in colder climates. Cotton, for instance, needs a warm growing season of at least 200 days, so it cannot be grown in the northern tier of states.

Precipitation. Water is available to plants mostly in the form of precipitation, by which we mean rain, sleet, snow, and ice. Dew and fog may also be important for some plants. Precipitation generally affects plant distribution by the amount of water that goes into the soil and is retained there. Some plants have become adapted to grow well in areas where there is little precipitation and therefore little soil moisture. Such plants are called **xerophytes**. Deserts, sand dunes, and other places that have light, sandy, well-drained soils best support xerophytes. Succulents, plants with thick fleshy stems or leaves, like cacti and stonecrops, are good examples of xerophytes. Conifers such as pines and cedars with very narrow needle- or scale-like leaves are also xerophytes. Cacti and other succulents often make good houseplants because of their low moisture needs, although since their habitats are often shadeless they have high light requirements. Other plants are adapted to grow in, or at least very near, water. These plants not only tolerate an abundance of water but usually cannot survive well without it. Such plants are called **hydrophytes**. Water lilies, pondweeds, and cattails are examples of such plants. Aquarium plants, naturally, fit into this category too. Still other plants have adapted their water needs to situations intermediate between deserts and lakes. These plants that have moderate water requirements are called **mesophytes**. The majority of plants belong in this category.

Soil. Soils, or **edaphic factors**, are equally important in affecting plant distribution. In addition to retaining moisture and thereby providing the direct source of water for absorption, soil also provides anchorage and, most importantly, nutrients for plants. As we have already discussed, most of the nutrients plants need are provided by and absorbed from the soil. The quantity of any of the necessary nutrient elements in the soil would affect the occurrence and distribution of plants. Soil fertility is also directly related to the physical nature of the soil (the size of the soil particles) and to the amount of precipitation. If the soil particles are fairly large and water drains through the soil quickly (sandy soils), then the nutrients, which dissolve in soil water, may drain (**leach**) away with the water, and the soil will be nutrient-poor and will not support good plant growth. If on the other hand the soil particles are very small (clay) and water cannot drain through the soil quickly, then the soil particles may become saturated with water and therefore have a low oxygen content. This condition would also affect the growth of plants, particularly those with high oxygen requirements in the roots. The acidity or alkalinity (pH) of the soil is also very important (see Chap. 2 for more discussion on pH). Most plants grow best at a pH between 6 and 7, which is slightly acid. Some plants, such as rhododendrons, azaleas, and blueberries, prefer much more acid soils (pH 3-5). Some plants, such as pines and oaks, contribute to the acidity of the soil when their leaves decay, releasing the high concentration of organic acids in them. Other plants are adapted to more alkaline soils (pH 8-9), such as some ferns and other plants that grow on limestone-derived soils. Thus as you can see environmental or abiotic factors play significant roles in affecting growth and distribution of plants.

There are several ways to examine the biotic factors, the plants and animals, in an ecosystem. One of the most significant ways is to examine them nutritionally—"who eats whom"—or, more specifically, to see how energy flows through them. Let us again examine the diagram on pg. 40, which very nicely summarized the energy flow or nutritional relationships occurring in any community or ecosystem. We have already established that green plants are the only organisms capable of utilizing light energy, carbon dioxide, and water to produce complex organic compounds—carbohydrates, proteins, fats, and so forth. In ecological terms the green plants are called **producers**, and either directly or indirectly all other living organisms (animals including people, fungi and bacteria) are completely dependent upon such plants as an energy (food) source. Animals may then be considered **consumers**. We may further categorize consumers into **herbivores** or **primary consumers**—those that feed directly upon plants, such as rabbits, and cattle who may eat clover and corn; carnivores or secondary consumers—those that feed upon the herbivores, like coyotes or hawks; and omnivores—those that utilize plants directly as well as feeding upon the herbivores and carnivores. People are a good example of omnivores. Both producers and consumers give off waste products and, in time, die. These waste products and dead organisms are then broken down into simpler, inorganic components such as carbon dioxide, oxygen, and nitrates by the bacteria and fungi which are the **decomposers.**

The above interaction between producers and consumers is called a **food chain**. It provides an excellent way to look at the flow or transfer of energy from producers to consumers. Food chains are usually diagrammed as several nutritional levels, usually three to five, that when stacked on top of each other form a pyramid. The width and depth of each level represents the amount of available energy at that level. It has been estimated that about 90% of the available energy at each level is lost when it is transferred to the next. Most of this lost energy is in the form of heat which is given off through cellular respiration. Thus if we were to look at a very simple three-level food chain, like the clover-rabbit-hawk example mentioned above, clover, the producer, would be most abundant and would support a smaller population of rabbits which in turn would support a still smaller

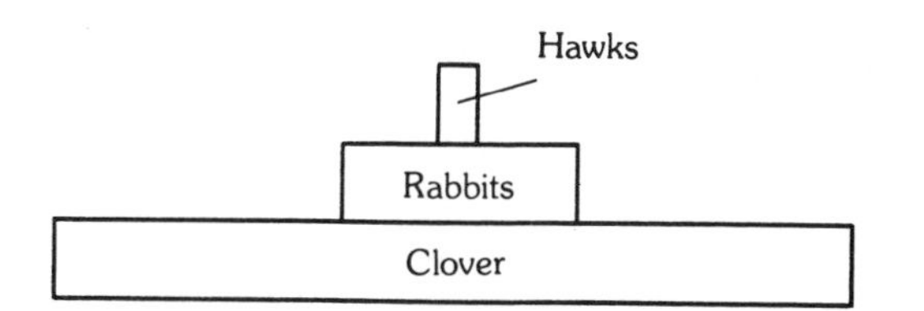

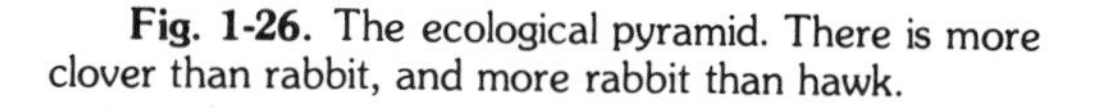

Fig. 1-26. The ecological pyramid. There is more clover than rabbit, and more rabbit than hawk.

population of hawks. (Fig. 1-26) Ultimately then, the producers determine the sizes of the populations of consumers.

Let us now look at the decomposing segment of this ecological cycle. If it were not for the action of fungi and bacteria which causes some of the raw materials needed by producers to be recycled, then the original source of raw material would probably have been used up long ago and we would not even be here to think about it. Let us consider just a couple of the elements needed by living organisms and see how they cycle and recycle through the system.

Carbon is the most important such element because it forms the skeleton-structure of all of the complex organic compounds of which organisms are composed. All of the carbon incorporated into land plants by photosynthesis comes from carbon dioxide (CO_2) in the air which gets there from (1) carbonates in lakes and streams, (2) combustion such as explosions of volcanoes, burning of wood, oil, or gas, and (3) cellular respiration by living organisms, particularly the decomposers. In fact, the decomposers account for a major portion of the CO_2 put into the system, since ultimately all of the organic matter in an ecosystem is channeled through the decomposers. (See Fig. 1-27.)

Nitrogen is another vital element. Of the elements needed by plants (besides carbon, hydrogen, and oxygen) nitrogen is probably needed most. It is necessary for the synthesis of proteins, nucleic acids, and other vital organic chemicals. At first glance a source of nitrogen would not appear to be a problem, since approximately 79% of the atmosphere is pure, molecular nitrogen (N_2), but, oddly enough, green plants are not able to use nitrogen in this pure form! Most plants absorb nitrogen from the soil in the form of nitrate salts (NO_3) such as potassium nitrate (KNO_3). There are several natural ways that nitrates may get into the soil (other than from the application of commercial fertilizers). Lightning converts nitrogen into nitrate. During a lightning flash enough heat is formed to break apart the nitrogen molecules in the atmosphere and recombine them with oxygen to form nitrates, which are absorbed by water molecules in the air and then fall to the ground in the rain. A far more important method is the biological conversion of atmospheric nitrogen into nitrates by certain bacteria and blue-green algae, a process known as **nitrogen fixation**, whereby they obtain chemical energy for their vital life processes. Some of these nitrogen-fixing bacteria are free-living in the soil, but most are found in association with green plants, usually in special nodules on the roots. The bacteria and the green plants form a **symbiotic** (mutually beneficial) relationship wherein the bacteria convert nitrogen into a form usable by green plants, and the green plants provide a home, organic food, and water for the bacteria. This arrangement allows those green plants that can associate with these bacteria to grow in soils that are low in nitrogen and thus compete better with other plants. Members of the bean family, such as clover and alfalfa, are well known for this symbiotic relationship with nitrogen-fixing bacteria. Another important biological source of nitrates in the soil is the breakdown of nitrogenous waste products from living and dead organisms by decomposers, particularly bacteria. In contrast, there is one special group of bacteria (denitrifying bacteria) that actually convert nitrates back into nitrogen which is then released into the atmosphere. This complex cycle is illustrated in Fig. 1-28.

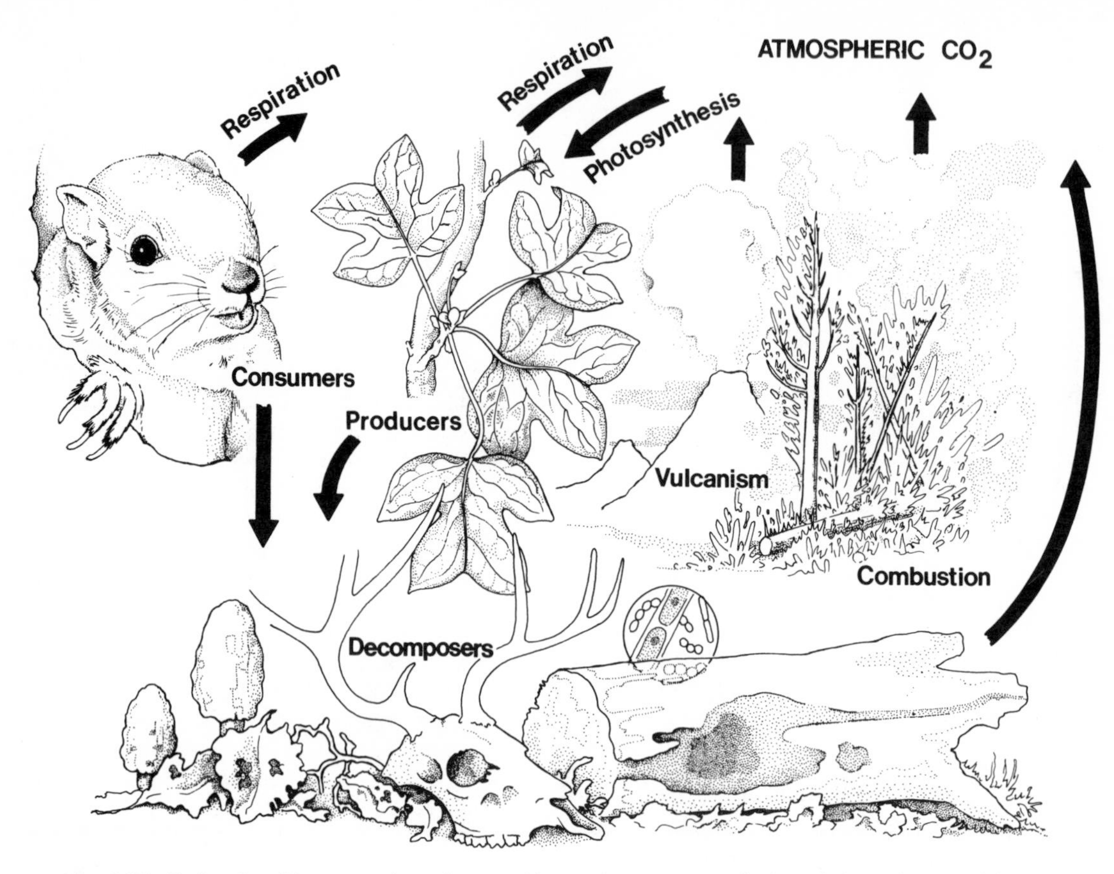

Fig. 1-27. Carbon is neither created nor destroyed but cycles systematically through living things and the inorganic environment.

Social interaction provides another way to look at the community. Let us briefly examine the social structure of the plants in a forest community to illustrate this. Social structure is determined by how plants live together in a community and compete successfully for light, nutrients, and moisture. There are several social levels or strata in the forest. Trees form the highest stratum with their branches and leaves forming a **canopy** at the top of the forest, thereby receiving maximum available light. Canopy trees must expend significant energy to grow to the top of the forest by forming a substantial trunk and root system to support the plant and to provide adequate nutrients and water. Beneath the canopy smaller trees and shrubs form the **understory**. These plants are also woody and must expend a considerable amount of energy to grow up to get the best of the light filtering through the canopy. Beneath the understory, herbaceous plants, forming the **ground cover**, absorb what little light trickles through the canopy and understory. These plants, lacking the ability to grow up to the bright-light zone, are either shade-tolerant or have other adaptations for getting enough light to carry on their life purposes. Examples of an alternate approach are the spring ephemerals so noticeable in the deciduous forests of the

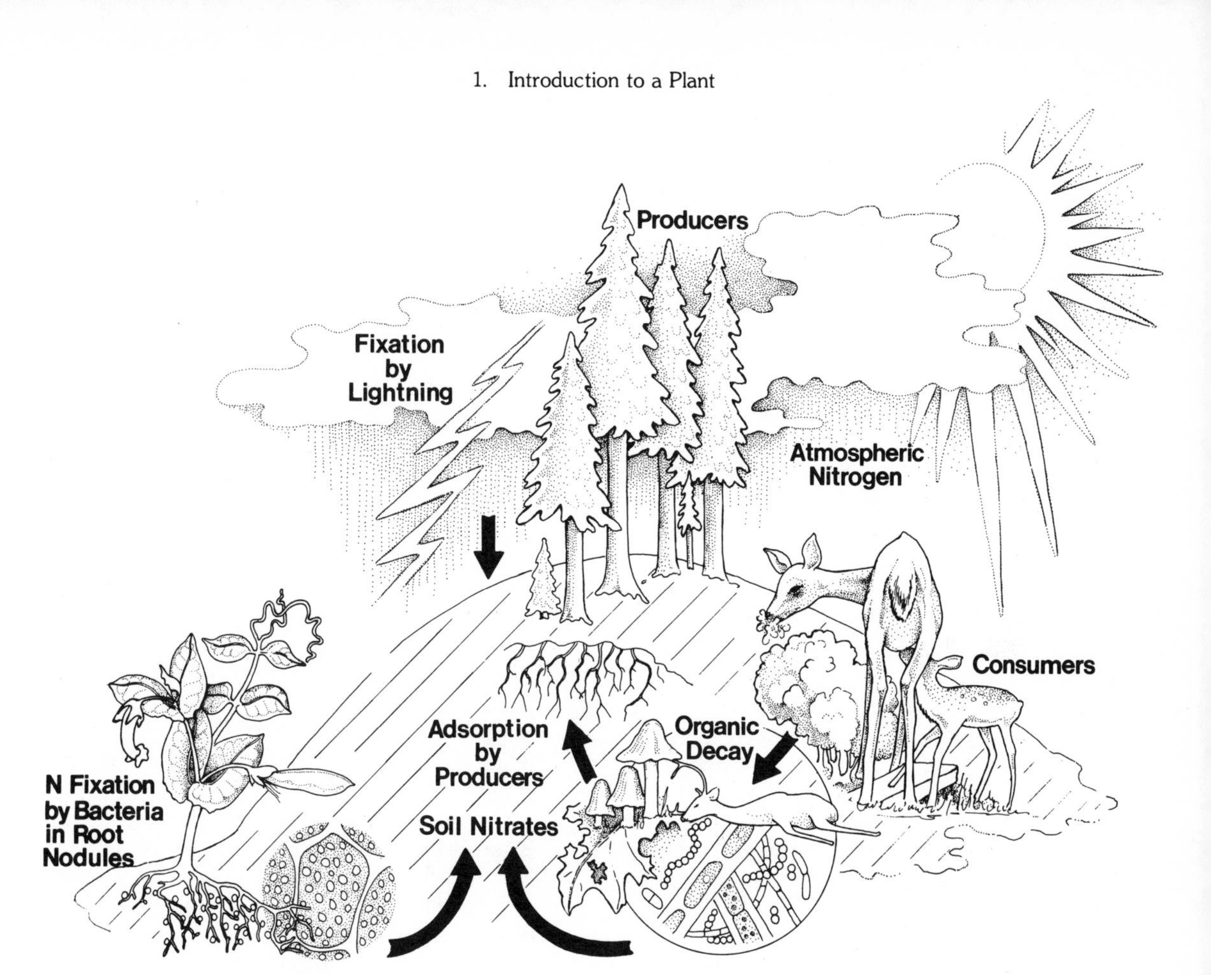

Fig. 1-28. Nitrogen is neither created nor destroyed but cycles systematically through living things and the inorganic environment.

eastern United States. Early in spring, as soon as the growing season begins, a myriad of small, herbaceous wildflowers like spring beauty, Dutchman's-breeches, harbinger-of-spring, and trout lilies push out of the ground from dormant bulbs and corms, leaf out, flower, form seeds, and store up new reserves for the next season before the forest trees have fully leafed out and closed the canopy. In this way these plants take advantage of the full sunlight that hits the forest floor during April and May. By June and July the leaves of most of these plants have already begun to wither and disappear.

Vines form another social stratum. These plants are mechanically dependent upon other plants for support. In some forests, such as flood-plain forests of the eastern United States or in tropical forests, woody vines are very important. Instead of forming massive self-supporting trunks, these plants grow very quickly, lean on or "crawl" up trunks of trees, and extend into the canopy thereby reaching full sunlight. This strategy is not without problems though. Instead of being 100 or so feet (30+ m.) high, as many trees might be, the vines, with continuous growth, may often be 700-1000 feet (200-300 m.) long, draped over the branches of trees. Think of the transpiration-cohesion pull necessary to get water from their roots!

Another distinct social stratum in some forests is formed by **epiphytes**. These are plants that are not rooted in the soil but rather use their roots to cling to other plants as a means of support. Epiphytes may be attached to the trunks of trees or to upper branches in the canopy. These plants have made another type of energy trade-off. In order to maximize light availability they have sacrificed a ready source of water and nutrients, the soil, and instead cling tenaciously to other plants closer to the light source than they could grow by themselves. Epiphytes obtain water from the atmosphere. To do this they have evolved unusual adaptations such as rosettes of leaves with cup-shaped bases to catch and hold water, or very thick, leathery leaves to cut down water loss. Nutrients are obtained either from water in the atmosphere or from the organic debris that accumulates around their roots. Some members of the pineapple family (bromeliads), for example, Spanish moss, are such epiphytes. Many orchids, especially in the tropics and subtropics, and even certain cacti have adopted the epiphytic way of life. Some mosses and algae also grow as epiphytes on the stems and leaves of plants, especially in the moist tropics. Examining the social structure of a plant community is a very useful way to interpret the growth habits of plants—why trees are trees, vines are vines, and so forth.

An ecosystem is dynamic. There are many interactions and lots of activity as we have seen in the above discussion. In such a dynamic situation, where organisms interact with other organisms and affect or are affected by abiotic factors, change is inevitable. Some organisms are able to outcompete others, and those outcompeted may disappear from the ecosystem while other organisms may move in. Such changes in the kinds of organisms present in the ecosystem in turn modify the abiotic environment of ecosystem. When we examine these changes, especially in the plant community, we find that they occur in an orderly and predictable manner. This orderly series of changes over a period of time is called **succession**.

A classical example of plant succession is seen in an old field in the northeastern United States (Deciduous Forest Zone) when it has been abandoned from agricultural use. (Fig. 1-29.) The abandoned field usually starts out as essentially bare soil. It is exposed to full sunlight and rapid drying. During the first year a few herbaceous plants may be seen growing. These are mostly annuals, including grasses. These plants can tolerate full sunlight and the dry

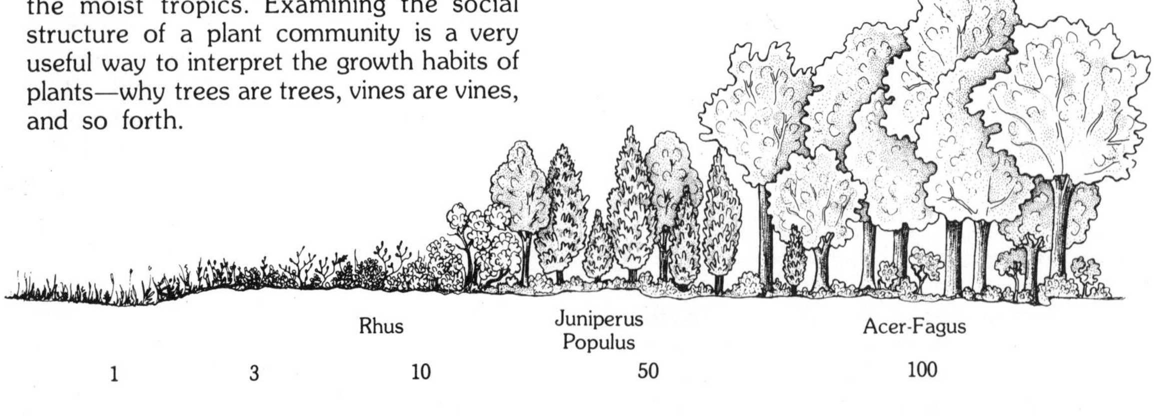

Fig. 1-29. The concept of succession presented as a sequence in space. See text (also pp. 55 in Ch. 2).

conditions. They complete their life cycle in one season and over-winter as seeds. In fact, the plants probably first get to the field as seeds blown in or carried in by animals. These plants may be considered **pioneers** since they can tolerate these rigorous, harsh conditions. This sequence will repeat itself for the first few years. Gradually as these annuals sequentially live and die, their stubble builds up some organic matter or humus that helps retain moisture. Over the next several (3-5) years, as more annuals continue to grow, a few herbaceous biennials and perennials become established, and the grasses begin to form dense mats of turf. These additional plants help to retain moisture, accumulate organic matter, and even provide some shade. The perennials die back at the end of the growing season but come up again every spring. They continue to add organic humus and increase the water-holding capacity of the soil and the amount of shade.

Gradually over the next 10-50 years a few shrubs and even some trees become established. In the seedling stage these woody plants need the shade and moisture provided by the herbaceous perennials. As the shade increases, the annuals can no longer find a suitable place to grow because the ground cover is too thick and there is too much shade and moisture, so they are gradually phased out. Some of the earliest shrubs and trees that get established are junipers and pines. These plants with their narrow needle-or scale-like leaves are able to survive in this still relatively dry habitat. As they extract nutrients and shed their leaves they make the soil more acidic. They also continue to provide more shade. Eventually some broad-leaved, acid-loving woody plants become established, such as oaks and hickories, their seeds carried in by squirrels from nearby woodlots. Gradually as more organic material accumulates from the build-up and decay of plant materials on the ground, more nutrients become available, more moisture is retained and more shade is provided, and eventually the pines and other conifers are replaced by the broad-leaved hardwoods. As more time passes and these processes continue, these plants further modify the community. The shade in the summer from the broad-leaved trees is greater than that provided by the conifers, so that less and less light reaches the forest floor. Many of the perennial plants that had grown there can no longer get sufficient light, and they die out. With the increased shade, other more shade-tolerant, moisture-loving trees such as beech and sugar maple become established. These plants can tolerate low light intensities as seedlings and young trees. Eventually, after about 200 years, the field will become a forest with a canopy formed by mostly beech and sugar maple trees, maybe with a few old oaks or other trees. The understory will be composed of a few shade-tolerant plants like flowering dogwood and redbud, and the ground cover will be a few shade-tolerant herbaceous plants or the spring ephemerals mentioned earlier. However, the dominant plants in both the understory and ground cover will be young beech and sugar maple trees and seedlings, because they are the only canopy-forming trees that in their youth can tolerate the dense shade on the forest floor.

Once the woods reaches the point where the dominant tree seedlings and saplings are the same species that make up the mature trees, we call it a **climax community** because it is now stable and will maintain itself for an indefinite period of time.

2. SOILS AND PLANTS

One of the unique aspects of planet Earth is the layer of life-giving soil over its surface. Soil is one of the important factors in growing plants successfully. All plants need basically three things to survive: light, mineral salts, and water.

Light is required as an energy source for all green, photosynthesizing plants. With this energy the plant can synthesize organic compounds needed to build plant tissues, but only if it has the necessary elements. These elements are in mineral salts that dissolve in soil water and yield nutrient **ions** (electrically charged particles). They are absorbed into a plant by its many root hairs. In addition, the water itself provides another element, hydrogen, which is a necessary component of all organic molecules that make up living organisms. Such molecules compose cells and tissues. Water also keeps the plant **turgid** or firm so that leaf surfaces are exposed to the essential sunlight. Soil therefore is a necessary medium for providing the minerals and water needed as well as providing anchoring material in which plants grow.

WHAT SOIL IS

By definition soil is the loose, unconsolidated material that covers the earth's surface. Dirt on the other hand is defined in the dictionary as a foul or filthy substance and the word is derived from ancient words for excrement. Soil is therefore the preferred word for our plant-growing substance, but one is always pleasantly impressed by the colorfulness of phrases such as "the old dirt gardener" as we strive to achieve that envied implied skill with plants.

In good, productive soils, roots thrive and provide the plant with essential minerals and water. Roots that grow deep into the soil and extend for long distances laterally can reach and absorb adequate nutrients and provide good mechanical support for the plant as well. Sturdily anchored plants, whether they are large trees or small annuals, will be wind-firm and drought-resistant.

Soils with compacted soil layers, too much or too little water, infertile ingredients, or toxic substances retard or prevent adequate root growth. Soil water passes readily into the tiny, fragile root hairs from which it moves across the root into the xylem tissue. The root hairs are short-lived and easily damaged and perform most efficiently in loose, friable soils with highly nutrient soil solutions.

HOW SOIL IS FORMED

Soil scientists generally recognize five major factors that contribute to soil formation: the parent rock that provides the inorganic portion of a soil, the climate or long-term weather effects on the parent rock and on the organisms present, the relief or topography of the area in which the

Fig. 2-1. The "Old Dirt Gardener" applying his expertise.

soil is forming, the living plants and animals whose activities affect in certain ways the progress and nature of soil formation, and, finally, the amount of time available and necessary for the interaction of these factors.

Parent Rock

Soil formation begins with the weathering and breaking down of rocks to produce a loose, granular material. Three major kinds of rock exist, each of which produces different kinds of soil. These are: 1) **igneous rocks**, formed from molten materials from deep within the earth that solidify as they cool when they arrive near or break through the earth's crust; 2) **sedimentary rocks**, formed by the massive accumulations of eroded rock particles (of any kind) or, in some cases, mollusk or other shells, that, under the effects of pressure and resulting heat, solidify into usually layered or stratified rocks; and 3) **metamorphic rocks**, which are forms of either igneous or sedimentary rocks that have been subjected to such extremes of heat and pressure that the original rock structure is modified beyond recognition.

Granite, which is a slow-weathering igneous rock, is the most abundant rock on the earth's surface. Its eroded form as a soil base produces rather acid soils of low productivity. Sedimentary rocks such as limestone produce neutral to alkaline soils that are very productive.

It is always helpful to know what kind of parent soil material is present so that modifications or improvements may be made, although such recognition is sometimes difficult. The underlying rocks of an area may be the source of the parent material, but many soils have been removed far from their source of origin by various agents such as wind, water, gravity, or even glacial ice, so that the home gardener must be careful not to jump to conclusions in assuming a possible mineral content of a soil. For example, in the glacial areas identification is complicated by the fact that glacial soils often represent a grand mixture of several kinds of soil materials depending on just what the ancient glacier picked up on its way south. Equally difficult to analyze could be alluvium deposited by water in a river delta.

In some situations soils will lack certain elements even though the parent materials are rich in those elements, if the soils are subjected to severe **leaching**, a process in which certain substances are dissolved in waters which flow away from the surface soils to layers deep below root levels.

In any event, rocks must be weathered until sufficient nutrients are available to support plants. Soils rich in certain essential elements are derived from rocks rich in these same elements, with the exception of nitrogen which is derived from the atmosphere.

Climate

Precipitation and temperature are the two climatic factors that directly affect soil formation. Because of its extensive effect on both the physical and biological processes in an area, climate is usually recognized as the dominant factor in producing almost any soil formation. An increase in either temperature or moisture results in more rapid soil formation by accelerating both chemical and physical weathering. In tropical climates we can find weathering effects 160 feet (53 meters) below the soil surface. The tropics contrast sharply with cold or frozen climates where there may be only a few inches of chemical and weathering activity.

Soil colors are also influenced by climate. The oxidation of iron in well-aerated soils can produce red soils, while light-colored soils either lack iron or are poorly drained and low in oxygen. It is interesting to note that dark-colored soils are often assumed to be high in organic matter and therefore more productive than lighter-colored soils. They often are, but some organic soils are so acidic that they must be heavily limed if they are to be usable. Fig. 2-2 shows areas of the United States that benefit from lime applications. Tropical soils too often look deceptively rich because of their dark colors. In these soils decomposition and leaching occur so rapidly and continuously that organic matter is depleted of its nutrients almost as fast as it accumulates.

High humidity and heavy rainfall lead to intense leaching. Our southeastern states are most likely to be severely leached, as shown in the accompanying map. (Fig. 2-3). The overworked soils have a low water-holding capacity and, without either freezing temperatures or abundant vegetation to slow the flow of water, especially in winter, the organic matter and soil nutrients are carried downward leaving poor, thin, acid soils. Under such conditions, even limestone soils, which are notably alkaline, become acid as the lime is leached away.

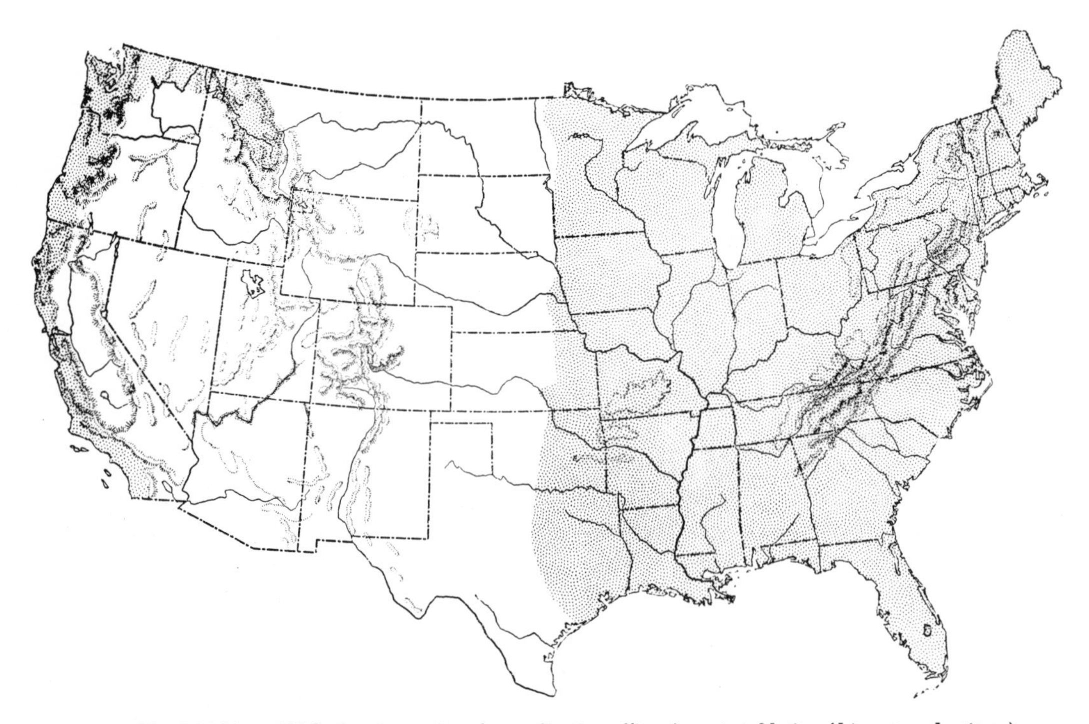

Fig. 2-2. Map of U.S. showing regions for application of lime (courtesy National Limestone Institute). Most of the soils in the shaded portions are acid and require lime for maximum plant growth.

Heavy winter rains on these soils can cause further damage by excessive erosion.

In arid regions where rainfall is reduced or scanty, little or no leaching occurs and salts tend to accumulate. Soils in such areas are usually alkaline. Soils in the southwestern states are of this type.

Relief or Topography

Relief affects erosion, drainage, and depth of soil and is effective primarily through its association with water. If one can imagine progressing from a steep hillside with thin soils and weak vegetation to gentler slopes with deeper soils and stronger plant growth and then to landlocked depressions with the greatest vegetative mass and thick, humusy soils containing decomposing plant material (see Fig. 2-4), the importance of relief is impressive.

Terracing, such as we see in many European and Asian countries, and contour plowing, common in this country (see Fig. 2-5), are attempts to cope with the influence of strong relief on the growing of field crops. Most home owners are not faced with this degree of effect of topography but must nevertheless be aware of the problems it can cause even on a small scale.

The Activities of Plants and Animals

There are literally millions of tiny organisms in a single spoonful of soil. These microorganisms, such as bacteria, molds, actinomycetes, and protozoans, along with the larger life forms, contribute significantly

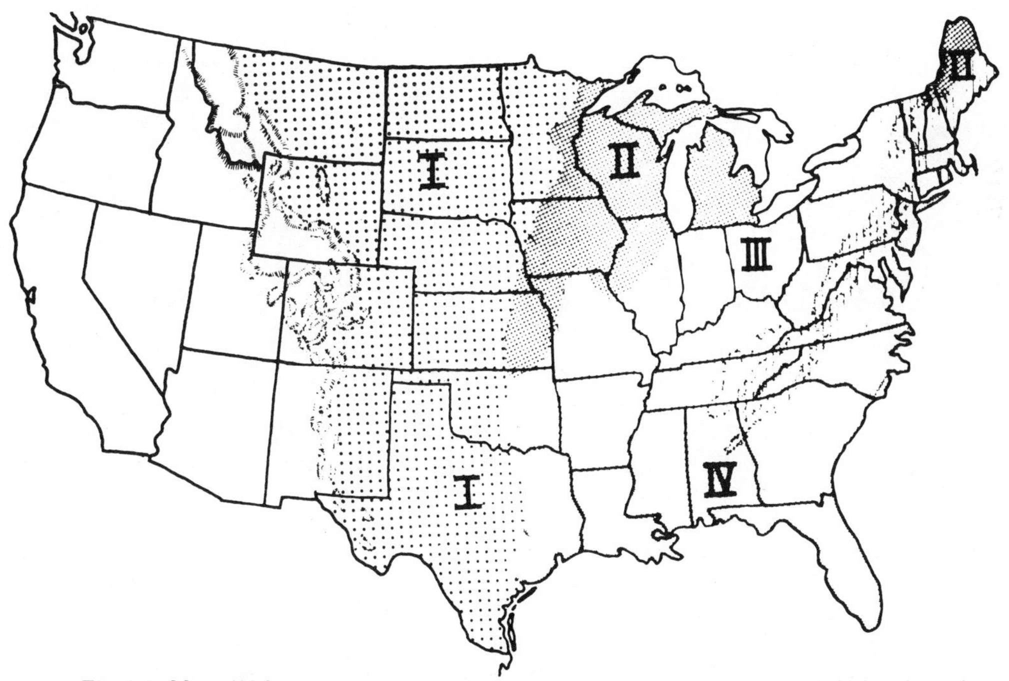

Fig. 2-3. Map of U.S. showing leached areas (courtesy Soil Science Soc. America). I, slight hazard of plant nutrient losses by leaching; II moderate hazard; III severe hazard; IV very severe hazard.

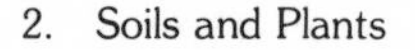

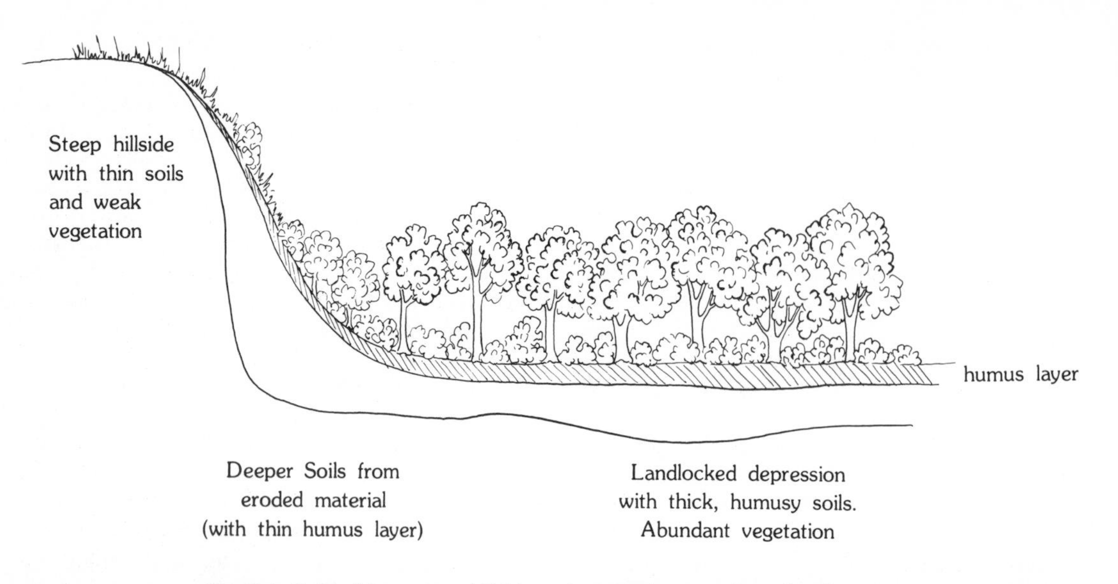

Fig. 2-4. Soil is thinner on a hillside and accumulates on the valley floor.

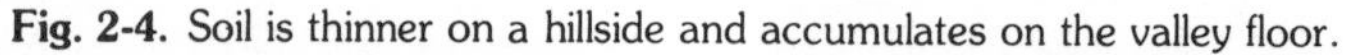

Fig. 2-5. Contour plowing protects the field from erosion while growing a cultivated crop. (USDA photo)

Fig. 2-6. A large boulder with its surface covered with a variety of lichens, each contributing to the erosion of the rock.

to soil building. Even on bare rock, simple lichens (see Fig. 2-6) can hasten the erosion of the surface by causing disintegration of the rock upon which they grow. As these simple plants die, their organic matter becomes mixed with the now-disintegrated rock particles to form the first tiny bits of soil. Soil permits the growth of more complex plants as well as bacteria and other microorganisms. After a few years, bits and pieces of plant material, both living and in states of decay, become mixed with rock particles, dust captured from the air, drops of moisture, and tiny animal forms, so that soil formation is well on its way.

Even in established soils, plants and animals continue to play important roles. Succession results in a change not only in predominant plant life but in the soils beneath the plants as well. And as the plant life changes so too does the animal life it attracts.

Within soils, roots interrupt the mechanical sifting out of soil particles according to size, and animals burrowing about interfere with expected soil-building processes by causing mixing within the soil material. These activities result in better, more homogeneous soils through which water disperses readily and in which large air pockets are minimized.

Time

All of the foregoing contributors to soil are dependent on the element of time. Time is needed for all of them to play their individual roles and to react to and interact with each other. Few things are achieved instantly, and certainly a productive soil is an achievement that requires a great deal of time. In well-watered temperate regions,

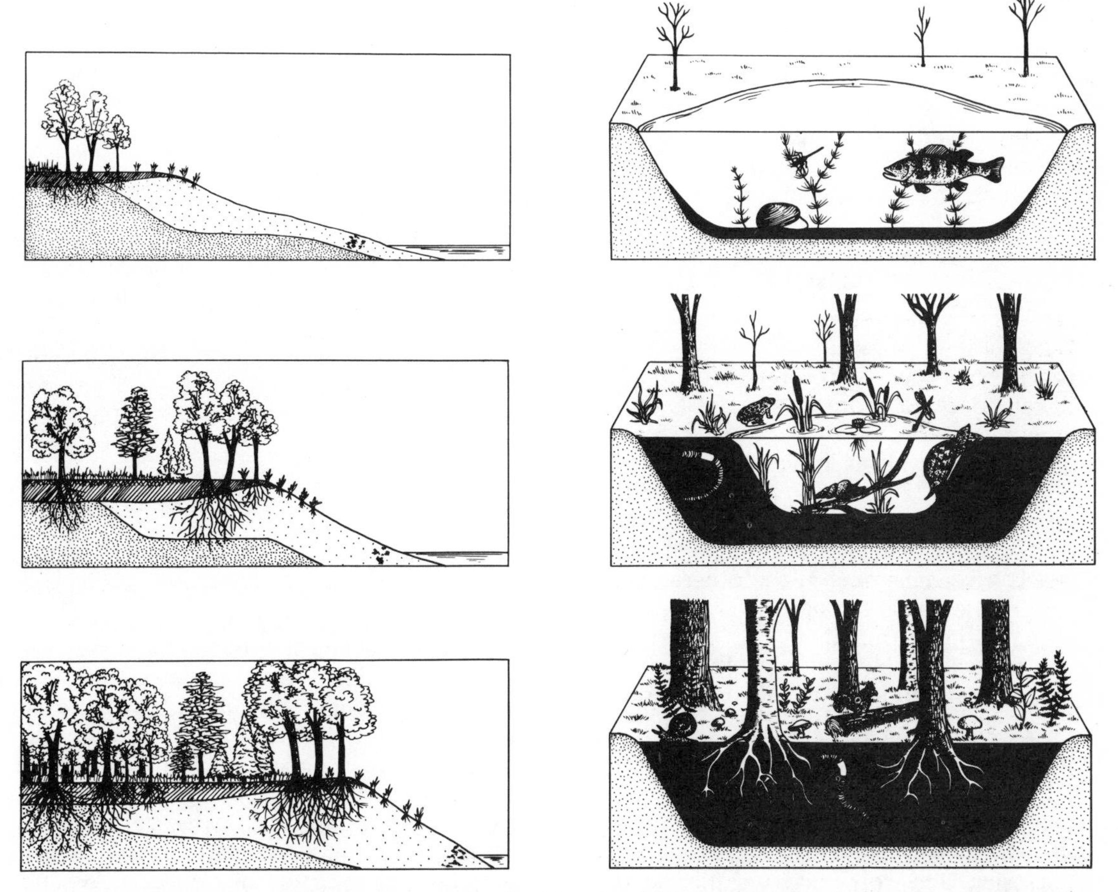

The role of succession in the development of soils. (left column) The sandy shore of a lake is colonized by a few specialized plants, which begin to build up soil. As soil accumulates, larger and larger plants are enabled to grow. Eventually the thick soil of forest floor overlies the original sand. (right column) The floor of a pond is covered with silt and organic sediments. As these accumulate, the pond gradually fills in, and an increasing diversity of plants grows in its margin, in their turn contributing to the soil. Eventually the pond disappears, and the forest grows in its rich soil.

well-developed soil layers will be established in about 200 years in an undisturbed situation. Under poorer conditions where rainfall is scant or temperatures are extremely high or low, considerably more time will be required, perhaps up to several thousand years.

In general, when the vegetation of an area reaches climax, so too does the soil beneath it, and all factors of the environment are in balance. However, just as time can be beneficial in allowing for the erosion of rock and the proper interaction of all the other soil factors, so too can it be detrimental when, with time, particle after particle of good soil is eroded away or important chemicals are leached down to inaccessible lower soil layers. It is important that the environment provide replacements if soil quality is to be maintained.

THE COMPOSITION OF SOIL

Established soils are generally considered to have four interacting parts: 1) soil minerals derived from the parent rock material, 2) decaying organic matter (**humus**), 3) organisms living within the soil, and 4) the air and water in soil spaces. There is inevitable interaction between these components, and a change in one compels a readjustment in the others. The nature of the soil is strongly determined by the part that dominates. For instance, where the water table is close to the surface a marsh can appear; where vegetation decays beneath stagnant waters a swamp may form; and if sands predominate, dunes or desert-like soils may develop.

Soil Minerals

Erosion and continued weathering of rock produce particles of various degrees of fineness from which many essential plant nutrients are derived. These soil particles can be classified by size: (Fig. 2-7).

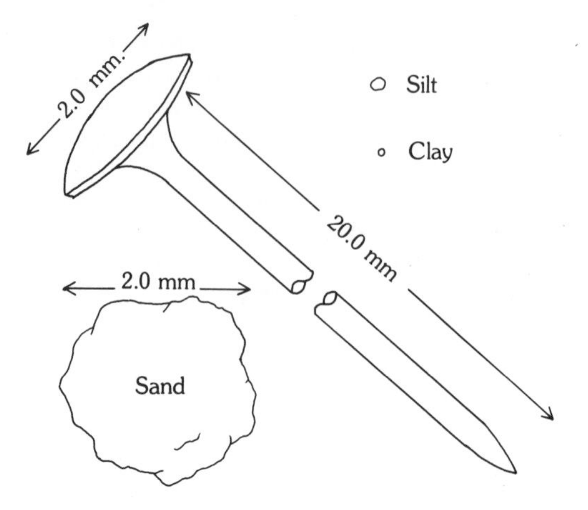

Fig. 2-7. Size comparisons for particles of sand, silt, and clay against the more familiar pinhead.

SAND—2.0 to 0.05 mm. in diameter
SILT—0.05 to 0.002 mm. in diameter
CLAY—less than 0.002 mm. in diameter
Particles larger than 2.0 mm. are called gravel and are not considered to be soil particles.

Established soils contain a mixture of these particles, and the relative amounts of sand, silt, and clay determine the **soil texture** (Fig. 2-8). Texture, in turn, determines the depth and rate of water penetration. (Fig. 2-9). This permeability of soil to water is related to the size of the soil particles and the spaces that exist between them in a soil. Coarse (or **light**) soils are those through which water passes rapidly. They are easily worked and are high in sand particles and other coarse fragments. Fine (or **heavy**) soils have very small pores between particles, so that water moves very slowly through them and retention is high. These soils consist primarily of clay particles and have little or no sand.

Soil can be further examined for **soil structure** which is the arrangement or grouping of soil particles. Sands are com-

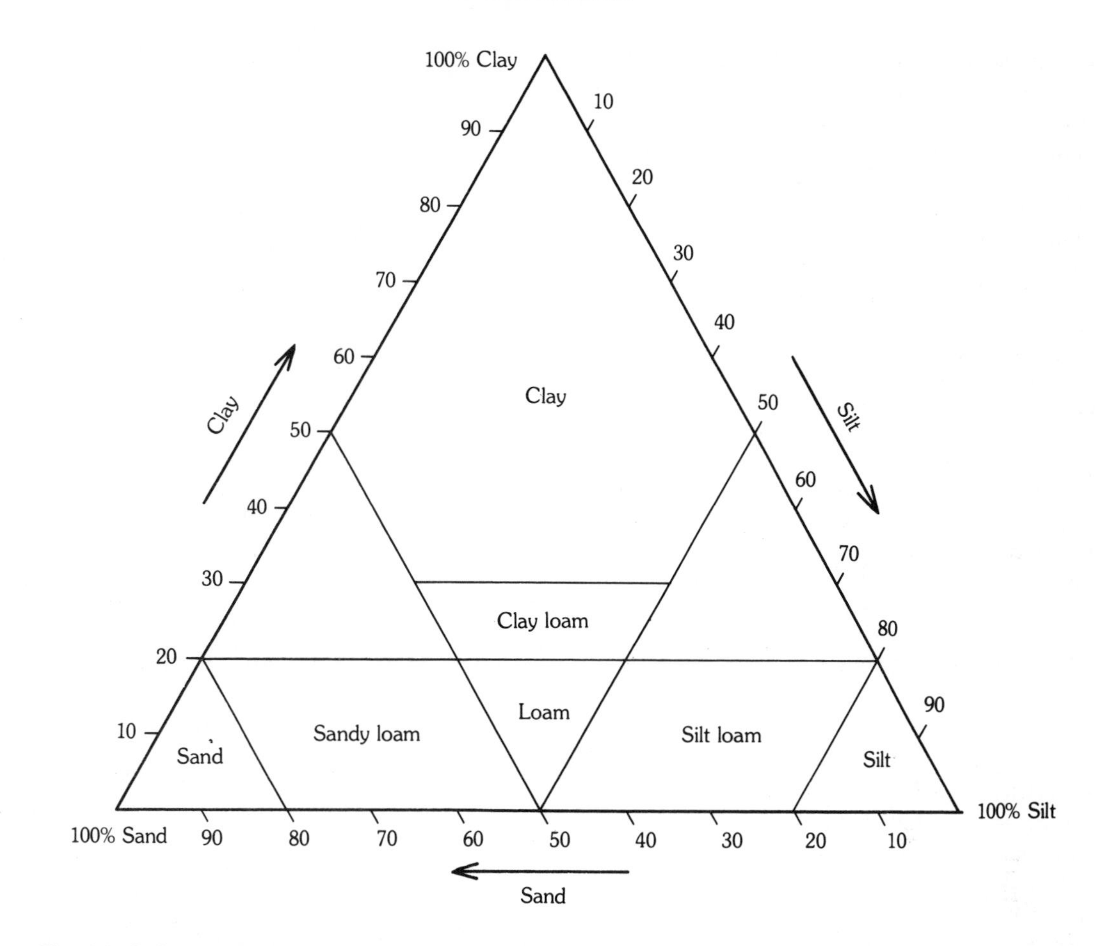

Fig. 2-8. Soil scientists use this system to illustrate the proportions of clay, sand, and silt that are associated with various soils. The soils named here are often further classified into additional subdivisions. (A soil which is 70% sand, 10% clay, and 20% silt is a sandy loam.)

posed of simple grains (usually of granite) that do not cling together and are highly resistant to further weathering. Smaller soil particles, however, gather into spherical globules called **aggregates**. These generally have a more complex porosity and therefore better water-holding capacity and thus form soils that are more habitable for plants than sandy soils. If humus is added this improves the structure of the soil but has no effect on the texture.

Naturally-formed aggregates are fairly water-stable, but when fine soils that are

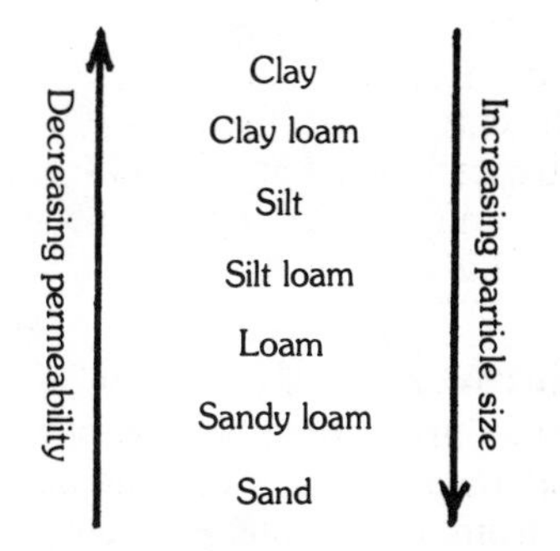

Fig. 2-9. A comparison of the main classes of soils in relation to their permeability and particle size.

very wet are turned over or plowed, artificial aggregates are formed. These artificial aggregates become compact masses of soil called clods in which the natural structure is disrupted and the water-holding ability is destroyed.

Sandy Soils. These large-particled soils are well aerated and easily worked, and in the spring they warm up quickly. On the other hand, because of their hard, crystalline (mostly silicate) nature, they dry out too fast and leach too easily. Under a summer sun they can get so hot that roots can be burned. The addition of clay and/or humus improves sandy soils immediately.

Silt Soils. Silt is fine-grained sedimentary material deposited by wind or water. Water-deposited silt often contains a large propostion of vegetable matter, but wind-deposited silt is usually free of any other materials. Like sand and clay, silt can be mixed with other components to become soil. A good garden soil, usually called **loam**, consists of about equal portions of sand and silt (about 40% each) and a lesser amount of clay (15-20%). Humus is then added to achieve the water-retentiveness required. Such a soil is highly desirable because it is easily cultivated and holds water well. If garden loams are properly turned or tilled so that they contain satisfactory amounts of air and water, they are said to be in good **tilth** and are ideal for plant growth.

Clay Soils. Clay particles qualify as **colloids** because they are of very small size and remain suspended in water for long periods of time. They also influence chemical reactions in soil and are one of the potential sources of nutrients for plants. They are viscous and gelatinous when wet and hard and cohesive when dry. A particle of sand is a single large entity, whereas a clay particle consists of submicroscopic particles with adsorbed water acting as a binding and lubricating agent within the particle.

Plants feed by releasing certain ions and absorbing other ions. These ions move through the soil dissolved in soil water. Clay soils have a high water-holding capacity. They also behave as negatively charged substances. This combination of features means that positively charged nutrient ions (cations) will be held by clay particles (acting as anions) and be available to plants as they require them for nutrients. This ionic-exchange process is discussed more fully at the end of this chapter (p. 68).

Clay soils are of such small particles and pack so tightly that when cold they warm slowly and when hot retain their heat for prolonged periods. They cake and harden so that water runs off the surface rather than being absorbed readily, and in the spring they dry out so slowly that planting must often be delayed. The addition of sand and/or humus will improve the structure and workability of clay soils.

Decaying Organic Matter (Humus)

Any form of decaying organic matter is a potential source of plant nutrients, and its addition improves the structure of almost any soil. It is the gardener's "cure-all" and is essential in a garden loam.

True humus must be distinguished from other organic material. It is the end product of decay brought about by the interacting physical and chemical activities of air, water, heat, and microorganisms. In nature, there is a progressive breakdown of organic matter. As true humus it usually combines with the inorganic soil minerals indigenous to the area. This mixture then becomes the top **layer** or **horizon** of a three-layered arrangement of soils called a **soil profile**. The sequence of layers in a soil profile overlain by organic matter in various stages of decay is especially evident beneath deciduous forests and appears as follows (Fig. 2-10):

TABLE 2-1. Layers of Soil Profile

LITTER	Freshly dropped surface material (i.e. leaves and twigs), undecayed.
DUFF	Partially decayed, but pieces of leaves, insects, etc. still recognizable.
LEAF MOLD	Duff now matted with fungus mycelium and small and microscopic organisms. Leaf mold is limited to forest situations, never occurs in cultivated areas.
HUMUS	Totally decayed organic matter, unrecognizable and of small particles. It grades imperceptibly into the first soil horizon.
A LAYER	This is the top horizon of the soil profile. It is very black in color, consisting of a high percentage of humus with a small fraction of inorganic soil particles and active soil organisms. It is easily penetrated by plant roots and is very water-retentive and subject to leaching. Its depth reflects the age of the forest.
B LAYER	A lighter-colored soil consisting of a mixture of organic material sifting downward from above and mixing with the light-colored inorganic soil from below.
C LAYER (Parent Soil Material)	Light-colored soil of mineral particles derived from eroded bedrock. Color is dependent on the type of rock. It may contain some leached material from above.
PARENT ROCK	Underlying native rock from which soil parent material originates by decomposition, weathering, and erosion.

Fig. 2-10. A soil profile. The A layer (or horizon) is high in organic matter, the C layer is nearly pure eroded parent rock material, and the B layer represents the transition between them.

The quality of soil depends on the presence and functions of organic material. This increases the water-holding capacity of a soil. One experiment showed that:

100 lbs. of sand could hold
25 lbs. of water
100 lbs. of clay could hold
50 lbs. of water
but 100 lbs. of humus could hold
190 lbs. of water.

As organic matter decays, chemical activities occur which release nutrient ions into the soil water that may be taken into the roots.

Any form of organic matter is going to modify soil structure by affecting the aggregation of soil particles. Since humus is the most finely-grained form of organic matter, it is easily incorporated into the aggregates. In cultivated soils, clod formation is prevented by the addition of ample amounts of organic materials. Well-structured soils are loose, moist, and easily worked. They are readily penetrated by roots and soil organisms.

Finally, excessive leaching is controlled by the high water-holding properties of soils rich in organic matter, and nutrient ions are therefore more readily retained in the surface layers.

Sources of Organic Matter

Gardeners need to add organic matter to produce the humus necessary for fertile soils. Flowers are picked and vegetables and fruits are harvested in the course of a season of gardening. These products have incorporated much of the fertile portion of the soil within them and are now removed from the site. The soil must have these nutrients replaced if there is to be a harvest again next year. This replacement can be achieved by adding organic matter and/or commercial fertilizers.

Some of the usual sources of organic additions are:

1. **Peat** or peat moss or, more correctly, moss peat. This is surface material that is almost pure organic matter in a partially decayed state. An area of muck soils is abundant in peat and nearly or entirely lacking in mineral content. The two most common forms of peat available to gardeners are Canadian (or commercial) peat and Michigan peat. Canadian peat is moss peat, nearly pure, partially decayed sphagnum or hydnum moss. It is homogeneous in texture and highly water-retentive. It is of high quality and more expensive to purchase than Michigan peat.

   Michigan peat is a heterogenous mixture of mosses along with pieces of sedges, trees, and shrubs. It is crumbly and dark-colored but not highly water-retentive.

   A good-quality peat has an excellent water-holding capacity. Water will usually constitute 30-70% of the weight of the peat, depending on type. Since peat is almost pure organic matter, it has a high nitrogen content, usually 1-3.5% While it may be reasonably free of weed seeds and detrimental fungi because of its acid quality, it is not usually sterile and may need to be treated before being added to potting soils.
2. **Leaf mold.** This naturally decaying plant material from a forest floor is a long-time favorite of organic gardeners. In its usual state of decay it is very black and highly nutritive. A layer of leaf mold carefully dug into garden soil will loosen it structurally and add greatly to its water-holding capacity. However, if leaf mold is over-used it becomes heavy and prevents proper aeration. One of its drawbacks is that it is not at all sterile and usually contains an abundance of nematodes and potentially harmful fungi and bacteria. Leaf mold derived from pine or oak forests will be acidic while that from beech-maple forests is usually neutral. If leaf mold is applied to a garden the pH requirements of the plants should be

considered (see p. 64).

3. **Compost.** By definition compost is simply vegetative debris allowed to decay to a state where it can be usefully added to garden soil. Almost anything in the way of plant trimmings, grass clippings, raked-up leaves, and even selected table scraps can be composted. In a pure state, compost will decay enough in a year or less to be then mixed with soil and added to a garden. Since a pile of pure organic matter can be unsightly, bad-smelling, and attractive to undesirable insects and other animals, composting is usually achieved by alternating layers of vegetable matter with soil to which commercial fertilizers and lime may be added. Fertilizers will provide the decay-causing bacteria with extra nitrogen, which they require, and the lime will counteract the slightly acid condition produced by decomposition, also reducing the odors.

   This pile should be wetted down and turned over periodically to mix the ingredients together and promote a more uniform state of decay. Composting is an excellent and inexpensive way to add both humus and plant nutrients in a safe and convenient form. (See Chapter 9 for compost formulas.)

4. **Animal manure.** Fresh manure should never be used in a small garden because a great deal of heat is generated during decomposition which could burn growing plants, especially roots, Well-rotted manure is considerably safer to use but will be low in nitrogen which will have been used up during decomposition. Commercial preparations of cow and sheep manure have been dried and sterilized. They are good soil additions because they contain about 2% readily soluble nitrogen for plant use.

5. **Cover crops.** For commercial gardening or crop fields, plants may be grown and then turned under the soil to provide nutrients or to improve the condition of the soil. Although legumes, such as clover and alfalfa, are most often used because of their ability to add nitrogen through the nitrogen-fixing activities within their roots, other plants, such a rye or buckwheat, may also be used, especially if supplemented with nitrogen fertilizers. The turned-under plants are referred to as **green manure**.

Organisms that Live in Soil

Microorganisms, which are so abundant in active soils, play a major role in the composition and formation of a soil. Most of their activities take place only when soil temperatures reach 60° F. (15.5° C.) or higher, the organisms maintaining dormancy at lower temperatures.

Soil microorganisms are of several kinds:

1. **Bacteria**. These single-celled life-forms, that often occur as chains or clusters, are the organisms that we usually associate with the "decomposer" level in a food cycle (see Chap. 1, p. 40). They cause organic matter in soil to decay by secreting enzymes that reduce complex molecules into simpler forms that will serve as nutrients for themselves and incidentally for roots.

   Nitrogen-fixing bacteria are especially appreciated members of the soil community since they are able to convert (fix) atmospheric nitrogen (N_2) into nitrates and nitrites, compounds that serve as plant nutrients (see Chap. 1, p. 45).

   The filamentous bacteria called Actinomycetes have attracted attention in recent years as being especially active in the rapid decomposition of humus-forming materials.

2. **Fungi**. This large and diverse group of saprotrophs includes the molds, mildews, and mushrooms that live and

feed on organic materials. Their vast, thread-like mycelia traverse the spaces between soil particles and attack any organic matter that may be encountered (see Chap. 1, p. 39). Occasionally we see dramatic evidence of the presence of fungi when a large and colorful fruiting body (spore-producing structure) is produced (see Fig. 2-11). The fungal mycelia play an important role in the structure of soils since they tend to hold soil particles together to produce aggregates.

In a specialized association called **mycorrhiza** certain fungi are symbiotically associated with plant roots and assist in the absorption of nutrients. Current investigations indicate that these relationships may be essential for the normal development of many kinds of trees and shrubs.

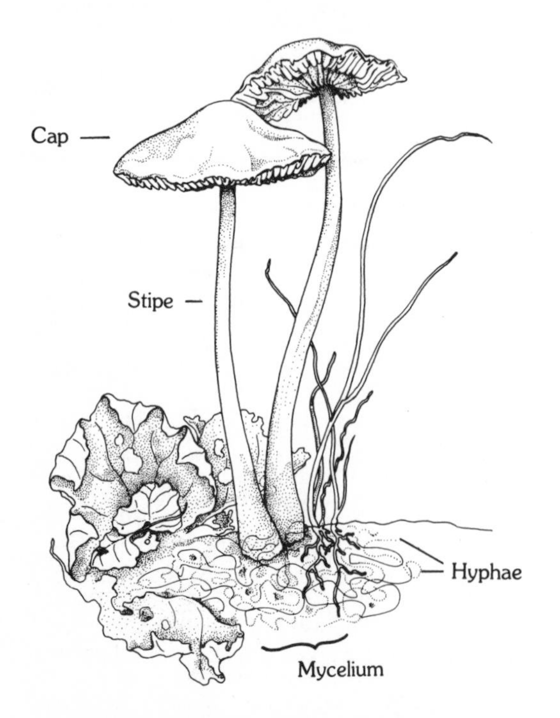

Fig. 2-11. Fungal mycelium in soil and the spore-producing mushroom produced by it.

3. **Protozoans**. These abundant one-celled organisms are of lesser importance in soils but by their very presence contribute to the organic portion of a soil and serve to convert certain molecules into more nutrient forms.
4. **Nematodes**. These nearly-microscopic roundworms are present in vast numbers in soils. Gardeners and turf managers are primarily concerned about the harmful species that feed on and injure root systems. Most of them, however, are free-living and non-harmful and contribute to soils by aiding in the decomposition of organic matter.

Macroscopic soil animals include ants, beetles, insect larvae (grubs), moles, and earthworms as well as passers-by such as mice and ground squirrels. Their main contribution is their ability to loosen soil and mix the various components into a more homogeneous medium.

Earthworms have long been recognized as indicators of soil quality. Loose, humus-rich moist soils have greater populations of earthworms than do tightly-packed, heavy, dry soils. However, their true benefits do not begin to match their highly magnified reputations. They do not contribute significantly to soil fertility, but their movements do break surface crusts and loosen and aerate soils. Their greatest contribution may be as a supply of organic matter when they die.

Roots, as soil organisms, play a significant role in the structure, fertility, and water-holding capacity of any soil. They are chemically active in removing nutrient ions and contribute to the supply of these ions when dead portions are acted upon by decay organisms. In their growth, they fill the spaces between soil particles, retarding water loss and leaching. In limited growing spaces, such as flower boxes and pots, they may become so abundant that they can be detrimental to normal soil processes (see Chapter 7, p. 187).

Soil Air and Soil Water

Close examination of a shovelful of soil will show that only about 50% of the volume is soil particles while the other half is porous space filled with air and water. (See Fig. 2-12). Both large spaces (**macropores**) and small spaces (**micropores**) will be present,

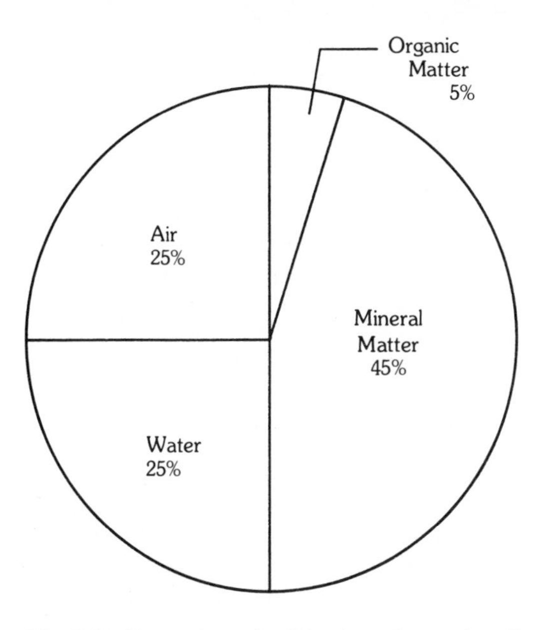

Fig. 2-12. Proportions of solids, air, and water in soil.

and they will be continuous or sporadic depending on the type of soil. A good soaking rain may completely fill all the pores with water for a few hours, but gravity will draw the water downward, especially from the macropores. As the water drains off it is replaced by air. More air replacement occurs as more water is removed by surface evaporation and plant transpiration.

Not all the water is lost, however, much of it being retained in the micropores. These pores are so small that the water in them is not affected by gravity, and random capillary movement occurs which involves the spontaneous flow of water due to the cohesion, or sticking together, of the water molecules.

When the macropores are filled with air and only the micropores retain water, a soil is said to be at **field capacity**. If the capillary water is also drawn off then the soil will not supply plant needs and it will be at its **wilting point**. When soils are **saturated** (soaked thoroughly so that water fills macropores) for prolonged periods, roots will be deprived of the oxygen essential for respiration and may die of suffocation.

Because of the close association between soil air and soil water, the air is nearly always saturated with water vapor. Occasionally a large air pocket will be present that will be filled with drier air. Roots in such pockets can be damaged by drying out. Special care should be taken when transplanting plants to firm the soil carefully and avoid possible air pockets around their roots.

Thus we see that both soil air and soil water are essential for plant growth and, in fact, complement each other. Each plays important roles in the physical environment and in the metabolic activities of every living plant during its entire life.

HELPING SOILS NOURISH PLANTS

Even casual reading of gardening literature will reveal that the soil is the source of many of the gardener's problems in growing healthy plants. Of these problems soil acidity and the availability of plant nutrients are of enough concern that tests have been devised for analyzing soils for these factors. Home testing kits, often quite expensive, are available for the do-it-yourselfers, but soil samples may also be sent to any state agricultural college, usually through the extension service, for a free or inexpensive expert soil analysis and a list of recommendations for improvement. Instructions for submitting soil samples are given in Chapter 9.

We'll take a look at these soil factors in the following discussion.

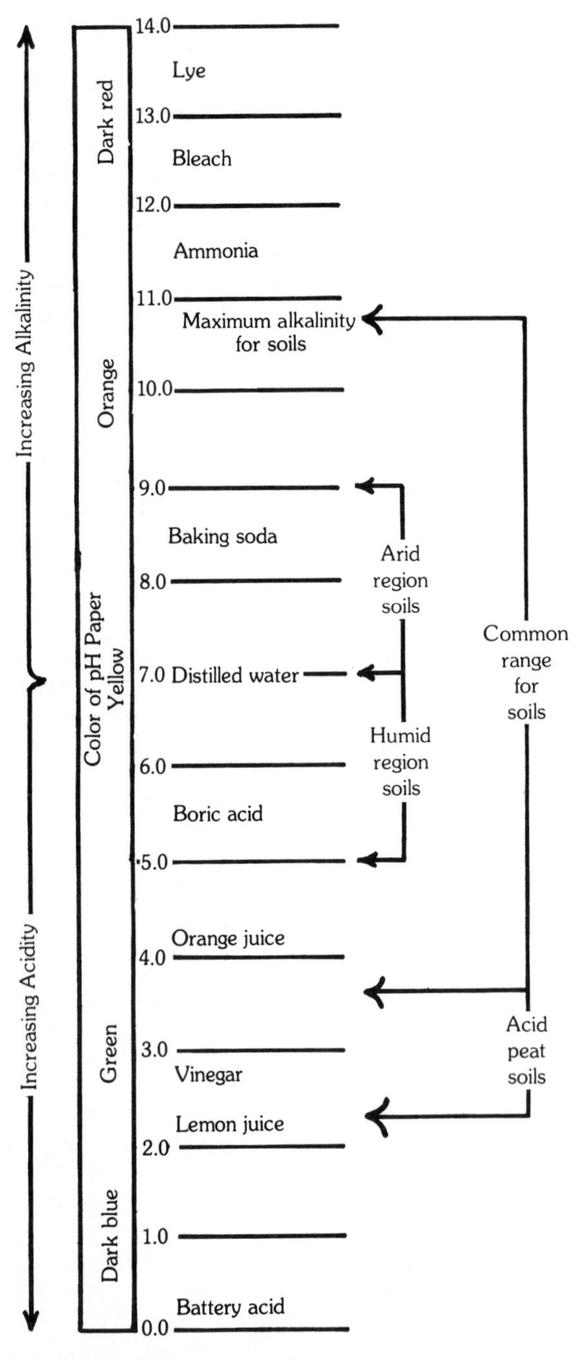

Fig. 2-13. The pH scale showing selected soil ranges and a number of common substances.

Soil pH

The availability of nutrients in soil is affected by the acidity of the soil solution. The acid or alkaline quality of the soil is determined by the concentration of hydrogen ions (H +) and hydroxy ions (OH -) present. If the two kinds of ions are in balance the soil is neutral. If H+ predominate the soil is acid, while those high in OH - are alkaline. Gardeners refer to acid soils as being **sour**; alkaline soils are said to be **sweet**.

Soil acidity frequently affects the growth of plants. Rhododendrons, azaleas, terrestrial orchids, and blueberries are plants which not only tolerate acid soils but grow more successfully in them than in neutral or alkaline soils. Celery, asparagus, mustards, and snapdragons grow best in slightly alkaline soils, and lime (a very alkaline substance) applications are often recommended to produce this quality in their soils. Most plants, however, grow best in soils near neutrality.

Plants, like animals, adapt genetically and physiologically through time to a variety of habitats. It would seem then that the demands of an environment have brought about tolerances or requirements for acid, alkaline, or neutral soils for plants, just as they have affected water and light needs or any others of a multitude of factors.

Soil acidity is expressed in pH units on a logarithmic scale of 14 gradations with #7 being the neutral point. Acid substances measure 0 to 6.9 while the alkaline range is 7.1 to 14. Technically pH is the logarithm of the reciprocal of the H + concentration in a substance in grams per liter. This simply means that at each smaller pH unit the H + concentration increases tenfold. A solution of pH 6 is therefore ten times more acid than one at pH 7, while a pH 5 is only one-tenth as concentrated as one at pH 4.

In a soil analysis it is important to determine the pH of the soil because the pH affects the solubility of plant nutrients in the soil, some of them becoming more soluble as the pH increases, others as the pH decreases.

To meet the needs of plants, there are a number of commercial preparations on the market to help the gardener adjust the pH of the soil. Oak leaves and pine needles have long been championed as soil acidifiers, but their actual effect is slight, and chemical additives such as aluminum sulfate ($AlSO_3$), flowers of sulfur (a powdered form of elemental sulfur), "micronized" iron, as well as a number of "name brand" preparations, are much more effective. For flowerpot volumes of soil, a few drops of vinegar added regularly to the water will suffice. Any of these substances will change the concentration of soil ions so that the $H+$ ions predominate and the soil becomes more acid.

To raise soil into the alkaline range almost any form of lime ($CaCO_3$) will be effective. It will also hasten the decomposition of organic material, by reducing the soil acidity to a more favorable level for soil bacteria, and make soil more crumbly, by changing soil aggregate structures. For houseplants in flower pots, crumbled egg shells or broken oyster shells mixed into the soil are adequate.

Soil Nutrients

As you read in Chapter 1, p. 24, we now know 16 chemical elements which are essential for good plant growth. Many of these elements are in salts that dissolve in soil water to form nutrient ions. See Table 2-2 for a list of nutrient ions usually used by plants (at end of chapter).

From the air and water the elements carbon, hydrogen, and oxygen are derived. Atmospheric carbon dioxide (CO_2) is incorporated into plant carbohydrates during photosynthesis. Water supplies the necessary hydrogen to complete such molecules. Oxygen enters plant cells via leaf stomates and root hairs to be used in respiration. (See Chapter 1, p. 19 for discussions of photosynthesis and respiration).

All the other 13 elements are derived from the soil. The decomposition of basic rock provides potassium (K), calcium (Ca), and magnesium (Mg). Humus and other organic materials release nitrogen (N), while phosporus (P) and sulfur (S) may be derived from either rock or humus. These six elements, in addition to carbon (C), hydrogen (H), and oxygen (O), are **macronutrients** required in large quantities for plant survival, and a deficiency in any one of them could retard growth or cause damage or death.

The remaining elements are **micronutrients** that are only needed in minute quantities. Iron (Fe), boron (B), copper (Cu), manganese (Mn) , molybdenum (Mo), chlorine (Cl), and zinc (Zn) are known to be essential although some other elements have also been detected in plant tissues.

Nutrients may be supplied to plants in many ways, and it matters not at all to a plant whether its nutrient ions are derived from the natural environment or whether the gardener applied supplementary materials in organic mulches or commercial fertilizers. In any event, a plant will absorb what is needed, when it is needed, and any excesses will remain in the soil.

Nitrogen, phosphorus, and potassium are standard components of fertilizers because they are the most likely to be deficient in garden soils. Their relative amounts are usually stated by a three-number series printed on the package. Thus a plant food labeled 7-6-19 means that it contains, by weight, 7% nitrogen, 6% phosphorus (calculated as P_2O_5), and 19% potassium (calculated as K_20). The rest of the fertilizer is inert ingredients. The nutrient elements are always listed in order: N, P, K.

The phosphorus-containing compound P_2O_5 and the potassium-containing compound K_2O are produced by heating samples of the fertilizers. These "ignition residues" are analyzed to determine the relative amounts of P and K in the fertilizer.

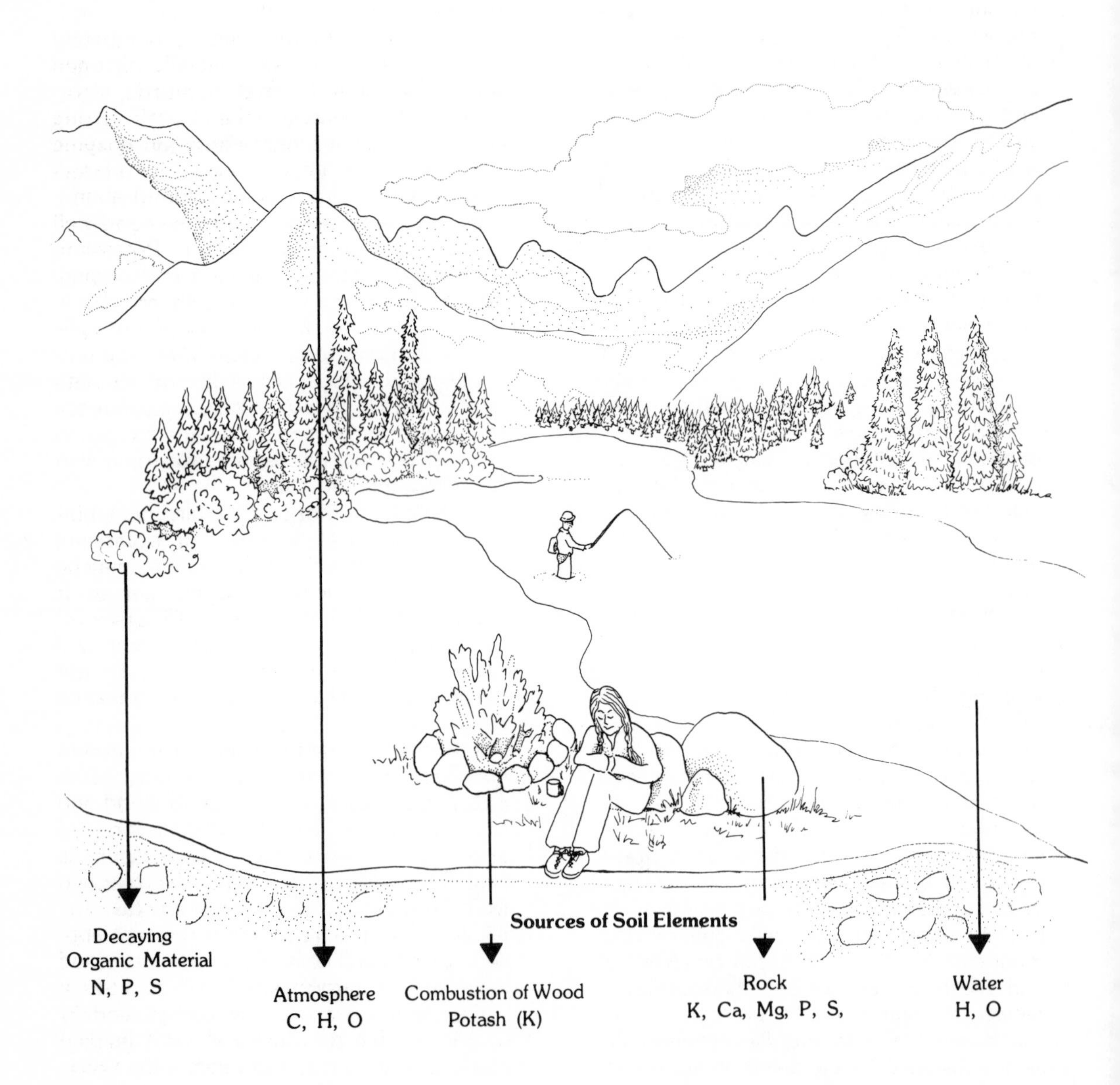

Fig. 2-14. The usual natural sources of essential soil elements.

Nitrogen is essential in proteins, the basic structural substance of living cells. Chlorophyll especially contains large amounts of nitrogen. Plants with adequate nitrogen produce large cells capable of absorbing large amounts of water. This absorption of water in turn influences the absorption of other nutrients by diffusion and osmosis (see Chapter 1, p. 21). Nitrogen thus is important in stimulating vegetative growth, especially of leaves and stems. It is available in a wide range of compounds, both organic and inorganic. If an excess of nitrogen is applied there will often be a luxuriant growth of leaves but little or no flower or fruit development. Fertilizers high in nitrogen benefit lawns but can produce undesirable results in a flower garden.

For good fruit and flower development and strong root systems, soils should have a good supply of phosphates, because phosphorus is essential in cell nuclei. It is also a part of the energy-transformation molecules, ATP and others. Accumulations of phosphorus have been found primarily in meristematic areas of stems and roots and in developing fruits and seeds, all of which have high energy requirements. Phosphorus compounds are not as readily soluble as other nutrients and so are not as subject to leaching. Because of their slow solubility they tend to accumulate near the soil surface where they are readily available to roots, and they are rarely absorbed by a plant in excessive amounts.

Garden stores sell a product called superphosphate to be used as a **top dressing** (to be spread thinly over a soil surface) or to supplement soilless potting mixes. By treating phosphate rock with sulfuric acid, grades of 16, 20, or 45% can be obtained, and these grades are usually specified in recipes for soil mixes.

Bone meal is frequently recommended as a substance to be dug into soils when transplanting perennials such as roses. It is derived from ground-up animal bones and decomposes to provide a little nitrogen and about 20%, by weight, phosphoric acid. However, it is so slowly available that it should be supplemented with superphosphate if a soil test indicates that phosphorus is needed.

Potassium interacts with a number of other plant nutrients, especially nitrogen and phosphorus, to produce sturdy, vigorous, disease-resistant plants with strong roots. It functions primarily as an enzyme activator and as such it aids in the production of sugars and starches, found abundantly in roots, and stimulates good cell growth and protein production. Potassium also affects stomatal opening and closing. When the stomatal guard cells are illuminated, potassium accumulates in the cells and the stomata open; in darkness the potassium leaves the cells and the stomata close. This function will, of course, influence the supply of carbon dioxide and oxygen in leaf tissue (see discussion on stomata in Chapter 1, p. 5).

Potassium is often in fertilizers as potassium carbonate, K_2CO_3, or potassium bicarbonate, $KHCO_3$, both of which may be called potash, or in other salt forms. It requires replenishing since these compounds dissolve readily in soil water and tend to leach from the surface layers and become unavailable to small, shallow-rooted plants.

Wood ashes, an organically derived source of potassium, will contain 2-10% potash depending on the type of wood and the exposure to leaching. Plant ashes are also high in calcium compounds such as lime. The negative ions which are released from both potash and lime will increase the alkalinity of the soils to which they are applied. Gardeners generally count on wood ashes (1) to supply potash, (2) to reduce soil acidity (by increasing alkalinity), and (3) to improve the structure and water-holding qualities of both clay and sandy soils. Overuse of wood ashes can be detrimental, both physically and chemically. Never use wood ashes near plants requiring acid soils.

The other nutrient elements are not normally applied artificially to a garden unless tests show deficiencies exist. Normal soils and water generally contain adequate supplies.

Soil Fertility

Sands and silts are mostly finely ground primary (unmodified) materials, usually quartz. Clay, however, is produced by the transformation of primary materials, mainly from sedimentary rocks such as shale, into complex secondary forms that are colloidal and microscopic and have negatively charged crystalline surfaces and thus behave as anions.

Soil organic matter (humus) consists of tiny amorphous particles whose surfaces also carry negative charges. Both clay and humus serve as potential sources of nutrients by attracting positively charged ions (**cations**) containing elements needed for growth. A cation already attracted to a clay or humus particle may be replaced by another cation and be pushed back into solution. The same type of activity may occur on a root-hair surface where a cation, produced by cell activities, is release into the soil solution and another cation is picked up and absorbed into the root.

Nutrient **anions** (negatively charged ions) also exist, and small quantities are usually in the soil solution. They may also be attached to chemically active sites, par-

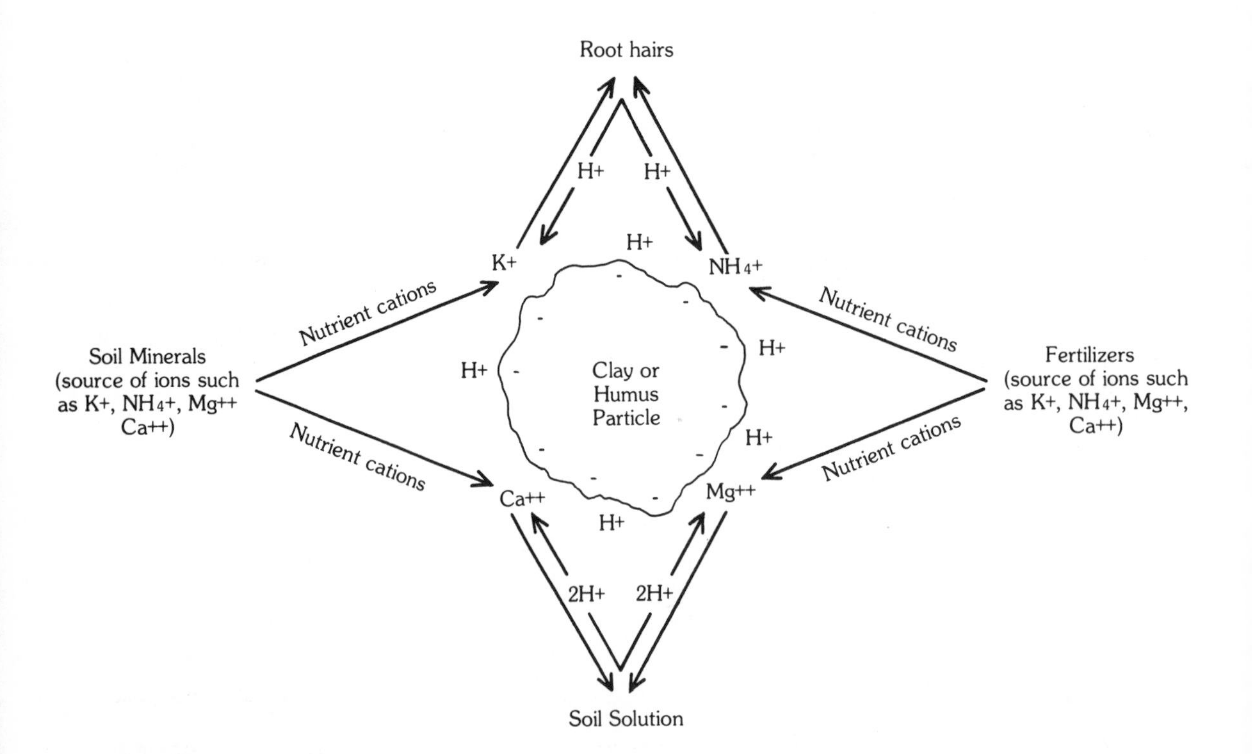

Fig. 2-15. Cation exchange in soils. The negatively charged clay or humus particles attract cations. Nutrient cations are derived from fertilizers and soil minerals. Respiration by root hairs and chemical reactions in the soil produce H ions. When the H ions are abundant, they replace or exchange with the nutrient ions, which then go into the soil solution or are attracted to the slightly negatively charged root hairs to be absorbed by the plant.

ticularly positively charged ones, on clay and humus particles. It is probable that, as with cations, a root cell releases an anion for each anion it uses.

If a fertilizer is added to a soil, it dissolves in the water and dissociates into its component anions and cations. The anions become available in the soil solution, while the negatively charged surfaces of the clay or humus in the soil attract the cations. Later other cations, such as H+ derived from water, will replace them so that they are then free to be picked up as nutrients by plant roots.

This explanation of cation (and anion) exchange is oversimplified since the actual interactions of cations and anions are a complex, tangled series of actions and reactions between soil particles, soil solutions, fertilizers, bacteria, and root hairs. All of these are potentially capable of producing and exchanging ions in the course of their normal functioning. (See diagram in Fig. 2-15.)

In general, the more clay there is in a soil, the higher the cationic-exchange capacity. Much more exchange capacity, however, is provided by the presence of even moderate amounts of organic materials, and they are beneficial in other ways as well.

Fig. 2-16. Soil management regions. Each region is based on a combination of climate, soils, and agriculture that make it different from the others. (See also Table 2-3.) KEY:

NUMBERS 1. North Pacific Valley Region. 2. Dry Mild-winter Region. 3. & 4. Rocky Mountain Regions. 5. Northern Great Plains Region. 6. Central Great Plains Region. 7. Southern Great Plains Region. 8. Mississippi Delta and Coastal Prairies Regions. 9. Midland Feed Region. 10. Lake States Cutover Region. 11. East-Central Uplands Region. 12. Southeastern Uplands Region. 13. Florida and Coastal Flatwoods Region. 14. Northeast Region. 15. Middle-Atlantic Coastal Plain Region (USDA).

TABLE 2-2. Information on Nutrient Elements Known to be Essential for Plant Growth.

Element	Symbol	Usual Natural Source	Ionic Form	Use in Plants	Signs of Deficiency in Plants
Carbon	C	Atmosphere	(CO_2)	Required for organic compounds.	These first three elements are components of the organic molecules that make up the plants themselves. They are unlikely to be destructively deficient since they must be available in the environment for plant life to exist at all.
Hydrogen	H	Water	H^+		
Oxygen	O	Atmosphere	O^{--}, OH^-, CO_3^{--}, SO_4^{--}		
Nitrogen	N	Organic matter	NH_4^+, NO_3^-	Required for many organic compounds, especially proteins. Deficiency common. General component of fertilizers.	Nitrogen must be in compound form to be usable to plants. Lack of this element results in stunted growth, leaves chlorotic between veins, noticeable first in older leaves then in younger ones. Veins reddish, shoots spindly, fruits small.
Phosphorus	P	Organic matter some soil minerals.	$H_2PO_4^-$	Essential for biochemical synthesis and energy transfer. Deficiency common. General component of fertilizers.	Stunted or dwarfed growth, leaves bluish or purplish, especially along edges and between veins. Veins reddish, necrotic, fruits slow to ripen.
Potassium	K	Soil minerals, some organic matter.	K^+	Principal cellular cation. Needed for sturdy plants. Deficiency common. General component of fertilizers.	Monocots pale green with yellow streaks, dicots with leaves bluish or spotted, margins chlorotic, often necrotic. Leaves wrinkled or corrugated or crinkled between veins. Lower leaves affected first, upper leaves later.

Calcium	Ca	Soil minerals	CA^{++}	Required for some enzymes and for strong cell walls. Deficiencies can occur. Lime applications helpful.	Leaves become chlorotic, curled or folled. Meristem break-down to point of death, roots poor, gelatinous, fruits poor or lacking.
Sulfur	S	Organic matter Some soil minerals	SO_4^{--}	Required for proteins and other organic compounds. Deficiencies may occur.	Leaves abnormally light-colored, stems slender.
Magnesium	Mg	Soil minerals	Mg^{++}	Component of chlorophyll, needed in phosphate transfer activities. Deficiencies may occur. Pulverized limestone applications helpful.	Mottled chlorosis between veins, margins yellowed, first appearing on older leaves. Leaves may wilt or be shed, often brittle.
Boron	B	Soil minerals	BO_3^{---}	Essential in some plants but function unknown. Unlikely to be deficient.	Shoot tips die, leaves curled or prematurely lost., internodes short, stems cracked or otherwise deformed, plants dwarfed. Flowers deformed or lacking, roots stubby. General break-down of meristematic tissue.
Zinc	Zn	Soil minerals	Zn^{++}	Required for activity of many enzymes, protein digestion. Unlikely to be deficient.	Leaves chlorotic and necrotic, young growth forming rosettes, leaves lost prematurely.
Iron	Fe	Soil Minerals	Fe^{++}, Fe^{++++}	Essential for many enzymes in photosynthesis and aerobic respiration of carbohydrates and for	Chlorosis between leaf veins first appearing on young leaves. Distinct bleached appearance and termination of growth.

Element	Symbol	Usual Natural Source	Ionic Form	Use in Plants	Signs of Deficiency in Plants
				electron transfer. Not usually deficient.*	
Chlorine	Cl	Soil minerals	Cl^-	Principal cellular and extracellular anion. Not usually deficient.	Deficiency never occurs under normal field or garden conditions.
Copper	Cu	Soil minerals	Cu^{++}	Required for some enzymes in photosynthesis and for iron utilization. Not usually deficient.	Wilted or dying leaf and branch tips, formation of side branches below dead tips on woody plants. Leaf color faded due to reduced pigments.
Manganese	Mn	Soil minerals	Mn^{++}	Required for activity of several enzymes and carbohydrate metabolism. Rarely deficient.	Mottled chlorosis between dark green veins, first on young leaves, later on older ones. Stems yellowish, woody, hard.
Molybdenum	Mo	Soil minerals	MoO_4^{--}	Required for activity of several enzymes and nitrate utilization. Needed by nitrogen-fixing micro-organisms. Rarely deficient.	Light-yellow chlorosis, unexpanded or reduced leaf blades.

*Iron, although present in soil, may not be available to a plant because it may be bound to alkaline components in soil. Soil acidifiers or chelating agents are helpful in such situations.

The following elements occur in plant tissues but are not believed to be essential in any direct way for plant growth: cobalt (Co), needed by nodule bacteria of legumes; silica (Si) and sodium (Na) which may have "supplemental" values for higher plants; fluorine (Fl), iodine (I), selenium (Se), aluminum (Al), arsenic (As), bromine (Br) and nickel (Ni), all of which may play roles as yet undetermined in plant metabolism.

Table 2-3 SOIL MA

Region	Factors Affecting Soil	Soil Conditions
1 North Pacific Valley Region	Affected by warm Japanese current modified by coastal mountains and Cascades. Native grasslands, rain forest on coast. Decomposition slow because of dry summers.	Low in organic matter, acidic sandy or silty, poorly drained. Low in water during growing season. Often lack N, P.
2 Dry Mild-winter Region	Pacific coastal area subject to warm moist winds. Periodic spring floods. Irrigation common.	Excellent alluvial soils in mountain valleys. Small areas of saline or alkaline soils where drainage is a problem. Good cropland except for deserts.
3 & 4 Rocky Mountain Region	Mountainous, heavily grazed. Pacific winds lose moisture before they reach inland areas.	Low in organic matter, lack N. Some soils alkaline. Plants should be salt- and alkali-tolerant.
5 Northern Great Plains Region	Northern extension of central plains. Native grasslands.	Soils deep and sandy. Basically fertile. Climate extremes make plant growing difficult.
6 Central Great Plains Region	Great flat plains subject to high winds, rapid evaporation. Native grasslands.	Soils fertile, but low water and excessive erosion have stripped land of good soil. Need N, P.
7 Southern Great Plains Region	Rarely forested, natural grassland and shrubs.	Slightly acidic to calcareous. Soils sandy or silty loams, sometimes with clay subsoils. Productive with good rainfall.
8 Mississippi Delta and Coastal Prairies Regions	Area formed from alluvial materials from Mississippi. Southern part extends along Gulf. North area consists of flood plain of Mississippi. Mostly low flat land.	Soils fertile but poorly drained. Vary from sand to clay to rich river silts and associated loams. Mostly deep, fine-textured. May lack P, N.

ɜEMENT REGIONS

Climate	Rainfall	Low Temperatures	Frost-Free Days (Growing Season)	*Hardiness Zones
Cool, humid	30-100″, mainly winter; summer droughts	Mostly mild, can freeze.	200	7, 8, 9 spots of 5, 6
Mild to semi-desert	30-60″ north, 3-12″ south; winter	Winters mild, but frost common in north, deserts.	200-300	9-10
Arid, cool to hot	nearly 0-50″ Irrigation necessary	Below freezing.	Variable, few to 250	4, 5, 6, 7, 8,
Dry, subhumid to semi-arid, extreme temperatures.	Irregular, 12-22″. Gardens must have supplemental water.	-20°F.	About 100	3-4
Cool, dry	Irregular, 14-38″. Gardens must have supplemental water.	Can go below 0°F.	140-210	5-6
Semi-arid west to humid east.	16-46″ but uncertain. May-Sept.	Extreme low temps. to mild.	180-300	7, 8, 9
Hot, humid to nearly tropical.	34-62″, sometimes deficient in summer	Sometimes below freezing, esp. in north.	180-300	7, 8, 9

***Fig. 9-7.** Hardiness Zone Map.

Region	Factors Affecting Soil	Soil Conditions
9 Midland Feed Region	Level to gently rolling land, glacial effects and deposits. Muck and peat in low areas. Soils formed under prairie vegetation.	Soils well-structured, fine, high water-holding capacity. Fertile, productive. Forest soils acidic, low in organic matter. Prairie soils deep, much humus.
10 Lake States Cutover Region	Extensively glaciated, many lakes, bogs. Cool moist forests.	Light-colored, acid, rather infertile, low in organic matter. Glacial soils gravely. Peat and muck in bog areas. Often need lime, P, K.
11 East-Central Uplands Region	Native vegetation was deciduous forest. Underlying materials mostly sedimentary. Limestones, shales abundant.	Subsoil with high clay content, slow permeability. Surface layer acidic, medium texture, low to moderate fertility, well drained. Stony soils in Appalachians. Limestone-derived in Kentucky, Tennessee. Likely to lack N, K.
12 Southeastern Uplands Region	Forested area, badly managed agricultural land in past. Land sloping to hilly, subject to erosion, hard rains.	Soils subject to extreme weathering and leaching. Acidic, low in nutrients, low water-holding capacity. Lack N, P, K.
13 Florida and Coastal Flatwoods Region	Low elevations, high water table. Underlying limestone.	Sandy, coral, limestone, peat-muck soils. Mostly acidic, poorly drained except in central Florida. Lack N, P sometimes Cu, Z, Bo, Mg.
14 Northeast Region	Widespread forests. Erosional materials from Applachians. Area glaciated, long cultivated. Parent materials crystalline-igneous or lake-plain sands and clays.	Alluvial deposits, many boulders. Soils acidic, thin top soil. Good to poor drainage. Lack lime, P, K.
15 Middle-Atlantic Coastal Plain Region	Coastal edge with alluvial deposits. Gentle slopes, some forests.	Soils mixture of marine, alluvial and glacial materials. Sandy loams have good drainage. Large areas of poorly drained soils. Swamps and marshes on coast. Most soils slightly acidic, badly leached. Need N, K.

Climate	Rainfall	Low Temperatures	Frost-Free Days (Growing Season)	Hardiness Zones
Cool, moderate humidity.	22-40″, summer	below 0°F.	140-180	4, 5, 6
Cold.	25-35″, summer	-20°F.	80-140	3, 4, 5
Humid, wide temperature range.	40-80″, summer; droughts frequent	-10 to 0°F.	120-210	6-7
Warm, humid.	35-60″. Oct.-March deficient in summer	10 to 20°F.	200-260	7-8
Warm, humid, mild winters.	46-64″, summer	20 to 40°F.	250-365 occasional cold waves	9-10
Cool, humid.	25-45″, spring-summer	-20°F.	90-200	3, 4, 5, 6
Mild, humid.	36-56″, summer	-20°F.	180-290	7-8

SOIL AND POTTING MIXES FOR PLANTS

There are probably as many "sure-fire" formulas for good growing media as there are gardeners. Everyone has a favorite, and many an argument has resulted for discussing the merits of this mix or that. With the introduction of the soilless mixes a whole new dimension was added and the converts to soilless growing took on the "I'll never-change" soil-mix supporters with enthusiasm. There are, of course, points in favor of soilless mixes just as there are for soil-based mixes, and some of them are discussed in Chapter 7. Below are some typical formulas for both soil and soilless mixes. Commercial products and chemical additives are available at most garden stores.

Soil Mixes*

These are based on combinations of topsoil or good garden loam, some form of humus, and sand for drainage. Additional materials may be helpful but are not usually essential.

Bush-Brown All Purpose Mix

1 part sand
2 parts fibrous loam
1 part humus
½ part well-rotted cow manure
Add a 5″ potful of bone meal per bushel.

Kitchen Sink Mix

6 cups loam or packaged potting soil
6 cups humus or peat moss
6 cups sand
3 cups vermiculite, perlite or combination of the two
1 cup gravel
1/3 cup packaged cow manure
1 tbsp. steamed bone meal
1 tbsp. limestone or crushed eggshell.

*All measurements by volume.

For succulents, double the sand or add an equal amount of perlite to either of the above recipes. For increased water retention for water-loving plants, double the humus portion. For more specific recipes try those below.

Succulent Mix

(for plants with thick, fleshy leaves, heavy stems and thorns)
1 part loam
1 part humus
3 parts sand
Add lime if acidic leaf mold or peat moss is used.

Light Gardener's Cactus Mix

Mix together equal parts of good garden soil and coarse sand.
Add very small amounts of leaf mold, bone meal, rotted manure and agricultural lime.
For Christmas Cactus and similar plants use more leaf mold and humus in the mix and substitute perlite, bark chips or osmunda fiber for sand.

Pelargonium Mix

(for plants that tolerate a loamy or heavier soil mix)
2 parts loam
1 part humus
1 part sand.
Add lime if acidic leaf mold or peat moss is used.

Many good gardening books are written by British authors who commonly refer to John Innes Mixes or variations of them.

John Innes Mix

Basic potting mix
7 parts loam
3 parts peat
2 parts coarse sand

For John Innes Mix #1
Add to basic mix, per bushel:

1/4 lb. complete fertilizer
1 oz. ground limestone
For John Innes Mix #2
Double above amounts for each bushel of basic mix.
For John Innes Mix #3 and #4
Triple or quadruple the amounts respectively.

Soilless Mixes

The University of California originated soilless mixes back in the '40's by combining sand and peat moss with nutrients. Following their lead Cornell has developed, since the '60's, similar but sandless formulas. Both the Cornell and U-C mixes must have fertilizers added regularly.

U-C Peat-Based Mix

Use equal volumes of sphagnum peat moss and sand. To each bushel add;
1 t. potassium sulfate
7 tbsp. dolomitic limestone
2 1/2 tbsp. agricultural limestone
2 1/2 tbsp. superphosphate-20%
1/2 t. chelated iron or fertilize regularly with high nitrogen fertilizer.

The Cornell Peat-Lite Mixes developed at the New York State College of Agriculture are being used for a wide variety of plants. The mixes are light weight, easy to handle, and water-retentive but drain readily and do not remain soggy. In addition, the formulations are more consistent with these materials than are possible with natural soils that often contain unknown ingredients.

Cornell Peat-Lite Mix

1/2 bushel sphagnum peat moss
1/4 bushel vermiculite
1/4 bushel perlite
Add: 5 tbsp. ground limestone
2 1/2 t. superphosphate-20%
7 1/2 t. 10-10-10 fertilizer
3 tbsp. granular wetting agent

Some formulations call for additional potassium nitrate (14-0-44) and iron sulfate for some plants (e.g. ferns) and seedlings.

A less precise but effective general soilless mixture is the following:

Soilless Humus Mix

1 part sphagnum peat moss
1 part perlite
2 parts leaf mold or rotted humus
To hold water longer or for a light weight hanging basket mix:
1 part sphagnum peat moss
1 part perlite
1 part vermiculite
Since terrariums are drainless, the soils need to be lightened and not over-watered:
1 part sphagnum peat moss
1 part sphagnum moss
1 part leaf mold or rotted humus
1 part sand and gravel

Epiphytic plants need water only occasionally and resent overly damp soils. Fir bark is the basis for these mixes:

Cornell Epiphytic Mix

(also recommended for light gardens)
1/3 bushel ground fir bark
1/3 bushel sphagnum peat moss
1/3 perlite
8 tbsp. dolomitic limestone
6 tbsp. superphosphate-20%
3 tbsp. 10-10-10 fertilizer
1 tbsp. iron sulfate
1 tbsp. potassium nitrate
3 tbsp. granular wetting agent.
Supplement with 20-20-20 fertilizer once a month.

Easy Epiphytic Mix

1 part sand
1 part chopped sphagnum moss
1 part medium grade fir bark
Fertilize lightly with each watering.

Plants growing under lights have special needs and the Indoor Light Gardening Society of America recommends the following three recipes for such plants. The proportions may be slightly varied for specific plants.

Tropical and Epiphytic Plant Mix
(for plants with coarse, tuberous or rhizomatous roots)
1 part sphagnum peat moss
1 part vermiculite
1 part perlite
For each 7 quarts of mix add 1 tbsp. ground limestone or dolomitic lime or 5+ cups crushed egg shells.
Fertilizers—alternate balanced type (20-20-20) with one high in phosphorus (10-30-10). Use 1/10 to 1/8 recommended amount with each watering.

African-Violet Mix
(for plants with fine root systems that require water-holding soils)
3 parts sphagnum peat moss
2 parts vermiculite
2 parts perlite
1 tbsp. lime for each 7 quarts mix
Slow-release fertilizer or fertilize with complete fertilizer on regular schedule.

-or-

2 parts sphagnum peat moss
1 part vermiculite
1 part perlite

Succulent and Cactus Mix
(for good drainage and low water-holding capacity)
1 part sphagnum peat moss
2 parts vermiculite
2 parts perlite
1 tbsp. lime for each 7 quarts mix.

Potting-Mix Components

Some of the soil recipes contain ingredients that sound very mysterious indeed. Herewith a glossary that should explain some of the terms and offer alternatives.

Topsoil, garden loam, packaged soil mix—All three are mostly natural soil materials. If garden soil is used be sure it has been composted and fertilized regularly in the past. Any old "backyard dirt" just will not do. Commercially packaged soil mix is the safest to use because it is sterilized and especially prepared. A bag of *soil* mix will be considerably heavier than an equal-sized bag of *potting* (soil-less) mix.

Sand—It is best to use "sharp builder's" sand available at lumber yards and garden stores or nurseries or from clean sand dunes. Particle diameters of 1.0-0.5mm. are much better than those of very fine sands that may be as small as 0.10-0.05 mm. Fine sands pack too tightly for plant roots. Ocean and lake sands are usually too salty and/or dirty to be used unless washed throughly and sterilized.

Gravel—Particle size should be just slightly larger than coarse sand. Bird-cage gravel or aquarium gravel are usually acceptable for use.

Perlite—A heat-expanded lava material that may be used in place of sand. It is light-weight, sterile, and neutral or slightly alkaline.

Vermiculite—Heat-expanded mica that may be used in place of leaf mold for water-holding ability. It is also light-weight and sterile, but avoid the very fine-sized type that packs too tightly. The "chunkier" (No. 2) type is preferred.

Humus — leaf mold — compost—All refer to organic materials that may be in various stages of decay. They provide good natural nutrients, improve soil structure and increase water-holding capacity. Any one of them will do.

Manure—Usually described well-rotted cow manure or commercially packaged cow manure. An organic source of nitrogen that is low in yield but offers some improvement to soil structure as well.

Sphagnum Peat Moss or Fibrous Peat Moss—A high-quality moss peat consisting of partially decayed sphagnum moss. It holds water well, but once dry it is hard to rewet. Slightly acid, so lime is usually added to neutralize.

Acid Peat or Humus Peat Moss—Similar to above but more decayed and finer textured. It will be more acid than fibrous peat because its density is about double. **Sphagnum Moss**—Comes in long fibers or in a milled form. Derived from living sphagnum moss plants that are collected and dried. Good water-holding capacity, neutral pH and nearly sterile.

Fir Bark—Chips or pieces or Douglas, white or red fir or redwood tree bark. Particle sizes range from fine (1/8-1/4") to coarse, a size suitable for walking paths. Packaged mixes often contain peat moss, osmunda or palm fiber and/or perlite along with the bark. Check the label. Usually acidic.

ADDITIVES:

Lime—Used to neutralize acid soils or counteract acidic ingredients and supply calcium. Agricultural ground limestone may be dolomitic or calcic or a combination of both. Dolomitic limestone is $CaMg(CO_3)_2$ and contains magnesium. Calcic limestone is $CACO_3$ only. Ground egg shells are a suitable alternative.

Gypsum—$CaSO_4$ will also supply calcium but has no influence on soil pH.

Superphosphate—Highly regarded as stimulant to strong root growth, etc.

Bonemeal—Yield 20% phosphorus. Slow to decay.

Iron Sulfate, Iron Chelate, & Potassium Nitrate—These are usually added to prevent chlorosis and produce strong growth. They are often found in soilless recipes, but a regular fertilizing program should make these separate additions unnecessary.

Borax—For boron deficiency and/or to neutralize alkaline soils. Plants are likely to suffer if too much is applied but will not really notice if it is left out altogether.

Mag-Amp and Osmocote—Slow-release fertilizers that are usually effective for 3-4 months. Replace with a new supply or substitute a regular fertilizing schedule at the end of the time period. A 4" pot will need 1 tablespoon of Mag-Amp but only 1 teaspoon of Osmocote.

Complete Fertilizer—A fertilizer containing all three essential elements—N, P, and K. Proportions of each will vary with the type and commercial brand.

Granular Wetting Agent—An additive for reducing water tension for increased penetration of water and nutrient solutions. Especially useful peat mixes since they dry out easily and are difficult to rewet.

3. SEEDS AND SEED PLANTS

Once upon a time there was a man named John Chapman (1774-1845) who grew seedling apple and peach trees on the frontier lands of Pennsylvania. Although apples were not a native plant, they became a staple of the pioneer families that moved into the American west through Pennsylvania and Ohio, and a producing apple orchard was often a prerequisite for even filing a land claim. No other fruit could be started so easily or be put to so many uses. They were eaten fresh, baked, or cooked or turned into apple butter, and late every autumn the excess was hauled by the wagon load to the cider mill to be crushed into a clear golden juice for cider, vinegar, and applejack.

Many an enterprising nurseryman recognized the value of seedling trees, and seeds were gathered from the cider presses and sown in nurseries so that the trees could be sold to the settlers as they established their homes in the new territories. John Chapman was one of these men who would collect a bag of seeds at a mill and return with them to a plot of land, plant them, enclose the area with a brush fence, and provide a meager bit of cultivation until

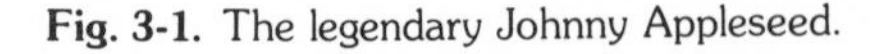

Fig. 3-1. The legendary Johnny Appleseed.

the trees were sold or given away or, in some instances, simply abandoned to become orchards. John moved his orchard nurseries with the frontier and along the way acquired a reputation that established him as one of America's folk heroes. He lived as the Indians did, rarely wore shoes, and shaded his face from the sun with a homemade pasteboard hat. The stories told by and about him became legend and he became the beloved Johnny Appleseed.

Although the life of John Chapman has been fantasized into colorful stories, his efforts still typify those made in the past to improve and disseminate fine food plants with seeds. It has been primarily through seeds that plants have found their way from their places of origin to distant places. Seeds have been responsible for feasts and famines, the rising and falling of dynasties and civilization. Johnny Appleseed, John Bartram, Marco Polo, and thousands of lesser-known people have been responsible for transporting and distributing seeds over the globe.

MENDEL AND DARWIN—A STARTING POINT

Another modest man played an even more significant role in seed history. Gregor Mendel (1822-1884) was an Austrian monk who discovered the basic laws of heredity. It was, in fact, because of Mendel's work that we now also understand how evolution, described by Charles Darwin (1809-1882) in 1859, is actually accomplished by plants and animals.

Briefly, evolution is simply a series of adaptations made over a very long period of time. Adaptations are the adjustments made by a population of plants or animals to survive in a particular way of life and reproduce in a particular environment. We must keep in mind that an organism must be adapted to the total environment in which it lives, and its survival depends on how well it fits that environment. The members of any population are slightly varied; that is, they are similar but not identical. The individuals that are endowed with the most favorable characteristics for living in the environment will generally survive the longest and produce the most progeny, and these will inherit the favorable characteristics. The less well-adapted individuals will have a more difficult time surviving and will, accordingly, leave fewer progeny, so that in time these variations will decrease in numbers or even be eliminated from the population. Thus a population adapts in the direction of those features that its environment demands for survival and fitness.

These were the kinds of observations that Darwin described in great detail to support his theory of **natural selection**, and today most of his observations and conclusions have been verified and are widely accepted. Natural selection, as the guiding force in evolution, was used by Darwin to explain *why* evolution proceeds as it does. He did not, however, have an explanation for the cause of the variations that occur in populations. It was Gregor Mendel and his studies of heredity that have provided the answers that Darwin did not have.

Mendel's experiments with the garden pea (*Pisum sativum*) demonstrated that characteristics are inherited in certain predictable ways by the transmission of "factors" from the parents to the progeny. Today, of course, we know that those "factors" are genes consisting of units of the molecule DNA contained in the cell chromosomes. In both plants and animals the genes in the egg and in the sperm establish the genetic or hereditary make-up of the new individual at the time of fertilization.

If we look at a few of the more significant changes that have occurred in plants through time, we see that they represent a series of adaptations to the changing environments of the geologic past. With the establishment of photosynthesis as a means of providing food in some primeval organisms, the related but separate worlds of plants and animals began. Animals continued to move about seeking their food and became generally more mobile. Plants, capable of making their own food, found it advantageous to stay in one place once they were in a light-providing locality rather than be subject to being washed or carried away to an unfavorable place. The ability of some tissues to anchor the plant and to absorb water and minerals led eventually to the development of roots. Some of the upper parts evolved to become more efficient food-making structures with expanded light-absorbing surfaces, which became leaves. Sturdy supporting stalks became stems that transported water and food and also lifted the leaves high enough for good light exposure and carbon dioxide and oxygen exchange.

Just as animals developed many diverse means of obtaining and utilizing food, withstanding various environmental conditions, and reproducing their own kind, so also have plants succeeded in developing specialized structures to enable them to survive in a wide variety of habitats. Plants, because of their immobility, have made structural adaptations especially in their stems and roots because these are the most permanent or longest-lasting plant organs and will be more subject than flowers or leaves to the rigors of the habitat. For instance, cacti exhibit many structural modifications that clearly allow them to survive in the challenging environment of a desert (Fig. 3-2). The most notable is a vast, fibrous root system close to the surface for rapid absorption of the infrequent rains. In addition, leaves are transformed into spines which are incapable of transpiration, and this reduction of surface exposure prevents excessive water loss. Thick, green, fleshy stems carry on photosynthesis and provide storage tissue for both food and water. Stomata occur on these stem surfaces. Leaves of many other plants often show interesting adaptations also, including being easily shed and replaced more or less seasonally. Flowers are the least resistant to the physical environment, since they need to make only a brief appearance. If the flower can be maintained just long enough to set seed, the goal is accomplished. Thus we find that many plants, though not at all closely related, often show similarities of root, stem, and occasionally leaf when they share the same environment, but their flowers reflect adaptations typical of their own particular family of plants (Fig. 3-3.)

Fig. 3-2. A cactus has fleshy stems and an extensive fibrous root system. The roots spread close to the surface of the soil where they rapidly absord any moisture available.

All of these adaptations have occurred by the accumulation of many slight genetic changes through many, many generations. These variations may be due to new combinations of genes in some offspring or to genetic mutations, which are slight but inheritable changes in the chemical structure of the genes. Mutations are caused by environmental factors such as radiation, chemicals, extremes of temperature, or by spontaneous "molecular instability" of DNA, a convenient phrase for naming what we cannot at present explain. Some of these changes are beneficial to the organism in its present habitat while others may be detrimental, and although almost any characteristic may be significantly influenced by the environment (Fig. 3-4), no plant can adjust itself to its environmental conditions more than its inherited genes permit. For example, a plant may grow

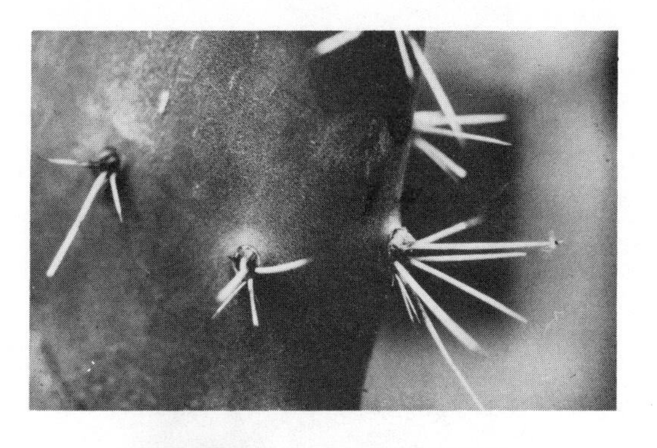

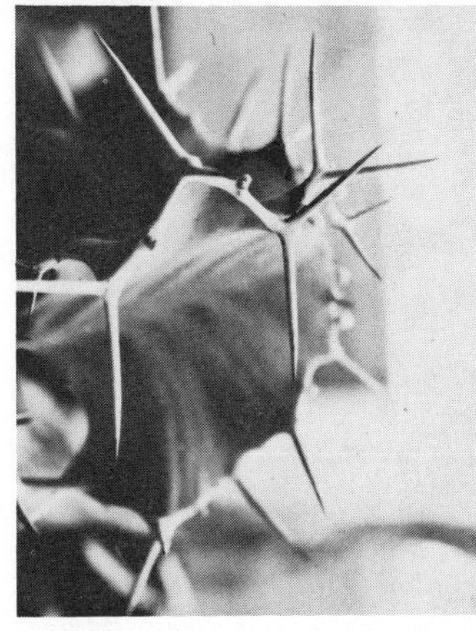

Fig. 3-3. Above, *Opuntia* sp. of the family Cactaceae sprouts spines from cushion-like areoles on its flat green stems. Below, *Euphorbia grandicornis* of the family Euphorbiaceae has succulent stems edged with long spines. Members of both families have made numerous adaptations to survive in hot, dry habitats.

Fig. 3-4. Two *Peperomia* plants that show the effect of soil on plant growth. The plant on the left is growing in well-watered but sandy, low-humus soil, while the plant on the right is in a rich, loamy soil.

larger and be healthier if its water supply is plentiful, but it does not follow that even more water will be better; the plant may succumb to a surfeit of water. The genes thus will establish a basic structure for a plant that can be, within limits, modified and/or influenced by the environment.

By selecting plants with tendencies toward or possession of desirable characteristics, from our point of view, and propagating these plants in preference to those with less satisfactory traits, we have turned our mid-continent grasslands into the "breadbasket of the world," we have fungus-resistant potatoes, an unbelievable array of delicious vegetables and fruits, rainbows of delightful flowers, and sturdy, well-formed trees and shrubs. And most of this progress has been accomplished since 1900 when Mendel's work on inheritance became known.

FLOWERING PLANTS AND THEIR GROWTH CYCLES

Seeds are in simple cones in most gymnosperms but are in many kinds of fruits in flowering plants. Similarly, the flowering plants have more variation in their growth cycles and forms than do the gymnosperms, which are all woody perennials, and this allows the flowering plants to live in all but the most forbidding climates, such as those of the poles. Plants are basically annuals, biennials, or perennials according to their growth cycles, and the differences in the cycles of angiosperms are discussed in the following sections (also see Fig. 3-6.)

Annuals

The seeds of annuals produce herbaceous plants that survive just one growing season. Seed is the only natural means by which these plants continue as a species over the years. The growing season for an annual varies from a brief six weeks or so, as in high alpine areas, to almost a full year as in some sub-tropical regions, and the

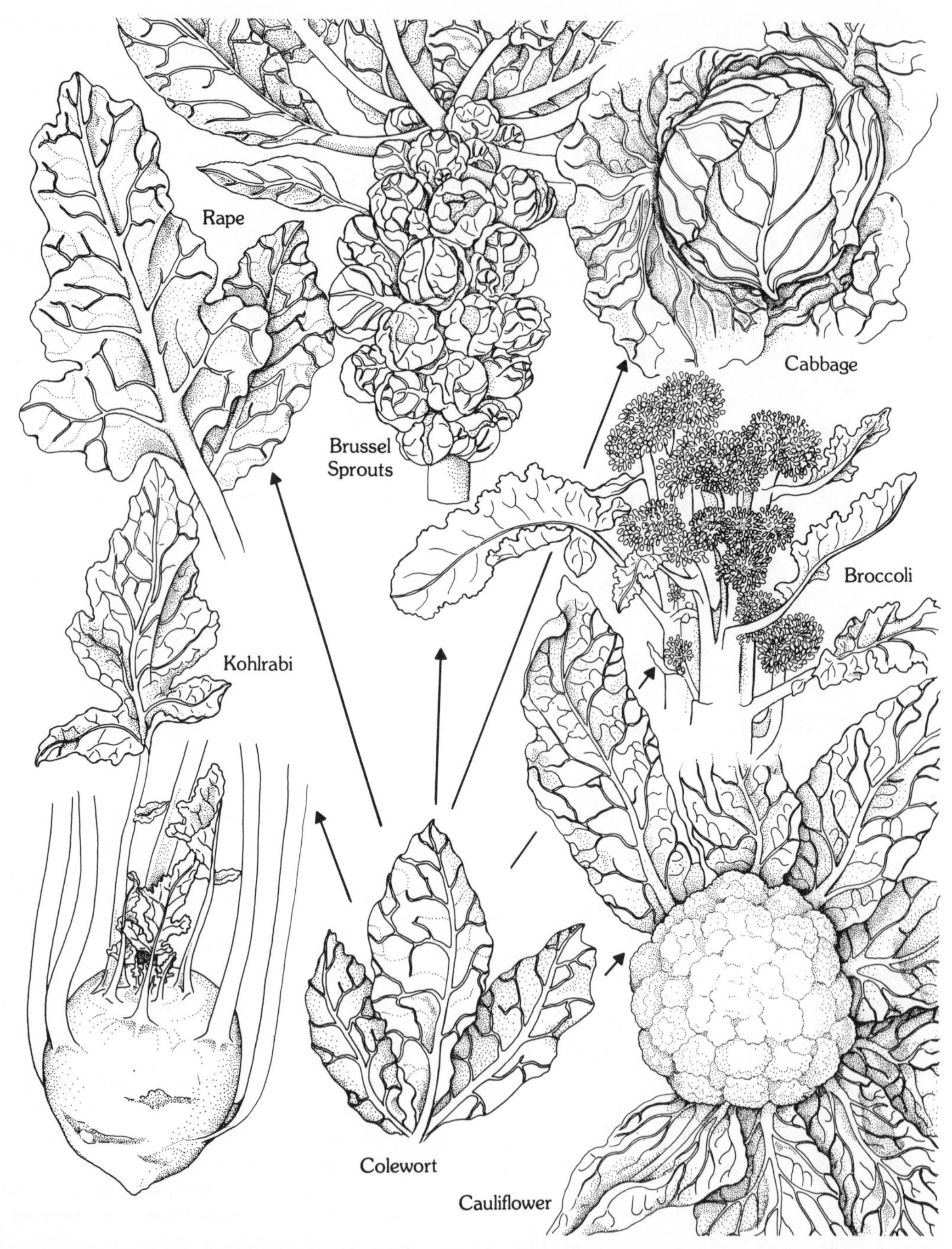

Fig. 3-5. Artificial selection over the centuries has modified the original wild cabbage *Brassica oleracea* into a wide variety of productive and delicious vegetables.

seeds and their requirements for germination reflect the demands of the environment in which the plants grow.

Many annual species evolved as early succession plants adapted to grow quickly and prolifically wherever and whenever spaces occurred that were unoccupied by the more long-lived species. Many of them are common today as the weedy pioneer colonizers of vacant lots and abandoned fields. Their seeds are adapted for easy dispersal and germination, and the plants grow rapidly since they must establish themselves before the larger, slower-growing plants arrive to compete with them. Certain other species of annuals are highly adapted plants in especially challenging environments, such as those that are notably hot, cold, or dry. Unlike the long-lived perennial plants that have evolved special vegetative features for existing in these environments, annuals produce seeds that have special structures or requirements that prevent germination, but maintain viability, until environmental conditions are totally favorable. Where conditions occur only rarely and/or briefly, a fast-growing annual will sprout, flower, and produce seed in the short time available before the environment becomes intolerable again.

Seeds of some of our common garden annuals such as radishes, lettuce, and certain legumes can be planted so that more than one crop can be produced in a single growing season if planting is scheduled carefully. Cold-sensitive perennials, such as snapdragons, grown outdoors are often treated as annuals in northern regions, since they are killed by winter's cold to which they are not adapted. Such plants are natives of warm climates where dormancy is not required and thus they have not developed the specializations necessary for surviving cold periods. The annuals that evolved in warm climates produce seed that cannot survive cold weather, whereas the seed of cold-climate annuals can lie dormant in the soil during the cold winter protected by tough seed coats and other specialized features.

Biennials

Biennial plants generally do not produce seed until their second year, after which the parent plant usually dies. Foxglove and hollyhock, two garden favorites for their colorful flowers, are biennials. Cabbage, celery, and beets are naturally biennial, but the home gardener rarely allows them to complete their growing span. Instead, they are picked when they are vegetatively tasty, and seed production occurs only in the commercial crop fields of the seed companies.

Perennials

A perennial is any plant that continues to grow and flower for several to many years after germination. Herbaceous perennials are found in most flower gardens and bloom rewardingly year after year if given good care. Woody perennials include all trees, shrubs, and woody vines. Most perennials have a specific yearly blooming period. However, a few, like roses, can be relied upon to rebloom one or more times during the summer.

Perennials have made many adaptations to survive the changing conditions to which they are subjected throughout the year. Changing daylengths, for example, trigger special light-sensitive chemicals in the plants to direct the activities leading to dormancy in the fall and toward leaf and flower production during the growing season. Similarly, cooling or rising temperatures and increased or lessening precipitation will gradually affect the physiological responses of the plants so that they prepare themselves for the coming season. There is further discussion on structural adaptations in Chapter 4.

The physiological and structural adjustments made by perennials will depend largely on the environments in which they originally evolved. Some plants, natives of

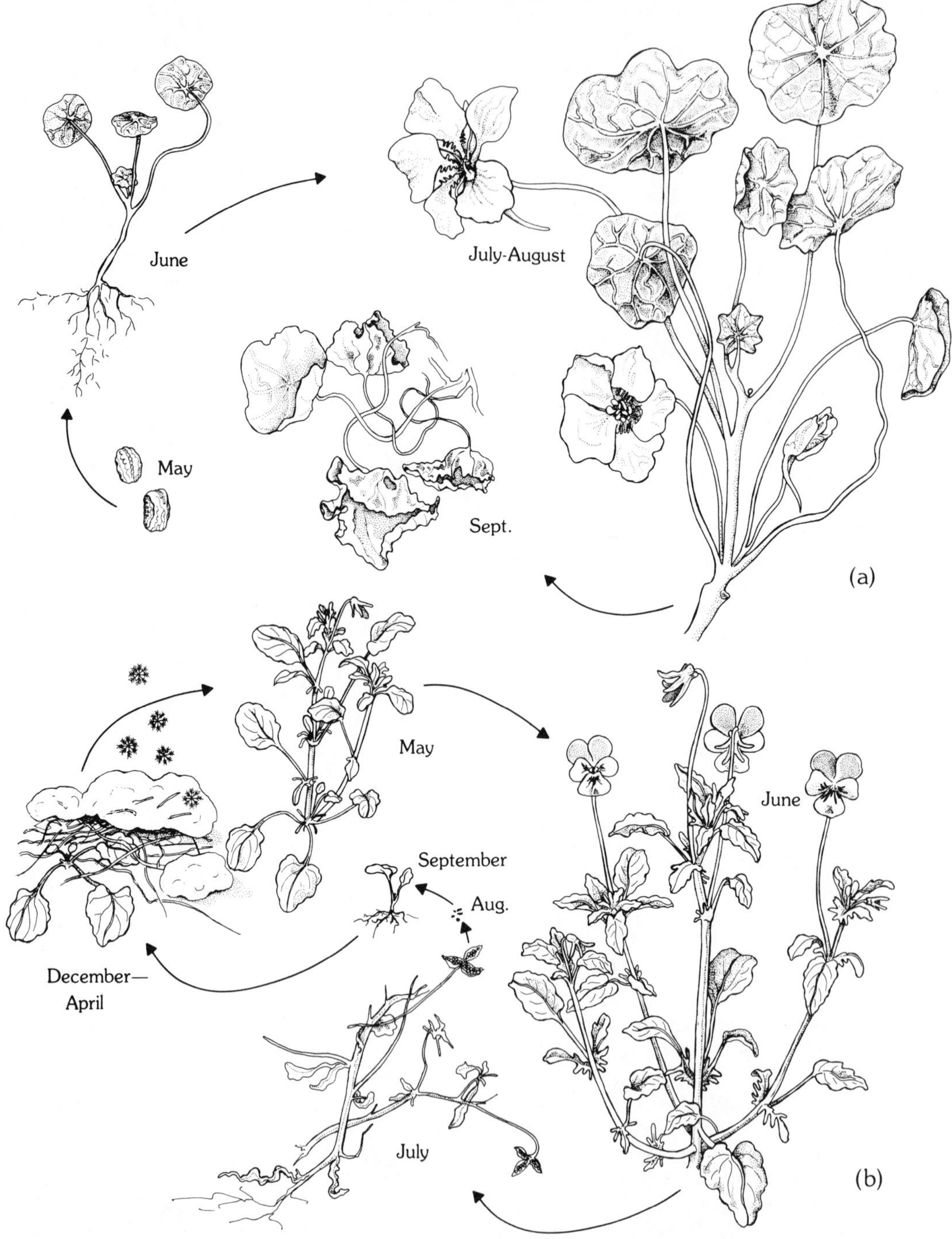

Fig. 3-6. Life cycles of annual, biennial, and perennial plants. (a) Nasturtium is an annual in most of the U.S. (b) Johnny-jump-up is a biennial, surviving into a second summer. (c) *Iris* is perennial.

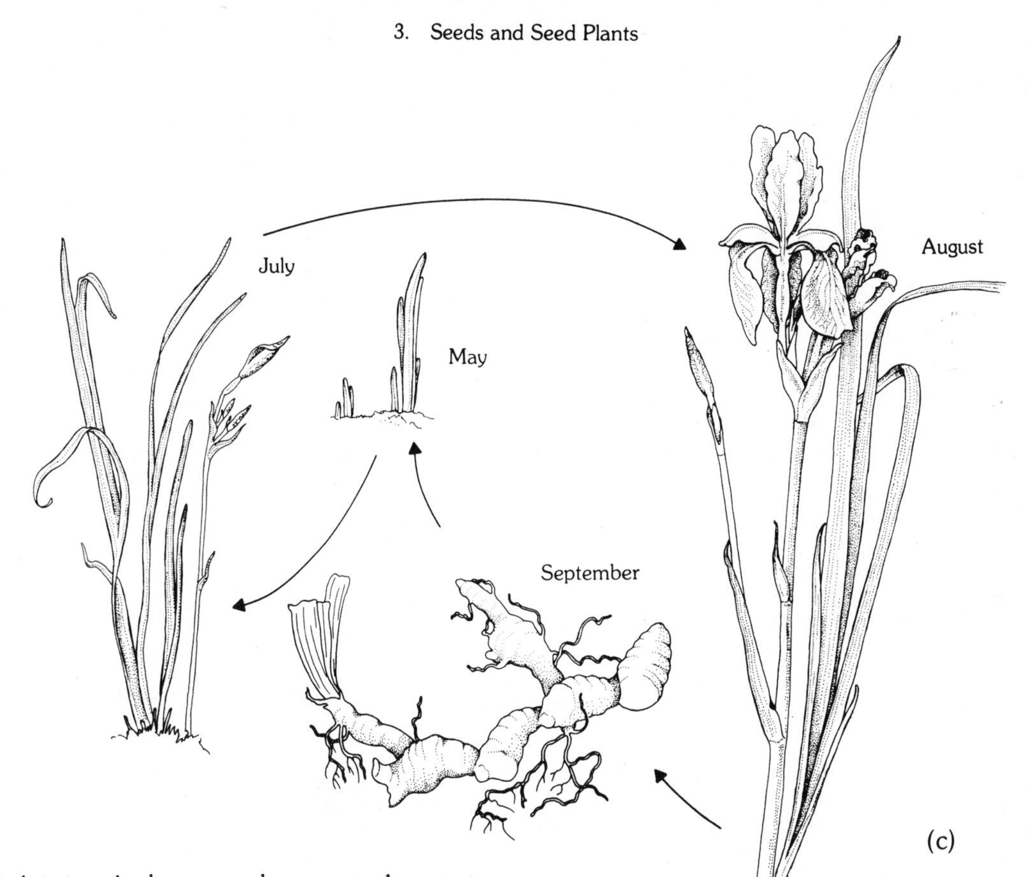

moist tropical areas, have no dormant period and stay green and leafy the year around. Other plants adjust to dryness but mild temperatures by dropping some or all of their leaves and easing into a period of near or complete dormancy. The perennials facing cold winters, whether herbaceous or woody, have had to make extensive adjustments to the lack of usable water, which is either absent or frozen, and to temperatures that reach the freezing point or below. Some of the many ways these plants have modified themselves in order to survive these conditions will be discussed in more detail later.

SEED DORMANCY

Most seeds are dormant when they are shed from the plant and will remain that way for a period of time prescribed by the plant's native environment. This dormancy prevents seeds, especially in temperate climates, from germinating at a time unfavorable for continued growth, such as in fall when winter is approaching. Some seeds may even require some drying or exposure to light before they will germinate, another adaptation to prevent germination until the seed is free of the parent plant and surrounding fruit. In addition, seeds will remain inactive if external conditions do not favor germination, even though the seed may, at this time, have completed its dormancy and now be able to germinate. Dormancy in seeds has been an important adaptation of plants to a variety of habitats.

Role of Water in Dormancy

When the seed is first formed it is composed of living, active tissues. It draws nourishment from the parent plant, to which it is still attached. When the embryo in the seed is formed and the fruit is ready to be shed, the water content of the seed drops drastically from about 70% to a level considerably below that, depending on species. This drop is the first necessary step toward a state of dormancy for the seed.

To maintain life during its dormant period, the seed must maintain minimum levels of water content and supplies of CO_2, O_2, and enzymes. Seeds normally subjected to lengthy over-wintering periods have the lowest levels of water within them, while those that germinate after a briefer, less severe period will maintain higher water levels. For instance, sugar-maple (*Acer saccharum*) seeds that are produced in the fall will be found to have a water content as low as 5%, but silver-maple (*A. saccharinum*) seeds produced in the spring for summer germination will average around 30% water. The low water content typical of seeds facing severe cold is an adaptive feature that prevents the formation of damaging ice crystals within the seed.

Other Features of Dormancy

Analyses have also shown that cold-resistant seeds have a high fat content, while seeds of milder climates have a higher water content, a low fat content, and a high proportion of carbohydrates. Since the chemical processes required for obtaining the energy from fats are more complex than those required for carbohydrates, the winter-subjected seeds are protected by the presence of fats from premature germination stimulated by brief, but not lasting, favorable conditions. Seeds in warmer climates, however, can take advantage of good conditions with their supply of readily metabolized carbohydrates.

Enzymes play an essential role in all types of plant growth, but they are inactivated at low temperatures. For storing seeds for future garden use, a range of 32-41° F. (0-5° C.) is recommended, since in this range tissues do not freeze but enzyme activity is nearly stopped. High temperatures will induce premature germination and may even kill seed tissue if high enough, especially if the water content is high.

Oxidation of stored molecules provides energy for a germinating seed. Thus dormancy can be better maintained during storage by lowering oxygen levels and/or increasing the carbon dioxide supply of the atmosphere surrounding the seed. Air-tight packages achieve such suitable levels and retard cellular activity.

SEED LONGEVITY

Since even in nature conditions for germination do not always exist when a seed is physiologically ready to germinate, most seeds have features that enable them to survive periods of unfavorable conditions. Thus we speak of a seed's longevity or period over which it can survive and still germinate. Seed-producing companies usually classify seeds into three groups according to longevity:

1. Microbiotic seeds. Those seeds able to survive less than 3 years in storage. Most garden annuals are in this group.
2. Mesobiotic seeds. Seeds capable of surviving 3-15 years. Herbaceous of both field and garden as well as thin-coated tree seeds are usually in this group.
3. Macrobiotic seeds. Seeds capable of surviving more than 15 years.

Seeds of marked longevity usually share a set of characteristics. They are likely to be larger and heavier than other seeds. They have hard, thick, smooth or even polished seed coats. The smoothness makes it difficult for fungus spores or bacteria to attach well enough to cause damage.

Finally, these seeds will not plump (expand) if soaked in water. Seeds of this type are produced by leguminous trees such as Kentucky coffeetree (*Gymnocladus dioica*) and honeylocust (*Gleditsia triacanthos*).

Longevity in any seed can be maximized with proper storage in dry, airtight containers, preferably under moderate refrigeration.

An interesting experiment was begun by Prof. W.J. Beal at Michigan State University in 1878. Into 20 pint bottles he put 50 seeds of each of 23 species of common Michigan weeds mixed in moist sand. The bottles were then buried in a sandy knoll. Every 10 years one bottle is opened and the seeds planted to test for germination potential. The 1970 sample, the 90th year, still yielded viable seeds of one species, moth mullein (*Verbascum blattaria*), with 20% germination. All plants appeared normal.

More recently, the U.S. Dept. of Agriculture has established the National Seed-Storage Laboratory at Fort Collins, Colorado. The laboratory is intended to be a storehouse of seeds of every known plant variety, and it already has more than 90,000 varieties stored in tight cans in climate-controlled rooms (Fig. 3-7). These seeds represent a germ-plasm bank of successful genetic combinations. They are tested for viability every five years, and if germination is low, they are regrown and new seed collected to replace the old. These seeds may one day be the source of as-yet-unrecognized genes that may mean better living for all of us.

Some legume seeds have remained viable under natural conditions for 158 years, and a few water-lily seeds, recorded as 250 years old, were successfully germinated. An older record is that of a 2000-year-old lotus

Fig. 3-7 The National Seed Storage Laboratory at Fort Collins, Colorado has 10 seed vaults. Here, the supervisor of the laboratory selects a sample for germination testing.

Fig. 3-8. The seeds of the water lotus, *Nelumbium nelumbo*, are well known for longevity. The large size and smooth, hard seed coat help the seeds resist hazards in their aquatic environment.

seed (*Nelumbium nelumbo*) found in the 1960's in a Neolithic canoe which was buried beneath 18 feet (5.5 m.) of earth in a peat bog near Tokyo. Peat cutters found the canoe and turned it over to archeologists who, in turn, found the seed. Dr. Ichiro Ohga, Japan's lotus expert, who had grown some 1000-year-old lotus seeds, coaxed the newest find into germination, and a year and a half later a beautiful shell-pink blossom appeared. Dr. Ohga was not overly impressed since it was "not different from lotuses today". This, of course, was not surprising since even 2000 years is a very short period in evolutionary time and any significant changes would require the pressures of environmental changes exerted over a very long time. Scientific evidence indicates that lotuses have apparently existed in their unchanging watery habitats from the ancient past to the present and have, therefore, had no need or stimulus to change form.

The oldest known seeds ever to germinate were 10,000-year-old tundra lupines (*Lupinus arcticus*) found in 1967 in a Yukon lemming burrow. It took them only 48 hours after planting to sprout.

Reports of many other germinations of ancient seeds, however, cannot usually be verified. For example, the reported finds of wheat seeds in Egyptian tombs and their subsequent germinations have been arranged by over-zealous tour guides and headline-seeking news reporters and have no basis in fact.

SEED GERMINATION

Seeds can be encouraged to break their dormancy only if certain physical and chemical requirements are met. Generally, setting the stage for germination involves time, temperature, water, oxygen and carbon dioxide, light or dark, and the production and activation of plant enzymes and hormones. Many seeds will germinate under a wide range of conditions, while others are quite specific in their demands.

Time

Proper timing, which allows a period of dormancy, is an important factor to most seeds. Primroses, for example, will germinate immediately upon being shed, or if that is prevented, they must be allowed a six-month period of dormancy before they will show vital signs again. Many garden perennials will lie passively in soil showing no signs of activity until their required period of dormancy has passed, at which point they are capable of germination if their other requirements are met.

Temperature

Temperature requirements, as well as water requirements, of seeds are usually similar to those for normal plant growth, so that plants and seeds in the same natural environments usually initiate growth at the same time. Some of our annuals and perennials that produce hardy (cold-resistant) seeds actually require the cool soils (sometimes as low as 40° F. [4.5°C.] of early spring to begin to grow. Their seeds, if collected, will not germinate unless given at least a brief period of chilling in the refrigerator (3 weeks to a few months depending on species). Most of the tender varieties prefer, if not require, warm soil temperatures of 70° F. (21° C.) or higher. A few fall-germinating perennials and biennials such as pansies are unusual in that they prefer the wide temperature ranges of fall for germinating, when warm days follow cool nights. However, it is difficult to define exact requirements for any one of the factors associated with germination, including temperatures, because they are all interrelated. To germinate, seeds need, for example, a certain period of time, even though temperatures may be ideal, for all the necessary physiological events to occur. This is an adpative advantage when it prevents premature germination during a time of unusual or unexpected, and short-term, warm temperatures, e.g., a "January thaw".

Many garden varieties of plants have resulted from hybridizing and selection so that the plants, under special cultural conditions, are esthetically or nutritionally satisfying. However, these plants or their seeds would be poorly adapted to natural habitats. While all seeds have certain needs in common, we circumvent some of these needs and adaptive safeguards, such as time and temperature, with special processing and packaging of the seeds of many of our garden favorites.

Water

Water must be available to any germinating seed. It softens the seed coat, making it possible for the seedling to emerge. For its initial growth the embryonic plant must depend on the nutrient reserves in the seed. These reserves can be transformed into usable molecules through the process of chemical respiration and transported to the seedling. Chemical respiration, as you will recall from Chapter 1, page 19, oxidizes (breaks down) complex reserve molecules and releases energy so that, in this case, a seedling plant can build its own tissues. Diagrammatically we can summarize these activities in the seed in this manner:

organic molecules in
storage tissue of seed
+
H_2O for fluid and changes
in chemical compounds
+
O_2 for oxidation
=
simple organic molecules to be
used in synthesizing molecules
for new growth
+
energy with which to synthesize
+
CO_2 and H_2O given off as waste

Carbon Dioxide and Oxygen

Seed germination cannot proceed, however, in water-saturated soils. Oxygen in soil air, as shown above, is required for the respiration of food reserves in the seed. Although water is needed to facilitate the diffusion of oxygen into the seed and carbon dioxide outward, if the water is overabundant there will be too little oxygen for the germination activities and perhaps too much carbon dioxide. Under these conditions the carbon dioxide can be toxic to the living cells of the seed. In the case of the long-lived lotus and lupine seeds mentioned earlier, germination was prevented by an almost total lack of air in their burial site. When the seeds were uncovered, oxygen became available and germination could occur.

Once the seedling has emerged above ground, atmospheric carbon dioxide is required for photosynthesis, and oxygen continues to be necessary for respiratory processes throughout the plant.

Light and Darkness

Some seeds have been shown to germinate best in the dark, others only if exposed to light (all other conditions being favorable), while others seem to be indifferent to light or dark. A number of experiments have been performed to discover how light influences germination and other kinds of plant growth. It is now known that some plant tissues, including seeds, may contain a light-sensitive pigment called **phytochrome.** Energy obtained by absorbing different wavelengths of light renders the phytochrome molecule active or inactive in promoting growth.

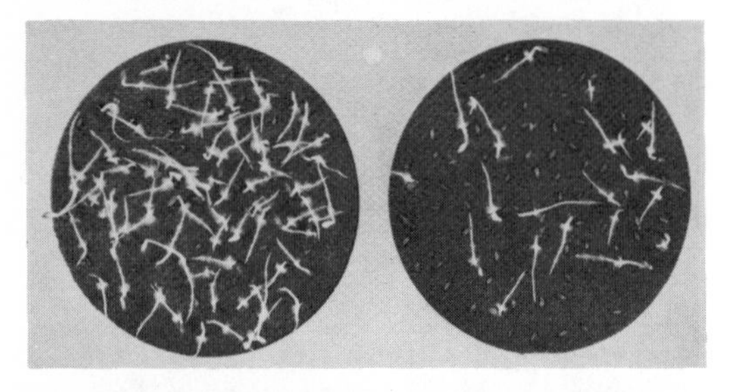

Fig. 3-9 Seeds of certain plants such as lettuce, bromeliads, birch, loblolly pine, and many common weeds germinate quickly when exposed to white light (left) but most remain dormant in darkness (right) even though favorable temperature and moisture are provided. (USDA photo.)

Many small seeds such as lettuce and common weed annuals will germinate only if exposed to daylight, which activates the phytochrome. When such seeds are buried deeply in the soil, the phytochrome reverts to its inactive state as a response to the darkness and, in effect, inhibits the germination of the seeds. (See further discussion on phytochrome activities in Chapter 6).

The gardener can benefit from this knowledge in recognizing that light-sensitive seeds must be planted close to the surface if germination is to occur. It will also explain why a sudden burst of new weed seedlings often appears in freshly-turned-over soils. Seeds that had remained dormant in the darkness beneath the soil are suddenly stimulated to germinate when brought to the surface and exposed to light.

Enzymes and Hormones

As explained in Chapter 1, page 19, all chemical activities involve a large number of enzymes that regulate the rate of such activities without becoming changed by them. The induction of enzymes for seed formation and germination is under the control of certain plant hormones.

Auxins are important to seeds in that they influence good fruit growth and development but are not directly involved in germination. They become important again when seedling growth occurs.

Gibberellins, produced by the embryo, play a central role in germination and in seed production as well. These hormones cause the endosperm (food reserves) to produce key enzymes important in producing usable molecules for embryonic growth and seed germination. Moist chilling or dry storage increases the levels of gibberellin in a seed.

Hormones of the cytokinin group complement the activities of the gibberellins by inducing cell division for embryonic development. However, another group of hormones, including abscisic acid, may inhibit or impede germination. After all, it is just as important for a seed to be prevented from germinating at the wrong time as it is to be stimulated to germinate when conditions are favorable. Abscisic acid and other inhibitors serve to oppose the effects of the hormones that promote growth. The presence or absence and the interaction of hormones can thus maintain dormancy or promote germination in a seed.

To summarize the interactions of these hormones briefly, a seed may germinate if:

1. Gibberellins are present.
   Seeds may fail to produce the hormones if required factors or conditions are not operating.
2. Inhibitors are absent or ineffective.
   Some seeds rely on a quantity of water to leach these hormones from the seed prior to germination.
3. Cytokinins are present.
   These hormones can inhibit the inhibitors, thus permitting the gibberellins to work.

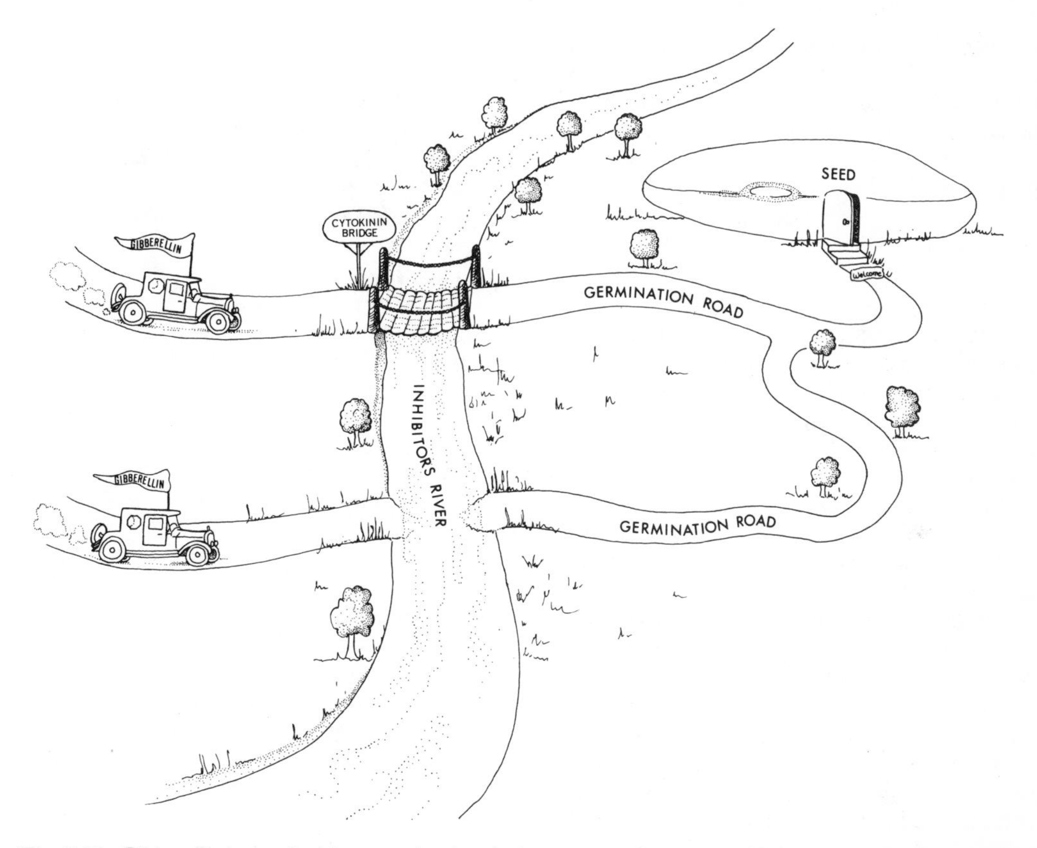

Fig. 3-10. Gibberellin is required for a seed to break dormancy and germinate. However, inhibitors, if present, will prevent gibberellin activity. The inhibitors may be absent or may be counteracted by the presence of cytokinin hormones.

Contrarily, seed dormancy is promoted by the absence or inactivity of gibberellins and cytokinins and the presence of inhibitory hormones.

Germination Inhibited by Soil Conditions

While it is true that some seeds are more particular about their conditions for germination that are others, they all, nevertheless, need a reasonable balance of all the above-discussed conditions and require a suitable medium in which to grow. In most cases this is, of course, soil, the composition of which can support or inhibit subsequent growth (see Chapter 2, Soils and Plants).

Interestingly, germination may occasionally be chemically inhibited by toxic by-products from other vegetation growing in the area. Tannins, phenolic compounds, or alkaloids from roots or leaf-and-stem litter are released into soil, where they can prevent seed germination and other plant growth. Such action may discourage competitor plants and reduce plant density in natural areas, but these toxins may cause difficulties for the gardener trying to germinate seeds in garden soils brought in from unknown sources. If such problems as poor seed germination or plant growth occur for no obvious reason, it may be that there are "chemical warfare" agents in the soil such as those produced by black walnut trees (*Juglans nigra*) or certain conifers. Flushing with water will remove these water-soluble agents.

Seed Coats

Covering every seed is a seed coat produced by the differentiation of the outer tissues of the ovule after fertilization occurs. This seed coat must be penetrated by the emerging seedling if the seed is to germinate successfully. Some seed coats offer little or no resistance and fall away easily after imbibing an amount of water necessary to swell the tissues and soften the seed coats. Other seeds have such tough seed coats that they must be abraded before water can penetrate. If the seeds roll down rocky slopes or tumble along in water from heavy rains, they will often be scraped enough for water to seep inside. Others are surrounded by acid fruits, the acid of which attacks the surfaces of the seeds.

Desert species often protect themselves from premature germination by producing inhibitory substances in their seed coats. Only heavy desert rainstorms will produce the large quantities of water that are required to leach out and carry away the inhibitors. Such rains usually provide enough water in the soil to permit germination and initial growth to establish the plants.

A package of seeds purchased by the gardener will occasionally give instructions to nick the seeds with a knife or sandpaper them. Geraniums (*Pelargonium*), morning glory (*Ipomoea*), and okra are typical examples. Such treatment followed by overnight soaking usually results in improved germination rates.

PACKAGING COMMERCIALLY PREPARED SEEDS

Seed growers realize a fairly good return on their investment of time and money only if they can sell seeds which can be relied upon to germinate well. Most aim for a germination rate of 85% or better for most garden and vegetable seeds. By protecting seeds against too much or too little moisture, freezing temperatures, fungi, mice or other rodents, as well as crushing or other forms of mishandling, the seed-producer tries to establish a reputation for reliable stock.

As a rule of thumb, a balance of humidity and temperature should be maintained in which the percent of relative humidity + the degrees of temperature (in F.) may equal but does not exceed 100. For example:

20% relative humidity + 80° F. = 100
60% relative humidity + 40° F. = 100

In either case germination will be retarded, and the seeds should remain viable, other conditions being suitable.

Some of the kinds of seed packaging available today are:

1. Paper packages coated with moisture-resistant substances or foil-lined. The interior of the package is often treated with a fungicide. Most seeds purchased today come in such packages. They are best used in the year they are produced.

2. Seed-tapes. Seeds are sealed within a water-soluble tape material so that seeds can be planted in a row and be properly spaced. Thinning should not be necessary. Since the sealing is done with water, it means that each seed has an atmosphere of water vapor around it which can be damaging to the seed and shorten shelf life. Tapes should be planted as soon as possible after purchase. (Fig. 3-11.)

3. Grow-sticks. These are wooden sticks that are the size of plant labels with a hole near the bottom in which a seed is encapsulated. **Theoretically you can plant** and label in one operation. This novelty suffers from the same problem as seed tapes.

4. Coated seed. Seeds of any size can be coated with hygroscopic (moisture-

Fig. 3-11. Planting seed tapes. (Photos courtesy Geo. W. Park Seed Co.).

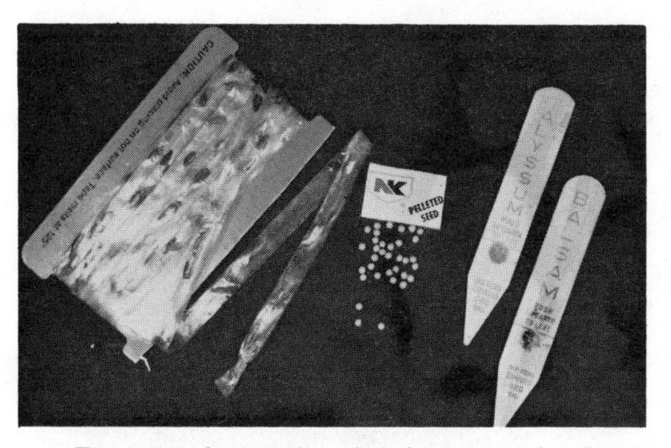

Fig. 3-12. Assorted seed packaging in color. Seed tape, pelleted seed, and grow-sticks.

attracting) materials to make them larger or more uniform for easier handling. Larger seeds are more easily spaced in the garden, and careful spacing will reduce the need for thinning. In addition, fertilizers and fungus- and insect-control agents can be added to the coating material. Coated seed has so far proved to be a popular and practical innovation in seed marketing. Because of the affinity of the coating for water, the seeds must be stored in a dry place until planted.

5. Canned seeds. This is the newest product and one which may prove more successful than any of the others. If seeds are kept at room temperature, a favorably low 25-30% humidity will satisfy the "rule of 100" and can be incorporated into the can at the time it is sealed and will remain at that level until the can is opened.

6. Freeze-dried seed. Recent tests have been conducted by the U.S.D.A. Research Service in which vegetable seeds were freeze-dried using equipment similar to that used for freeze-drying coffee. The freeze-dried seeds successfully survived storage conditions at very high temperatures that killed conventionally-treated seeds. Even at moderate temperatures the survival rate was far better for freeze-dried than for untreated seeds. The research results have been so encouraging that efforts will now be made to improve the freeze-drying techniques and determine optimum conditions for each kind of seed.

GROWING SEEDS INDOORS

Use a good-textured loam consisting of one part topsoil, one part mixed sand and perlite, and one part crumbled peat moss for most seeds. The perlite will prevent packing too tightly. The soil mixture should be pulverized or screened. Spread the soil evenly into a planting flat or flower pot with drainage holes and tamp down with a flat board. (See Appendix to Chapter 2 for recipes for soilless mixes and Chapter 4, pg. 125 for special peat seed-starting containers).

Seeds may be broadcast over the surface or planted in rows about 2" (5 cm.) apart. Overcrowding should be avoided to prevent weak, spindly plants. If the seeds are very fine, they will need little or no covering of soil. If the seeds are large, sift more soil over them to the depth recommended on the package, usually equal to the thickness of a seed. Press fine seeds into the surface, and firm soil over the larger seeds with a tamper or flat board.

Water evenly, preferably by setting the flat in a tub of water until the water has seeped its way to the surface. If this is inconvenient, dampen with a very fine spray so as not to disturb the seeds. The seeds must never be allowed to dry out, so a covering of some kind may be required. A pane of glass or sheet of plastic wrap extended over the top of the flat will keep moisture in, but it should be loosely fastened so as not to inhibit air circulation. A single sheet of newspaper, cut to size, and placed directly on the soil surface will keep the soil surface damp and can be wetted without displacing the seeds beneath.

The flats should be placed in a warm, shady place until the seeds have germinated, at which point coverings must be

Step 1

Step 2

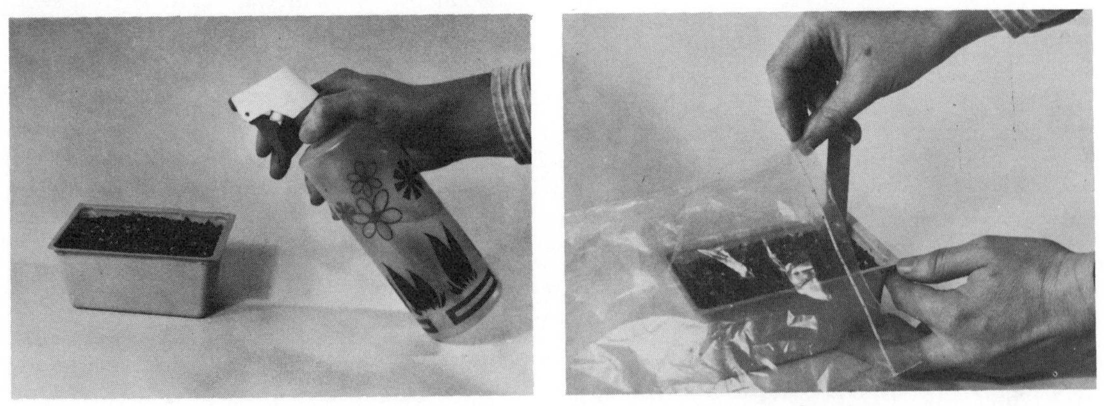

Step 3

Step 4

Fig. 3-13. Planting seeds indoors. Step 1. Spread the soil flat or pot. Step 2. Broadcast seeds or set in row 2" apart. Step 3. Water evenly by setting flat or pot in tub of water or water with a fine spray. Step 4. Cover with plastic wrap or single sheet of newspaper until seeds germinate. Later thin out excess seedlings.

removed and the seedlings can be placed in a bright sunny spot to encourage sturdiness. See Fig. 3-13 for pictured instructions.

Thinning out is an absolute necessity . Overcrowding is a major cause of poor plants. Plants growing closely together compete for water and ions in the soil as well as for light for photosynthesis and for space to grow. If the excess plants are not removed, all the plants will be weakened, as they will produce elongated, thin-walled cells as they attempt to "reach" beyond their competitors. When thinning, never pull plants from the soil if the roots of remaining near-by plants will be damaged. Instead, cut off these unwanted plants at soil level. Their roots will quickly die without the support of the upper portion of the plant.

If plants are being grown for outdoor use they should be started about 6-10 weeks before they are to be transplanted into the garden. When planting times are being planned, frost dates and other outdoor conditions should be considered (see Chapter 9).

For pot plants, the seedlings should be

Step 1

Step 2

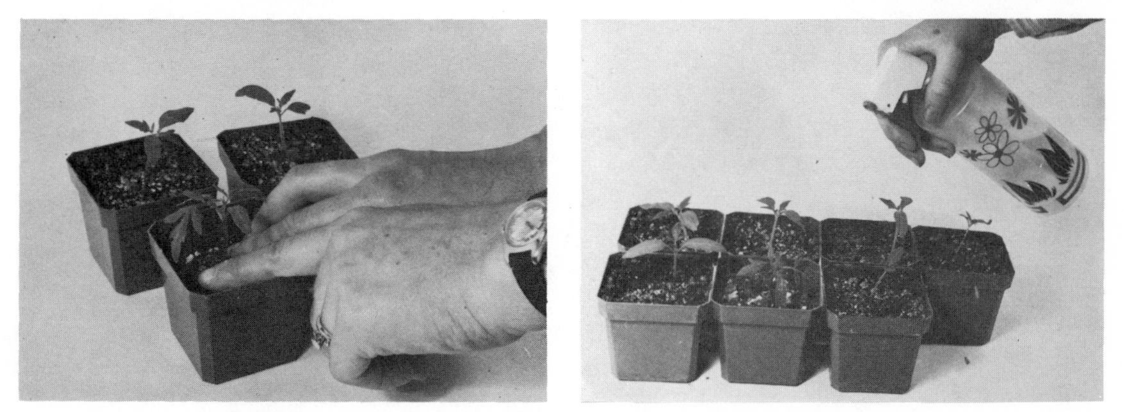

Step 3

Step 4

Fig. 3-14. Transplanting seedlings. Step 1. Allow seedlings to grow in flat until they show a second set of true leaves. Step 2. Lift seedlings carefully from soil and pinch off excessive straggly roots. Step 3. Place seedling into soil in pot just a bit deeper than it was in the flat. Firm soil. Step 4. Water with a fine spray; shield from full sun for a few days.

transplanted as soon as they are large enough to handle easily without damage. This is usually when they have a second pair of true leaves. Most flats are shallow, and root growth may be restricted if plants are kept too long in the flat.

When being transplanted the plants should be removed only a few at a time to avoid drying the roots out. If straggly, excessive root growth has occured, pinch back the excess. The plant should be planted into a hole in the soil so that it will be just a little deeper than it was in the flat. Firm the soil around the plant with your fingers and water with a fine spray. Potted plants may be bottom-soaked by setting the pots in shallow water. Shield the plants from bright sunlight for a day or two to prevent wilting and to allow the seedling to recover from the shock of transplanting. Increased humidity achieved with a plastic-bag tent over the plant may be especially helpful for a few days. See Fig. 3-14 for pictured instructions.

There are many media that may be used to germinate seeds successfully. Shredded or milled sphagnum moss, vermiculite, perlite, and the soilless mixes are most commonly used. Sand alone packs too tightly for satisfactory growth of most seeds. The roots can be easily removed from the other substances without damage at the time of transplanting. However, since there is little or no nourishment in these materials, the seedlings should be transplanted as soon as first or second leaves are formed, or if not, a balanced fertilizer solution suitable for houseplants should be applied every two weeks for continued growth.

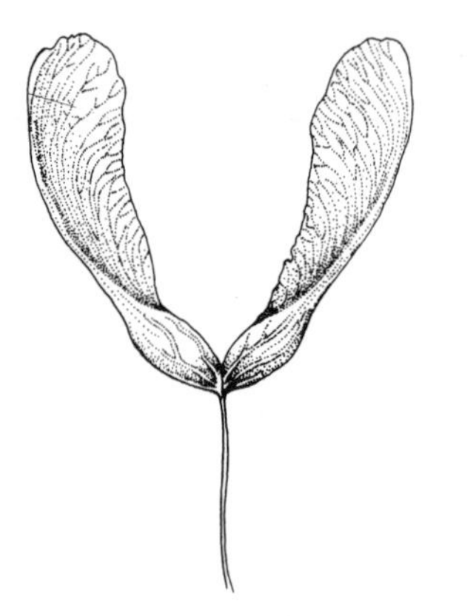

4. VEGETATIVE PROPAGATION

Seeds are an inexpensive way to obtain a large number of plants, but the little seedlings face an uncertain future because of their their susceptibility to disease and other environmental hazards. The resulting crop also will be genetically variable, including some plants that are not especially desirable, and seed production requires fertile pollen and ovules and a means of accomplishing pollination.

Vegetative propagation, in which a part of a plant is removed and allowed to grow into a new plant, offers some obvious advantages, not the least of which is that only one plant is needed to start new ones. As explained in Chap. 1, p. 27, this allows for the propagation of plants that may be infertile, seedless, or non-flowering, as are most cultivated foliage plants, or for the perpetuation of a particularly nice plant selected from a group of seedlings.

Many plants, even in their natural habitats, have developed means of vegetative propagation, often nearly dispensing with the sexual process altogether. Clumps of sumac and aspen derived from buds on far-spreading roots, clusters of daylilies and daffodils surrounding the parent plant from which they sprout, and new plants grown from fragments of willow branches or from buds that form an drop from *Kalanchoe* leaves represent some of the kinds of vegetative propagation. The plants produced by these natural methods of propagation constitute a clone of genetically identical plants derived from a single parent. A clone may be produced artificially if several cuttings or plantlets are removed from a houseplant and potted up. A characteristic feature of these vegetative propagations is the production of adventitious stems or roots, and, as we shall see, plants are remarkably adept at providing themselves with new stems and roots under a variety of conditions (Fig. 4-1.)

Fig. 4-1. Adventitious roots from the base of the stem of a large *Agave* plant.

UNDERGROUND STEMS, ROOTS, AND LEAVES

The fossil record reveals to us that the first angiosperms were small trees that existed in the mild climates of the Mesozoic era well over 100 million years ago. From these trees our present-day herbaceous plants evolved along many evolutionary pathways in response to the pressures and demands of a continually changing world environment. Trees, as a plant form, had disadvantages in some habitats. For example, such large plants require long growing periods before seed production can occur; they require a considerable amount of growing space; and they produce a lot of (energy-requiring) tissue that is extraneous to the primary function of producing seeds. As world climates became cooler and drier, as geologic evidence tells us they did, plants had to develop tissues that were resistant to the dryness and cold. As seasons were established, periods that were unfavorable for growth led to the evolutionary development of dormancy in plants. Dormancy became an important feature for survival, but other modifications provided solutions to the disadvantageous characteristics of trees and led to herbaceous plant forms.

The reduction of overall size and amount of tissues in the evolving herbaceous plants had some immediate ecological advantages. Many could now grow where only a few grew before, and the reproductive rates increased. However, since winter conditions made it difficult for some of the increasingly delicate stems to survive, some of these plants developed specialized organs underground. Thus protected, these organs stored food and water and were also capable of dormancy after the above-ground portions died back. When conditions became favorable again, buds on the underground structures produced new growth above ground utilizing the stored reserves of food. The herbaceous angiosperms that did not evolve the necessary protective and nourishing structures found their habitats limited to those that were less challenging, and they often produced only seeds as the means of survival during adverse periods.

Some interesting studies have recently shown that temperature, as one of many factors, is and probably always has been important in the evolution of specialized plant tissues. Along with light, water, and CO_2, it influences photosynthesis and the distribution of its products (sugars) into roots, stems, and leaves. When daytime temperatures are low the sugars tend to accumulate in roots and leaves. When daytime temperatures are high, rates of both photosynthesis and respiration increase, and the sugars are primarily utilized in the production of new leaves. If the night temperatures that follow are also high, reserve foods are used in respiration, and (since photosynthesis is not occurring during the dark hours) organic reserves are greatly reduced and the production of new tissue is accordingly decreased. If, however, cool nights follow warm days, respiration and other processes are slowed, with the result that the sugar content of the plant tissues remains high, often accumulating in specialized tissues such as thickened leaves or fleshy underground structures.

Since, in temperate regions, day and night temperatures fluctuate more in spring and fall than they do in summer, growth and/or storage rates are high at these times, lower in summer. At the equator, where both day and night temperatures remain high nearly year around, the plants have had to adapt to such conditions and they have become so specialized that they are now unable to flourish outside this selected range. Correspondingly, plants accustomed to fluctuating temperatures will not grow well and certainly not produce some of their typical specialized organs if subjected to uniformly high temperatures. Thus we see that adaptations of both plant form and physiology contributed to the

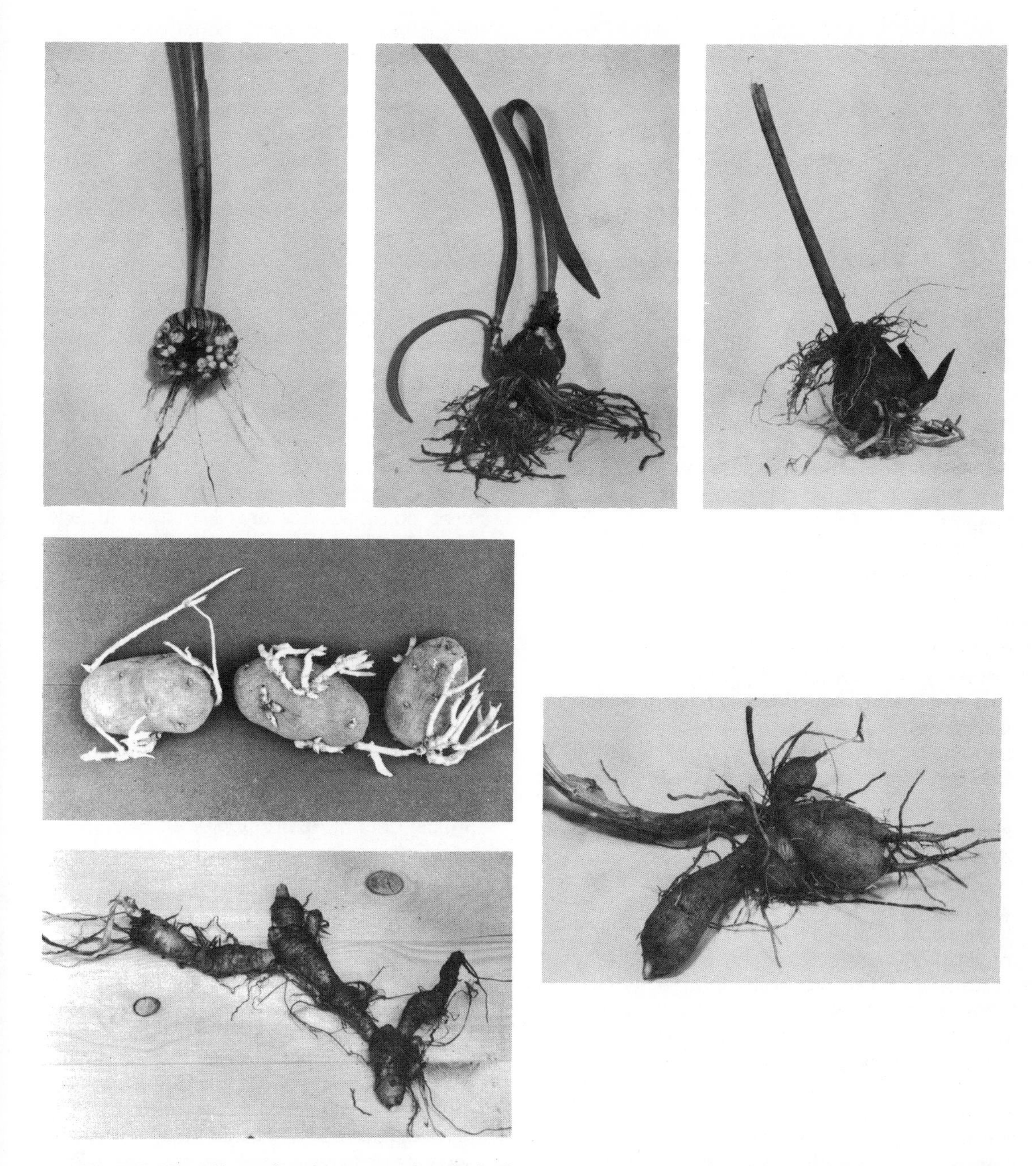

Fig. 4-2. Naturally produced structures by which plants may vegetatively propagate themselves. (top, row, left) *Gladiolus* corm with cormels. (top, center) Tunicate bulb of *Amaryllis* with bulblet. (top, right) Scaly bulb of a lily (adventitious roots are on stem above bulb). (center, left) Sprouting stem tubers of potato. (center right) A cluster of *Dahlia* root tubers. (bottom, left) Branching rhizome of *Iris*.

establishment and evolution of underground storage structures in temperate-zone plants.

The meristematic growing tips of plants, i.e.,the root and stem tips and lateral buds, benefitted by the development of underground structures. These tips are very sensitive to cold and dryness, so herbaceous plants evolved a variety of modifications to protect them. In some plants, underground creeping stems, called **rhizomes**, already a feature of the more primitive ferns, appeared. Grasses today are well known for their rhizomes which spread extensively just beneath the soil surface. Further modifications produced fleshier, food-storing rhizomes such as we see in *Iris* or *Canna*. **Stem tubers** are produced at the ends of otherwise slender rhizomes as terminal storage structures and are seen in a number of commonly-grown plants, most notably the Irish potato *(Solanum tuberosum)*. In each case these underground organs can be identified as stems by their nodes, marked by buds (sometimes called "eyes") or branches, the internodes between them, and the terminal bud that gives rise to new growth. The delicate lateral and terminal buds are well protected by the surrounding fleshy, moisture-providing tissues, the covering of soil overall, and in some climates, a protective blanket of snow as well.

Corms are also underground stems, and they probably originated in soils that were not as suitable to the spreading nature of rhizomes and tubers. These short, fleshy storage stems are covered by the basal portions of the leaves. Each year one to several new corms are produced on top of the old one, which then shrivels and dies.

Stems, however, were not the only parts modified for underground storage. **Bulbs**, of which there are many varieties, consist of modified, scale-like leaves containing ample supplies of food reserves such as sugar, starch, and protein. These fleshy leaves are attached to hardened stem tissue called the **basal plate**. When growth starts, roots are produced from the outer edges of the basal plate, and leaves and flowers emerge above the ground from growing tissue within the protective and nourishing bulb. Bulbs generally increase in size over the years and multiply in number. Monocots of the families Liliaceae, Amaryllidaceae, and Iridaceae most commonly produce bulbs.

Although stems and leaves require more extreme modifications, roots too have developed features so that they are more protective and sustaining of the aerial portions during times of stress. A large, fleshy tap root acting as a storage organ occurs in many plants, especially those that survive for many years. **Tuberous roots**, such as we see in dahlias, elephant's eyes, and sweet potatoes, are usually fleshy storage growths to which buds and stems are attached at the proximal end. There are no buds or "eyes" on the surfaces of root tubers as there are on stem tubers.

Unlike animals with predictable lifespans, these kinds of plants discussed above do not usually die of "old age" but most often succumb to an unfavorable environmental event. Allowed to grow without disease or injury, they grow and multiply vegetatively over the years and almost seem capable of immortality (Fig. 4-2).

The Gardener Steps In

The various types of vegetative structures discussed thus far have been evolved by plants in their wild and native habitats. People, however, entered the world of plants by finding ways to use their natural methods of reproduction to human advantage in developing additional plants for pots and gardens as well as in prolonging the life spans or extending the habitats of many of them.

Each of the types of underground structures, i.e., rhizome, stem and root tuber, corm, and bulb, can be found in any

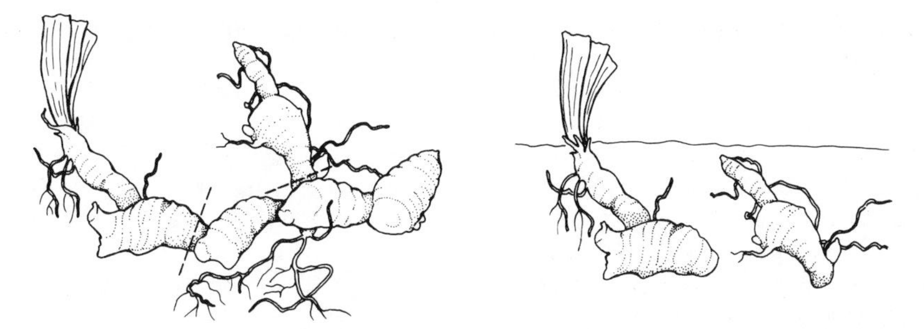

Fig. 4-3. Dividing an *Iris* rhizome. 1) After flowering, cut leaves off at about 6″ height. 2) Cut between segments as shown by dashed lines to remove young growth from older segments. 3) Replant young segments, discard old.

number of plants that brighten our gardens and window sills, and each of them requires particular care if it is to be maintained and/or propagated successfully.

Among the plants that produce rhizomes are many of the *Iris* and orchids, the cannas, lilies-of-the valley, and most ferns. These are plants capable of living for many years, but, just as the oldest stems of most plants become less productive, so also do the older rhizome segments. Also, when rhizomes become overcrowded, new segments are restricted in their growth. This is particularly evident when flower production is reduced or absent and usually occurs in *Iris* after a period of four to five years. To revitalize irises, lift the rhizomes out of the soil soon after blooming is completed. Break or cut the youngest rhizome segments apart from the older sections, which may be discarded. For ease of handling, and because many of the roots will be unavoidably broken off reducing their absorptive abilities, the leaves may be cut down to about 6" (15 cm.) in height. They will be replaced with the summer's growth. The young rhizomes should contain at least one node from which new growth can be initiated. Plant the rhizomes just beneath the soil surface, positioning them horizontally so that they can grow forward without being obstructed by such things as fence posts or neighboring plants. (See Fig. 4-3.)

Fig. 4-4. Dividing a potato stem tuber. 1) Select a firm, healthy potato and section it into several pieces. Each piece should have at least one "eye". 2) The pieces would be planted in loose, well-drained soil.

Both stem and root tubers have a rich supply of reserve food for new plant growth, but stem tubers are handled quite differently from root tubers when used in propagation. The stem tuber, such as a potato or *Caladium*, will have a number of incipient buds (eyes), each of which may give rise to a shoot. The tuber should be sectioned so that each piece has at least one eye. Plant the sections about an inch (2.5 cm.) deep in a loose, well-drained soil, and the eyes will soon grow into new plants. (See Fig. 4-4.)

Although all stem tubers can be divided, many of them, if not divided, will become more productive of flowers as they increase in size over the years. Tuberous begonias and *Gloxinia* each produce rounded tubers with a cluster of buds on top that are easily injured. As an alterntive to sectioning the tuber, you can take stem or leaf cuttings from the plants during the growing season as a means of obtaining new plants (see section on cuttings in this chapter).

Dahlias are common garden perennials that form clusters of root tubers that require dividing from time to time. Adventitious buds that will produce the new stems can be seen on dormant plants if you look carefully around the base of an old stem where it attaches to the cluster of tubers. To separate a tuber from the cluster, make clean cuts at the base of the old stem and between the buds so that the tuber will contain at least one stem bud. The tubers should be separated in the spring when it is easy to see the lively new buds around the base of the old stem. Plant each tuber on its side with the bud end upward in a hole deep enough to cover the tuber. (See Fig. 4-5.)

Sweet-potato tubers occur singly and, for propagation, are planted horizontally close to the surface of the soil. They produce adventitious shoots on the upper surface and fine roots from the lower surface. When the shoots have several leaves and roots of their own they may be removed from the old potato and planted in fields for further growth.

In the popular gardening literature you will often find the term "bulb" used to refer to almost any form of underground storage organ. However, there are only two types of true bulbs and both consist of thick scale leaves containing stored food attached to a modified stem called a basal plate. Plants of **tunicate bulbs** consist of tightly layered fleshy scales underground and long, straplike leaves above ground. Flowers are produced at the top of a tall green stalk (**scape**) arising from between the leaves. Tunicate bulbs are the source of most early spring woodland flowers as well as *Narcis-*

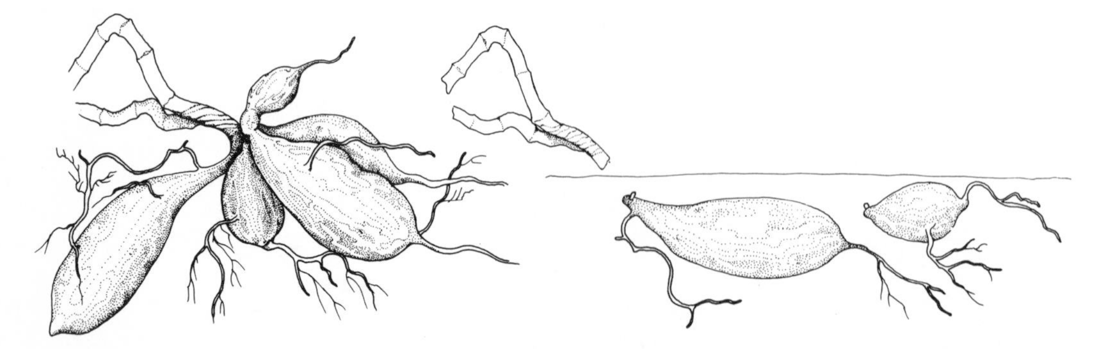

Fig. 4-5. Dividing *Dahlia* root tubers. 1) Tubers should be divided after winter storage, before replanting in the spring. 2) Divide tubers between buds where old stems and tubers join. Each tuber should have at least one bud. 3) Replant individual tubers horizontally just beneath the soil surface.

sus, tulips, *Amaryllis*, hyacinths, onions and ornamental *Allium*. Bulblets, which are really small branch buds, often form in the axils of the outer scale leaves. Adventitious roots appear from the base of the developing bulblet, and soon afterward it may be separated from the parent bulb to become a new plant. These bulblets are often referred to as **offsets**, a fairly descriptive term for these vegetative productions. (See Fig. 4-6.)

Scaly bulbs are made up of a cluster of small, loosely bound scales attached to a common basal plate. A tall stem, bearing leaves arranged alternately or in a whorled fashion for most of its length, emerges from the center of the bulb, and one to many

Fig. 4-6. Dividing a tunicate bulb of *Amaryllis*. 1) After winter storage gently pull bulblets from the large parent bulb. They are connected at the basal plate. 2) Replant all bulbs and bulblets so that the apex is protruding above the soil surface. 3) Bulbs of hardy plants, i.e., tulips, daffodils, etc., should be planted outdoors in the fall well below the soil surface.

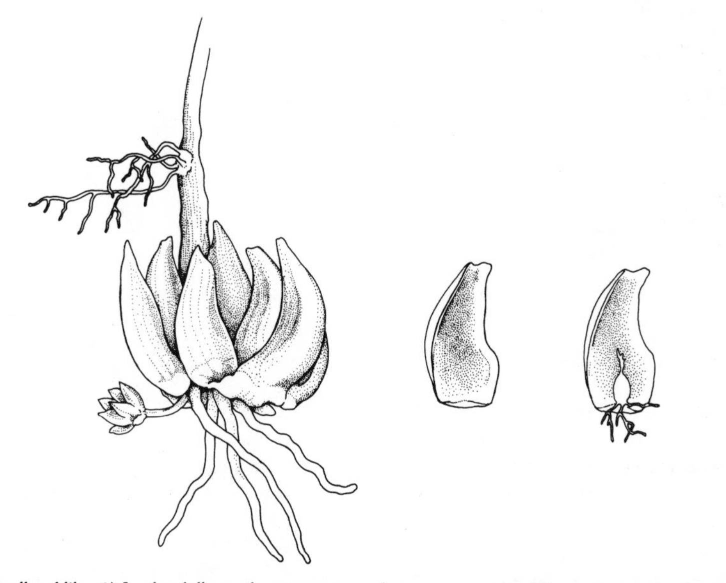

Dividing a scaly bulb of lily. 1) In the fall gently separate and remove some of the outer scales form the parent bulb. 2) Plant the scales base down in a shallow sandy mix or put into plastic bag containing moist sphagnum moss. 3) New bulblets will appear on the scales. 4) Plant bulblets in a pot of well-draining soil and place in a cold frame until spring.

large showy flowers are produced at the apex. True lilies such as the favorite Easter lily (*Lilium longiflorum*) are produced from scaly bulbs. Tiny dark adventitious buds called **bulbils** often form in the axils of the leaves on the stem in some lilies (see Fig. 4-7), and bulblets may form on the scales of the underground bulb. Any of these may be removed and planted to grow into new plants.

Fig. 4-7. Bulbils in the leaf axils of a lily.

Individual scale leaves of large, mature scaly bulbs may also be used for propagation by gently pulling them off at the base. They should then be set base down into a shallow flat of well-drained sandy soil. Young bulbs will form about the edges of the scales. About a dozen scales can be removed from a bulb without interfering with its blooming potential.

Bulbs may be **hardy** (able to withstand freezing), but some are **tender** (not resistant to freezing weather) and must be lifted for storage each fall. However even in warm climates some of the tender bulbs can be maintained more successfully as garden plants if they are stored for the winter season. Even though the temperatures may not exceed lower limits, winter rains or dryness can be detrimental to the bulbs when left in the soil year around. To maintain bulbous plants (and almost any other cultivated plant) in good size and vigor it is important to remove seed pods at their earliest stage unless seeds are wanted. Seed and fruit development deprive bulbs of nourishment that will otherwise be available to increase bulb size and form bulblets.

Corms produce a small, pellet-like form of offset called a **cormel**. These cormels will be found around the base of each new corm. They may be removed and sown the following spring to grow into new corms, although it will take two growing seasons before they bloom (see Fig. 4-8.)

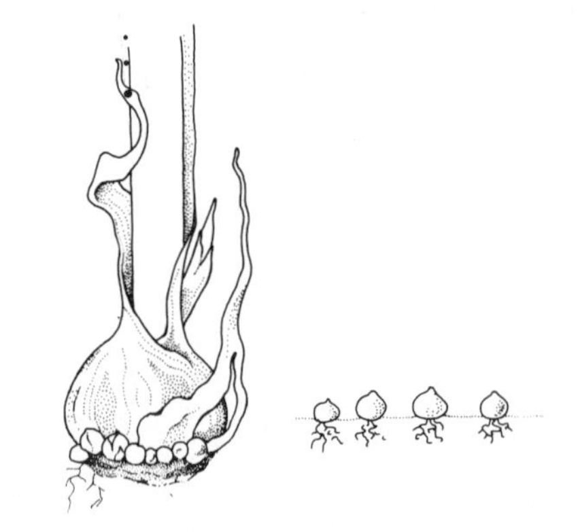

Fig. 4-8. Dividing corms of *Gladiolus*. 1) Lift tender corms with cormels in the fall and store. 2) In the spring, gently remove small cormels from the parent corm. 3) Plant cormels in the garden just below the soil surface and spaced so that increases in diameters will not crowd them. 4) Corms of hardy plants such as crocus may be left in the garden where cormels will increase the number of plants over the years.

Fig. 4-9. Planting bulbs for forcing. Step 1. Select bulbs suitable for forcing (those pictured are hyacinth bulbs). Nearly fill pot with well-draining loamy soil. Step 2. Place bulbs so that tips will be above soil surface. Step 3. Fill spaces between bulbs with additional soil. Step 4. Water thoroughly. Label with plant name and date. Allow to grow in cold, but not freezing, location.

The corms of both spring- and fall-blooming crocuses are hardy. The corms of *Gladiolus*, like some other underground storage structures such as the tubers of *Dahlia* and the bulbs of *Amaryllis*, are tender and must be lifted each fall and stored. Tender bulbs, corms, rhizomes, and tubers are all similarly treated when being stored indoors for the winter season. First, they are gently removed from the garden soil and spread on newspapers to dry out a bit. When the tops die off, after about a week, cut them off close to their bases. Don't pull; you might tear into their lower portions. Don't remove any firm and healthy roots. They contain good stored food that will be needed when planted again. Store the corms, tubers, etc. in paper bags in as cool and as reasonably moist a place as you can manage in your house or garage, preferably about 50° F. (10° C.). Avoid using plastic bags since they reduce air circulation and encourage mold growth. If the air is overly dry, as it may be if your storage area is near the furnace, a packing of very slightly dampened peat moss or sawdust can be placed around the structures. *Gladiolus* corms are especially susceptible to tiny insects called thrips, but dusting the corms with a 5% malathion powder or sulfur before storing will control these pests.

Forcing Bulbs and Corms for Indoor Bloom

Hardy bulbs and corms of tulips, hyacinths, crocuses, and narcissus can be induced to bloom indoors during the bleak days of winter if given proper treatment (**forced**). This method simply involves exposing them to their required cold period artificially and then gradually bringing them into increased warmth and light (see Fig. 4-9).

1. In the fall select bulbs and corms especially noted for good forcing qualities. Garden-store displays and mail-order catalogues usually provide this information.
2. Use a soil mix of 1/3 garden soil, 1/3 sand and/or perlite, and 1/3 peatmoss. For each quart or liter of soil add a teaspoonful of 5-10-5 fertilizer. Be sure the pot has a drainhole in the bottom.
3. Plant several of the large bulbs in a pot so that they are about an inch (2.5 cm.) apart and the tops of the bulbs are above the soil line so that water will not collect on the growing tips and possibly cause rotting. Small bulbs and corms of crocus and grape hyacinth should be about an inch (2.5 cm.) below the soil surface.
4. Set the planted pot in a tray of water until the soil surface is wet.
5. Place the pot in a refrigerator or other cool place with the temperature between 40-50° F. (5-10° C.). Do not freeze! Using a refrigerator is a way for gardeners in warm climates, where these plants cannot be grown outdoors, to enjoy these spring flowers.
6. Wait at least 10 weeks for hyacinths, 12 weeks for most other bulbs, and up to 16 weeks for late-blooming tulips before removing them from the cold. During this time roots will grow and pale yellow leaves and perhaps a bud will emerge above the soil. Check occasionally to see if water is needed.
7. Bring out into subdued light and temperatures as close to 60° F. (16° C.) as possible until the leaves become green and the buds begin to expand.
8. Gradually provide more intense light for good flower color and development. Do not put into direct sun, however, because the heat and light intensity will age the flowers too rapidly.
9. Enjoy.
10. When flowers fade, remove flower stalks and continue needed watering until foliage yellows and dies naturally. Let soil dry.
11. Remove bulbs from pot and store in paper bags until fall. Being brought into bloom early drains the bulbs' reserves more than usual, so in the fall they should be planted in the garden, where they should remain for a year or two to recover.

Tender bulbs such as paper-white *Narcissus* and *Amaryllis* do not require cold-storage treatment since they are normally adapted to milder climates. *Amaryllis* bulbs can be maintained for years as blooming pot plants, but paper-whites are rarely good for more than one forced blooming period unless they can be allowed to recover in the garden where climate permits.

All plants of a "bulbous" nature store food in their various underground structures. To do this they must produce abundant green leaves to carry on the photosynthesis necessary for this food production. Cultivated plants will perform best if flower stalks and seed pods are removed before seed production progresses very far. *Leaves must never be cut off when the plant stops blooming.* Leaves should remain on the plant for the duration of its normal growing period, and appropriate light intensity should be provided. Check a gardening handbook for the fertilizer needs of specific plants and supply whatever will best produce sturdy growth and abundant food reserves. With good care these plants can survive for years, many of them becoming increasingly more productive of lovely flowers.

ABOVE-GROUND STRUCTURES

Many plants, both woody and herbaceous, have developed means of propagating themselves vegetatively with above-ground parts. Since these structures are generally exposed to year-around weather conditions, they must survive as well as any other aerial feature and be able to become dormant where cold or dry seasons exist.

Suckers, runners, stolons, and **layers** are the usual forms of above-ground modifications of stems for propagating. They all represent methods of producing new plants rapidly that are robustly competitive in habitats crowded with a wide variety of other plants. Since these vegetatively reproduced plants remain attached to the parent plant at least until they are well established and often for indefinite periods, supplementary support will be received from and may be eventually given to the parent plant and perhaps even other adjoining plants of the clone. These are plants that have found successful methods of invading and establishing themselves in new territories that plants without these features could not begin to challenge. It is evident that these structures are used primarily as a means of producing new plants vegetatively and not as storage organs since there is little food stored in these stems beyond what is required for day-to-day existence. In this respect they differ markedly from the underground rhizomes and tubers.

Suckers, Runners, and Stolons

Suckers are essentially ground-level vertical shoots that grow from the base of the plant, or from adventitious buds on roots just under the surface, and are capable of growing into full-sized plants. They are primarily responsible for increasing the size of the clumps of lilac, flowering quince, witch hazel, and hollygrape and are, in these cases, desirable up to a point. When the stems begin to crowd each other, some of them should be pruned out. If new plants are desired, individual suckers may be dug carefully, to minimize root damage, and transplanted to a new site. Genetically, of course, they will duplicate the original plant.

Suckers on some trees, such as mountain ash, locust, or golden rain, are not only unattractive but may draw some nourishment from the parent plant (see Fig. 4-10). Since we usually intend these trees to be single-trunked and as vigorous as possible it is best to remove any suckers as soon as they appear.

Fig. 4-10. Mountain ash (*Sorbus*) trees often produce suckers from the base of the tree which are unattractive and could be detrimental to the tree. They should be pruned off.

Grafted rose bushes frequently produce suckers from the rootstock. Plants used for such rootstocks are selected for their vigor and hardiness and rarely produce desirable flowers. Since suckers from these grafts will carry the genetic information from the rootstocks and not from the more attractive upper portion (scion), they will be worthless as flowering material and should be promptly removed.

Spontaneously produced plantlets similar to suckers, and (like bulblets and cormels) often referred to as **offsets**, are commonly produced by certain foliage houseplants. These are outgrowths of the parent plant that can, if desired, be removed to another pot after they have sufficient roots to survive on their own. *Agave, Aglaonema, Sansevieria, Rhoeo,* and the bromeliads produce easily separated offsets or divisions. (See Fig. 4-11).

Fig. 4-11. Bromeliads such as these produce offsets or divisions from the base of the parent plant.

The tight clusters of plantlets, sometimes called **crowns**, of African violets, *Asparagus*, *Aspidistra*, *Oxalis*, and *Streptocarpus* are somewhat more difficult to separate from each other but, like the more easily defined offsets, can be propagated into individual new plants.

The terms runner and stolon refer to stems naturally specialized for vegetative reproduction and are usually synonymously used. A **stolon**, by definition, is any basal branch that is disposed to root, while a **runner** is a very slender stolon. The plantlets produced at the tips of the stolons of strawberry, *Saxifraga*, hen-and-chicks, and *Chlorophytum* are familiar to plant growers as a source of easily obtained new plants. The hanging stolons of potted *Saxifraga* and spider-lily plants will often branch and produce leafy growths at several nodes of the stolon. Little or no root growth will occur unless the base of the plantlet contacts a suitable growing medium.

For plantlets to grow into new plants they must first make adventitious roots. Adventitious roots usually arise from the nodes of stems, and, in fact, most monocosts and some dicots have root-primordial tissues regularly included in their stem nodes. These tissued usually remain in a dormant state until environmental conditions encourge their growth. When the stems of such plants come into contact with moist soil, for instance, they root quickly. In their native habitats, this ability allows them to establish themselves competitively among the other plants. Most stolon-produced plants will root more quickly if they remain attached to the parent plant while roots form. The little plants can be rooted either in the same pot as the parent plant or in a pot placed close by. Gently push the plantlet into the soil surface. A hairpin or loop of twist-em can be used to anchor it to the soil. Since the stolon is living tissue, it contains food supplies that need replenishing, so, after the new plant is well established, cut the stolon off as close

Fig. 4-12. Stolons develop into vegetatively reproduced plantlets in many species. The stolons of *Episcia dianthiflora* (left) are thin, runner-like extensions. Plantlets of *Sempervivum tectorum* (center) are produced on short, thick stolons from the parent plants. *Chlorophytum* (right) is spectacular as a hanging plant.

to the new plant as possible so it doesn't burden the young plant with its support. If the plantlet is detached from the parent before planting in soil, leave a short length of the stolon attached, since under these conditions the stolon may provide some nourishment while the roots are forming (Fig. 4-12).

Layers and Layering

Layers are normal branches that root when they come in contact with soil and produce new plants. Some plants form layers very readily. Blackberry, raspberry, *Forsythia*, and *Rhododendron* produce adventitious roots when stems bend over far enough to touch the soil. A layered branch is in some respects like a stolon, and new plants can be produced at the stem apex or at nodes or other areas along its length.

Three descriptive names may be applied to the most common methods of layering. **Tip layering** is accomplished when a stem apex comes in contact with the soil, roots form, and new stem growth follows. If you try to walk through a patch of wild blackberry brambles, you will find that your feet tend to catch in loops of blackberry stems that have formed where branch tips have anchored themselves to the soil with adventitious roots. **Simple layering** is achieved when adventitious roots appear at any point or node along the stem wherever it touches the soil. **Serpentine** or **compound layering** is the development of roots at several points along the stem (see Fig. 4-13). In both of the latter cases the stem apex may extend above the soil level unless tip layering has occurred.

The gardener can take advantage of this ability of stems to produce roots by manually bending branches down, pegging them to the soil until the roots form, and then cutting the developing plant from the parent. Commercially grown berry bushes and many flowering shrubs may be layered to produce new plants. Nursery growers

Fig. 4-13. (left) Tip-layering in wild black raspberries. The branches, bent over in the snow so that their tips contact the soil, will produce adventitious roots from the apex by the time new leaves begin to appear in the spring. (right) Serpentine layering in wild grape vines. As the vines extend over the ground, adventitious roots grow from the nodes to anchor the plant firmly to the soil.

usually make a notched wound in the stem of a selected plant in the portion to be buried to stimulate faster growth of adventitious roots, since wounding disrupts the normal flow of materials in stems and results in an accumulation of root-stimulating plant hormone in the wound area. Stems may be anchored to the soil with forked sticks until roots form, after which the new plant can be separated from the parent and planted elsewhere.

Air Layering

Air layering is an ancient method of producing adventitious roots on the stems of plants that might be reluctant to do so or that can't be bent easily over to the ground, or it may be faster and surer than stem cuttings (see below, p. 116) for some plants. It is a popular technique with growers of foliage houseplants such as *Dracaena, Dieffenbachia,* and *Ficus* as well as certain broadleaved evergreens such as holly, *Rhododendron,* and *Gardenia.* It is best done just as new growth begins in the spring and the plants are emerging from their semidormant condition. (See Fig. 4-14.)

1. Select a branch end that looks as though it will make an attractive new plant.
2. Cut a V-shaped slit 1/3 of the way into the stem.
3. Prop open the slit with a match stick or toothpick.
4. Dust the wound surface and nearby area with commercial rooting-hormone powder.
5. Using water-soaked sphagnum moss (not moss peat, which is too acidic), squeeze out the excess water so that a packing of moist but not dripping moss can be placed firmly around the wounded area of the stem.
6. Cover the wad of sphagnum moss with plastic wrap to retain moisture. Tie securely to the stem with twist-ems at both ends.
7. Check the moisture of the moss packing periodically and re-wet if necessary.
8. When a mass of new roots protrudes through the moss, cut the stem off below the rooted area and gently remove the plastic wrap and the sphagnum moss. If some of the moss is entangled in the roots just leave it in place since removing it may damage roots.
9. Plant in suitable soil in pot or garden bed.

Fig. 4-14. Air layering a rubber plant (*Ficus elastica*). Step 1. Cut slit 1/3 way into stem. Step 2. Prop slit open with stick. Step 3. Pack wound area with moist sphagnum moss. Step 4. Cover sphagum moss with plastic wrapping.

THE GARDENER TAKES CONTROL

Artificial Vegetative Propagation

Cuttings are leaves or sections of roots and stems that are removed from a plant and induced to grow into whole new plants. Most plants can be propagated by some kind of cutting. Florists, farmers, greenhouse growers, and nursery growers utilize cuttings as a fast and inexpensive way to produce plants of known genetic quality. This is, in fact, the only reliable way to propagate a new seed-grown hybrid that shows desirable characteristics. Such plants may be patented under the U.S. government patent law of 1930 which protects them for a period of 17 years. Patented plants may be commercially propagated only if there is a legal agreement between the holder of the patent and the potential grower. Plant patents do not, of course, forbid the windowsill gardener from propagating a few plants for personal enjoyment.

Stem Cuttings

Cuttings taken from stems are probably the most widely used method of propagating plants vegetatively. Cuttings from the stems of many plants, especially those that are herbaceous, will produce adventitious roots quite readily if placed in water or moist soil. A dusting of root-inducing hormone powder will stimulate growth from woody stems or others that may be somewhat poor or slow in producing roots.

When a stem cutting is taken from a plant, hormones, especially auxins (see Chap. 1, p. 24) play a major role in wound healing and the production of adventitious roots, both of which are essential in the successful transition of a cutting into a complete new plant. In the wound area their presence may stimulate the cells to divide into an undifferentiated mass of tissue called a **callus** which is formed after the flow of auxin from the growing tip is disrupted and the hormone accumulates in the wound area. The auxin promotes the synthesis of an enzyme called cellulase which digests cellulose in cell walls. This loss from the cell walls softens them so that the cells enlarge, take in additional water, and begin dividing, quickly forming a protective callus. These same hormones, auxins, in different concentrations and interacting with other factors, will later bring about the differentiation of the callus tissue and other minimally specialized tissues, such as parenchyma, cambium, and pith of a stem, into appropriate new tissues for stem repair and for adventitious roots.

Hardwood Cuttings

Hardwood cuttings from the woody stems of shrubs and trees are cut from the current growth in the fall of the year. At this time the new tissues are well established but not fully mature and can still be easily induced to grow. If available material is not abundant, the cutting can be made with a simple diagonal cut just below a bud. Preferably, a mallet or hammer cutting will be used in which the cutting includes a portion of a main stem with a twig growing from it at right angles. Such a cutting will have a larger area from which adventitious roots may be produced. (See Fig. 4-15.)

Fig. 4-15. Hammer or mallet cuttings are often used for woody plants. Cuttings can be made on dormant plants (re: text instructions) or, if greenhouse plants are used, just prior to the resumption of active growth in the early spring.

Hardwood cuttings are best handled in the following manner:

1. Cut the dormant stems into 6-10" lengths (15-25 cm.), although results of recent tests show that many plants produced roots more reliably and abundantly from longer cuttings up to 24" (60 cm.).
2. The stouter stems from a plant should be used, since they usually root more vigorously than the thinner stems.
3. Tie the cuttings in bundles, pack in moist sphagnum moss, and seal in airtight plastic bags.
4. Store at 40-50° F. (5-10° C.) in a refrigerator or buried in sand outdoors. The cut ends will callus over during storage.
5. In early spring, plant the cuttings in sandy soil with only the tips of the cuttings exposed.
6. Water and fertilize with a general garden fertilizer.
7. By fall roots and shoots should be well formed, and the new plants can be transplanted where desired.

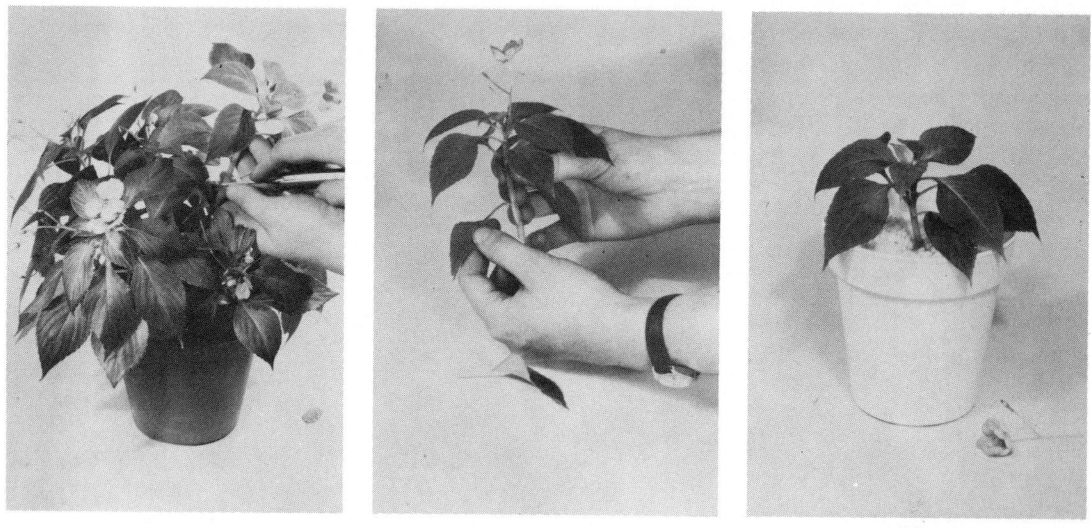

Step 1 Step 2 Step 3

Fig. 4-16. Making a softwood cutting of *Impatiens sultana*. Step 1. Cut terminal portion from stem. Step 2. Remove leaves at lower end and any flowers or flower buds. Step 3. Place in moist medium and protect from direct sunlight for first few days.

Softwood Cuttings

Softwood cuttings are taken from plants that are growing vigorously and generally consist of terminal new growth. Plants with green herbaceous stems are commonly used as softwood cutting material. Shrubs and trees also yield softwood cuttings from the tips of current shoots early in the growing season before they have become woody. Annuals and biennials as well as flowering garden perennials and herbaceous foliage plants produce young shoots, often referred to as **slips**, that can be removed from the plant for propagating. Older, fibrous stems should, in all cases, be avoided, since new growth is not easily initiated in fully specialized tissues and the results can be expected to be poor.

Softwood cuttings of most broad-leaved evergreens, many flowering shrubs, and vines are usually cut in mid- to late summer after blooming. However, since there are a few, such as lilac, that should be cut while in bloom or at other special times, it is best to look up the requirements for your particular plants in a gardening handbook before attempting any cuttings.

The procedure for taking a softwood or herbaceous cutting (slip) is fairly standard for most suitable plants: (See Fig. 4-16.)

1. Early morning is the best time, when the plant's leaves are at maximum water capacity after a night of low transpiration loss and before the plant has to face the more rigorous daytime conditions.
2. The terminal portion of 2-6" (5-15 cm.) is best, although long stems of herbaceous plants such as *Coleus*, *Impatiens*, or *Begonia* can be cut into several sections, each of which will root successfully.
3. Cut the stem at a 45° angle, especially if thick, to increase the surface for water uptake and root formation.
4. Remove leaves from the lower end of

the cutting so that at least two nodes (for sufficient root formation) can be inserted into the planting medium.

5. Remove any flowers or flower buds, and continue to remove any that form while the cutting is rooting.
6. Remaining foliage may have to be slightly reduced to prevent excessive transpiration (water loss) from the leaves. Without roots the cut stem end cannot take up as much water as needed, and severe wilting may occur from which the cutting may not recover. If the leaves are few but large, leaf area can be reduced by cutting some of the terminal portion from the leaves.
7. Dip the portion of the cutting that will be inserted into the soil into a root-promoting hormone powder. To avoid over-applying this auxin-imitating powder, which may actually inhibit root growth if used in large amounts, mix one part powder with eight parts sterile sand. A quick dip into this mixture applies just the right amount for most stems. It may be necessary to dampen exceptionally smooth stems, and fuzzy stems may need to be brushed off a bit, but most stems are easily coated.
8. The planting medium should be a loose water-holding material from which the rooted cuttings can later be removed without damage to the new tender growth. Coarse sand, vermiculite, perlite, sphagnum moss, or combinations of these are commonly used. Moisten the planting medium and insert cuttings at an angle to assure good light exposure. Firm medium around the cuttings and moisten again if needed.
9. Place the cuttings in a shaded place for the first few days to reduce transpiration. Then, depending on the normal light requirements of the plants, move them into brighter light or direct sun if the leaves are firm and not wilted. It is sometimes advantageous to cover with plastic or glass to keep humidity high and prevent wilting. Gradually this covering should be removed as the cuttings strengthen. Good air circulation is important, but sharp drafts can damage fresh cuttings.
10. Keep the medium moist, not soaked. Do not allow to dry out. Provide warm temperatures, 65-75° F. (18-24° C.), and additional bottom heat if possible. Heating coils for this purpose are available from most garden supply stores. Root formation is faster if the planting medium is warmer than the surrounding air.
11. When the roots are about one inch (2.5 cm.) long, the cutting should be transplanted into suitable soil. If the cutting cannot be transplanted at this time, a fertilizer solution should be used to supply necessary nutrients for continued growth.
12. If cuttings must be held for a time before planting, wrap them in wet paper towels, place in a loose plastic bag, and refrigerate until planting time.
13. Some stems "bleed" when cut. Sugary phloem sap or milky-colored juices from special ducts exude from a cut surface. Moist succulents have these juicy stems. Such cuttings must be allowed to dry and the cells to seal off the flow, thus reducing the risk of bacterial invasion at the cut end that may cause rotting tissues.

Cuttings of the cane-like stems of plants of the family Araceae, which includes *Philodendron*, *Dieffenbachia*, and *Aglaonema*, deserve special comment. These sturdy green stems are often seen bearing a tuft of leaves at the apex but are bare below that to the soil. Such plants become increasingly unattractive as the leafy apex reaches ever higher and the bare stems become ever longer. To remedy this situation, the following procedure is recommended: (See Fig. 4-17).

Step 1

Step 2

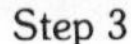

Step 3

Fig. 4-17. Making cane cuttings of *Dieffenbachia picta*. Step 1. Remove lanky cane close to base of plant. Step 2. Cut cane into several short sections and a leafy top section. Step 3. Plant top section vertically and other cane sections either vertically or horizontally in moist perlite. New shoots will emerge from protruding "dimples" on stem.

1. Cut off the leafy top, treat the base of the cutting with rooting-hormone and plant in damp perlite or vermiculite. When roots have formed, pot in loose, humusy soil.
2. Cut the remaining stem into 4-6″ (10-15 cm.) sections. Leave about a 6″ (10cm.) stump attached to the rooted portion in the soil.
3. The rooted stump will produce a new leafy aerial portion from buds on the stem if kept watered and supplied with light.
4. The other stem portions may be planted either vertically or horizontally in a suitable medium, i.e., sand, perlite, or vermiculite, or even soil. If planting vertically, dip the lower portion of the

cutting in rooting powder and plant at least one node deep in the medium. Stem sections planted horizontally should be buried halfway in the medium after rooting powder is dusted onto the lower side. Adventitious roots will arise from the areas below the surface of the medium in both cases, and new shoots will arise from buds on the stem above ground. Before they begin to grow, these buds have the appearance of slightly protruding dimples in the smooth stem area between node joints.

5. A humid atmosphere can be maintained over the cuttings by covering the containers with a plastic-bag tent.
6. When the new shoots and roots are well established on the cane sections, they can be planted into a pot of well-draining soil. (See table of soil recommendations in Chap. 7, pg. 198).

Root Cuttings

Roots, unlike stems, do not have nodes from which new growth is usually easily initiated. Adventitious shoots, however, may form naturally on the roots of certain plants, a trait that, under normal growing conditions, seems to be limited to a few species, but some others may be induced to form them as a response to injury. Garden perennials that grow naturally in clumps, such as poppies, bleeding heart, *Phlox*, and *Delphinium* are often divided by separating a whole plantlet, with roots and stems, from the clump. These same plants may also be propagated from root cuttings alone, although this is often difficult for the amateur gardener to do. Since there are no nodes or other obvious initiating points on roots, experience is needed to learn to judge which portions of roots will be most likely to produce buds. Generally you should select a part of a root that is neither too old and woody nor too young and fragile. Cut the selected root into 3-4″ (7-9 cm.) lengths. Bury the sections horizontally just below the soil surface, and mulch for the winter if there is danger of freezing. In the spring they will (you hope) produce the adventitious buds from which new plants will grow. Fig. 4-18 illustrates a root cutting.

Leaf Cuttings

Leaf cuttings, as the name implies, utilize the leaf as a source of material for new plants. Leaves used in this way do not become a part of the new plant but serve only as an origin and source of early nour-

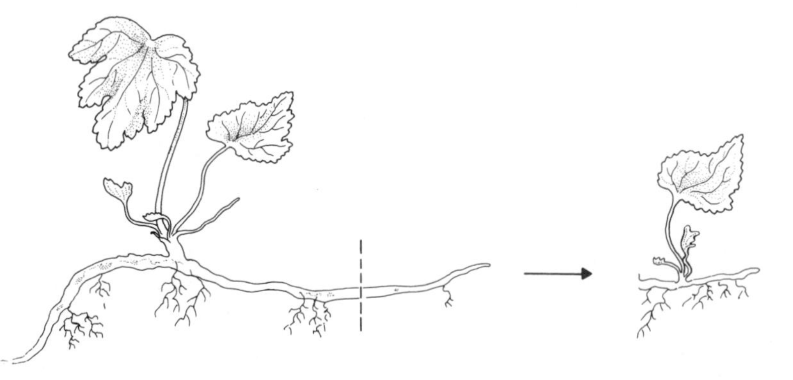

Fig. 4-18. Root cutting from Japanese anemone. 1) In the fall, remove several pieces from healthy roots. 2) Bury the piece horizontally in a shallow pit or cold frame. Mulch to protect from freezing. 3) New buds will appear by spring and the root pieces can be planted in the garden to produce new plants.

ishment for the developing plant. The old leaf often persists as an attachment long after the new plant is well along and may need to be removed if it detracts from the appearance of the new plant.

A surprisingly large number of plants can be induced to produce adventitious shoots from their leaves under ideal laboratory conditions and care, but in general the home gardener is most likely to be successful with the fleshier types of leaves rather than thin, papery ones. The thicker leaves have more closely packed cells with a greater amount of stored food and water and are usually heavily cutinized (see Chap. 1, pg. 5) and are less inclined to wilt severely. A coating of wooly or glandular hairs on a leaf will also retard water loss from leaf surfaces. In some plants whole leaves must be used; in others sections of leaves are sufficient. For best results use healthy, mature leaves that have not yet begun to age.

1. Leaf-and-petiole cuttings: The easiest type of leaf cuttings utilizes leaves that are borne on petioles such as those of African violet, *Gynura*, and some of the varieties of *Peperomia*. Detach the leaf with its petiole with a clean, smooth cut. Dip the petiole end in rooting powder and insert into the rooting medium. The leaf blade should protrude well above the surface. New plants will form at the end of the petiole (Fig. 4-19).

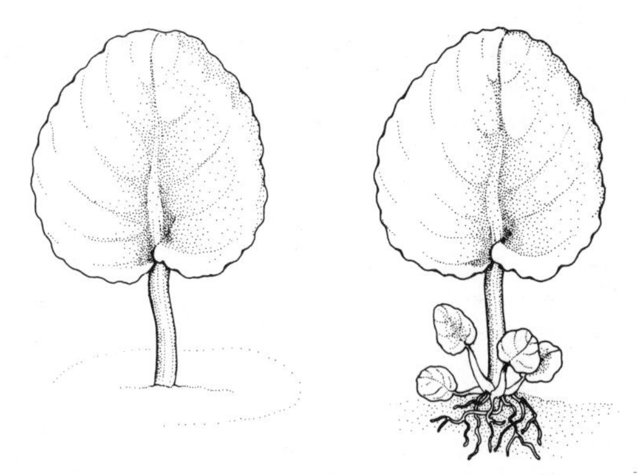

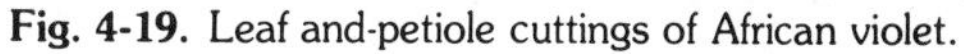

Fig. 4-19. Leaf and-petiole cuttings of African violet.

2. Leaf-blade cuttings: Whole leaves or sections of them may produce new plants even if they don't have petioles. Small, fleshy leaves of sedums (donkey tail) and crassulas (jade plant) can be plucked and planted base down in moist sand.

The terminal portion of the thick leaves of aloes and haworthias can be cut off, allowed to heal for a day, and planted base down in rooting medium.

The very long strap-like leaves of *Sansevieria* can be cut into several sections. Allow the cut ends to heal, dip in hormone powder, and again, plant base down in damp sand or other well-draining medium. When cutting a long leaf into several sections, take care to plant each with the terminal end uppermost. Plants do know which end is up and will not produce new plants if reversed (See Fig. 4-20).

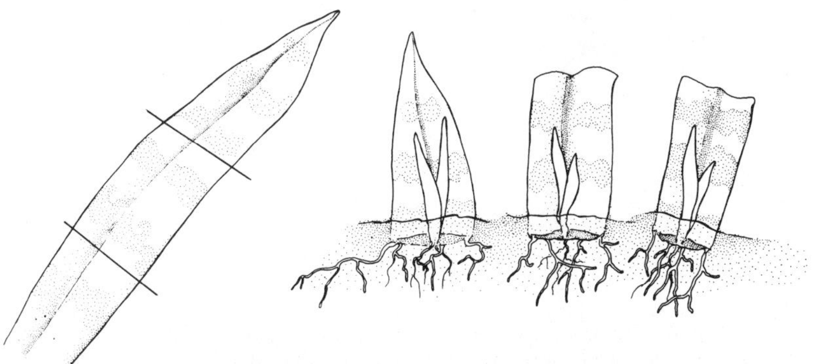

Fig. 4-20. Leaf-blade cuttings of *Sansevieria*.

These leaf cuttings will each produce a new plant from the base of the cutting. The old leaf portion may be removed after the new plant is well established.

An interesting phenomenon will be seen if variegated leaves of *Sansevieria* are used as cutting material. These leaves are a mottled green with creamy white edges. The new plantlets are derived from cells in the central green area of the leaf and, apparently, carry active DNA instructions for green leaf color only, since none of the leaves on the new plant will be bordered with white color. (See Fig. 4-21.)

Plants produced by vegetative propagation, or, in fact, any parts of a plant, that vary genetically from the parent plant are

Fig. 4-21. The parent plant of this *Sansevieria* is a yellow-edged variety. When a leaf cutting is taken, the new little plantlet will be derived from cells in the central area of the cutting and will lack the yellow border.

unusual and are called **sports**. While such aberrations occur infrequently, variegated plants seem to be especially subject to these genetic changes. Perhaps the genes responsible for variegation are in some way less stable and more likely to mutate than are the genes for more normal green colorations.

The large leaves of the Rex begonias with strong veins will give rise to new plants if several of the veins are slit crosswise and the leaf pressed firmly against a moist sand-and-peat medium. Small stones may help to hold them down. Roots will be produced at the wounds, followed by new shoots. Alternatively, large *Begonia* leaves may be cut into triangles, each bearing a portion of a main vein, and planted base down in a rooting medium. Roots and new shoots will then form at the basal end (See Fig. 4-22.)

3. Leaf-bud cuttings: These cuttings are especially useful if stem-cutting material is in short supply or if the plants have small or papery-thin leaves that do not produce adventitious growth easily. By this method new plants will be produced from the tiny bud at the base of the leaf. Remove a leaf and its auxillary bud by gouging slightly into the stem from which they grow (see Fig. 4-23). Cover the stem portion of the cutting with damp sphagnum moss until adventitious roots and shoots are produced. With this method a single branch of a *Rhododendron*, for example, that will produce only one stem cutting, will produce a dozen or more leaf-bud cuttings, a fine advantage if a number of plants is desired.

4. Spontaneous leaf propagations: Among the many oddities of the plant world, *Bryophyllum* and *Kalanchoe* have attracted considerable attention. These plants have leaves with natural indentations along the edge from which new little plants spontaneously form under almost any conditions, whether the leaves remain attached to the plant or not (see Fig. 4-24). *Bryophyllum* leaves are even sold in souvenir shops

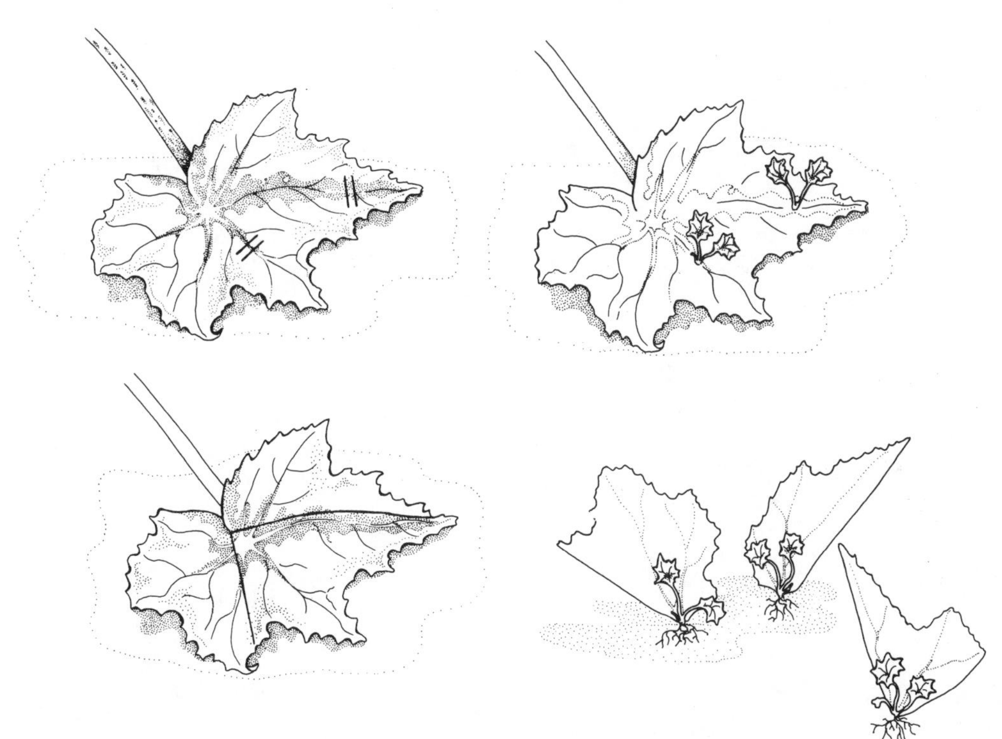

Fig. 4-22. Leaf-blade cuttings of *Begonia*.

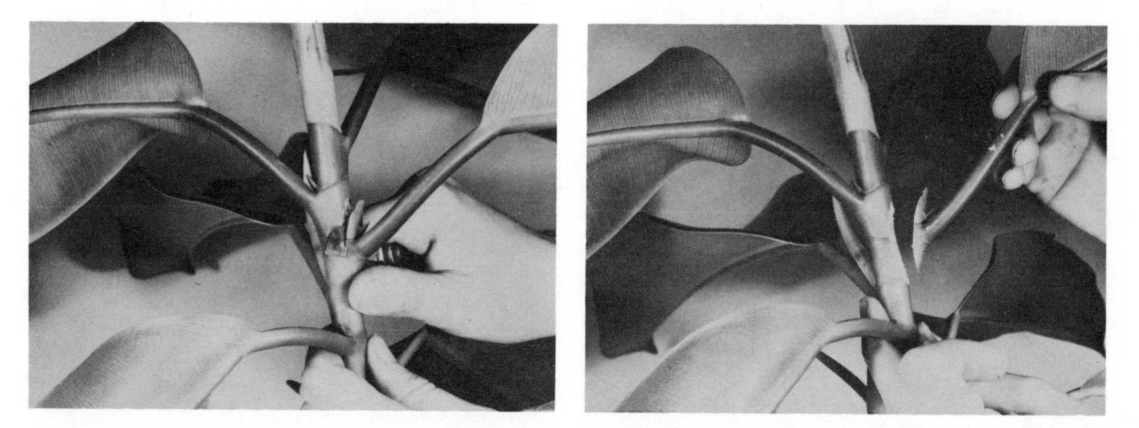

Fig. 4-23. Taking a leaf-bud cutting from *Ficus elastica*. Step 1. Select a leaf with a healthy tiny bud in the axil. Step 2. Remove leaf and bud from stem and place base down on moist sphagnum moss until adventitious roots form.

Fig. 4-24. *Kalanchoe daigremontiana* (above) and *K. tubiflora* are often called "maternity plants" because of the many tiny plantlets that form on the leaf margins and tips.

with instructions to "pin the leaf to your kitchen curtain and watch while new plants miraculously appear in a few days". The little plantlets often drop off and become established in the soil around the parent plant. These may be repotted into their own pots.

A few ferns such as mother fern (*Asplenium*), and the piggy-back plant (*Tolmeia*) are among the other plants that may also spontaneously form plantlets on their leaves.

Planting Media and Containers for Cuttings

Windowsill gardeners often find it so simple to start leaf or stem cuttings in water that one might wonder why solid rooting media are suggested at all. Many houseplants do indeed propagate well in water and, in fact, can be maintained in water culture almost indefinitely if the water is changed regularly and sometimes supplemented with additional nutrients. *Coleus*, *Impatiens*, the wandering Jews, *Scindapsis*, and *Begonia* are but a few that will thrive in water.

However, plant roots formed in water will be large, thin-walled, and fully turgid. When such roots are transplanted into a normal soil, they are easily damaged because of their thin-walled fragility and suffer shock when the turgidity of the cells drops rather rapidly when the supply of water is lessened. Cell functions will be somewhat disorganized until the roots adjust to this new environment. Some cells of the root may not survive at all. Because of the interactions of the root and shoot there will be a simultaneous effect on the shoot of the plant as well. The overall result is usually a setback for the plant from which it may never fully recover.

The recommended planting media for cuttings include sand, perlite, vermiculite, peat moss, and sphagnum moss, either alone or in various combinations. For cuttings of easily rooted plants such as *Impatiens*, *Begonia*, *Gynura*, etc. a 50-50 mixture of sand and perlite is quite adequate as long as the plants are transplanted to a more nourishing medium soon after they have rooted. A popular mixture for almost any cutting is:

2 parts shredded sphagnum moss
1 part sand or perlite.
Peat moss may be substituted for the sphagnum moss for acid-tolerant plants.

It is important to provide not only physical support for the cutting but also oxygen, water, and, after roots form, at least some nutrients which are obtainable from the organic material. The medium should be loose enough to allow the newly formed roots to be removed from it without damage when it is time to transplant.

The double-pot method is popular for starting cuttings, especially when just a few houseplant cutting are needed. A large pot is filled with perlite, sphagnum moss, or other medium suitable for cuttings, and a smaller clay pot, with its drain hole plugged, is inserted in the middle. The small clay pot is filled with water or wet sand, and the cuttings are placed in the medium in the large pot. The water in the small pot keeps the cutting medium moist and adds humidity around the cuttings. A plastic-bag tent will minimize early wilting. (See Fig. 4-25.)

When a cutting is well rooted it can be transplanted into its own pot containing a soil mixture suitable for the species. Some recommendations for appropriate containers will be found in Chap. 7, p. 180, and soil recipes are in the Appendix, Chap. 2.

One of the most exciting technical advances in modern gardening has been the peat pot, introduced in 1954. It is composed of 70% good European peat and 30% wood fiber. To this is added soluble fertilizer. Seedling plants and cuttings grown in Jiffy-Pots (also called ErinPots and Pullen Pots) are stronger and grow faster than comparable plants in clay or plastic pots. The Jiffy-

Fig. 4-25. A double pot arrangement in which to root cuttings. A plastic bag tent will help to retain humid conditions.

Pot is filled with a rooting medium and the cutting or seedling is inserted. When roots are abundant, the plant is transplanted while still in its Jiffy-Pot, thereby avoiding root disturbance and supplying degradable material for soil enrichment.

The idea was so appealing that other forms are now on the market. The Jiffy-7 appeared in 1967. This is a compressed peat disk encased in plastic netting. It has a pH of 5.5 to 6.0, which is a level acceptable to most plants, with nutrients added. When soaked a few minutes in water the Jiffy-7 expands to seven times its compressed size. The Jiffy-9 incorporates a gel binder in the peat to hold it together, and the plastic netting is eliminated. Jiffy-9's usually have preformed holes for planting. Kys-Kube is a variation that is a square block composed of peat and vermiculite with fertilizers added.

All of these products are sold under a number of different names as well as those cited. There are additional variations on the peat-pot idea produced by many garden-supply manufacturers.

5. MEETING THE ENEMY*

Life on planet Earth depends on the maintenance of a basic four-link food cycle. Green plants, through the process of photosynthesis, incorporate energy derived from the sun into food (sugar) molecules. These plants are eaten by herbivorous animals who in turn are eaten by carnivorous creatures, some of whom also manage to gulp down fellow carnivores as well. The dead and waste materials from these plants and animals are reduced to basic raw materials for recycling by the activities of the fungi and bacteria (see discussion on food chains and cycles in Chap. 1, p. 44).

NATURE'S DIVERSITY CREATES PROBLEMS FOR HUMANS

All ecosystems on earth depend on the food cycle. A garden is a small unbalanced ecosystem within a limited area and attracts and maintains its own food-cycle organisms. The gardener is, of course, providing as near ideal conditions as possible for the "green plants" part of the cycle and here is where the balance within the cycle gets disturbed.

Healthy, abundant plants attract herbivorous life forms that, in turn, provide food for carnivores. Our monocultural crop systems encourage outbreaks of insects and diseases, since there are not enough other kinds of plants around to harbor potential enemies of the organisms attacking the crops, and the pests can spread easily from plant to plant.

We till and cultivate the soil in both cropland and our gardens, producing habitats for weeds and pests while destroying organisms, or their habitats, that could help control them. Land is laid bare and robbed of its productive nutrients and animal homes and foods as it erodes away with time.

Travelers often return deliberately or inadvertently with new species of organisms that, lacking native controls, may flourish and become exceedingly destructive in our view.

In an effort to maintain control we often apply pesticides which, to our horror, make the situation worse by destroying not only the pest but its enemies, and invariably it is the pest that makes a successful comeback, usually in far greater numbers than before.

All is not futile, however, and the gardener will find that nature provides a good deal of help in producing a worthwhile garden. Insects are the most notable pests of the garden, and it should be encouraging to note that only about 2% of our known insects are pests. The other 98% are innocuous, many of them even being predators or parasites upon the pest species. Other predator-prey-parasite relationships exist among all the other known garden inhabitants.

*See Chapter 7, Indoor Gardening, for discussion of pests and diseases of houseplants.

Weather may help control some pests too when it becomes too hot, too cold, too wet, or too dry for a particular pest. Since food is important to garden friend and foe alike, the gardener can do much to control rampant pests by adding or eliminating certain plants to affect the pest population directly or by encouraging the increase of its enemies.

It has long been evident that no one approach is effective for most pest problems. Chemical pesticides were once hailed as being the answer to our pest problems, but over the years pests have become resistant to chemicals, their effectiveness has not been long-lasting enough, residues can remain in soils and crops, pesticides can disrupt natural controls, and, most seriously, they can harm non-target organisms including humans.

The best approach lies in what is called "integrated pest management". This is a "system of pest management that will bring the most benefits, at the most reasonable cost, on a long-term basis, to the farmer and to society" (C&EN News, April 23, 1973). The goal is to coordinate the best of available control techniques rather than to rely completely on chemical pesticides or on any one other method.

Above all it must be kept in mind that the goal is control or suppression of pests, not total eradication. If a total-eradication program is successful against a given pest, its natural enemies will be deprived of their food supply and, if they have not been eradicated by the program also, will either starve or leave the area in search of new prey. If the pest organism is reintroduced it will have the opportunity of multiplying unchecked until a new pesticide program is applied. Predator-prey relationships must be maintained for long-term control to avoid the overuse of pesticides.

If we adjust our thinking to operate on an ecological basis we should become aware of the fact that plants and pests exist together and damage often occurs. Is the damage tolerable? Will there be economic loss? Is the damage esthetically disturbing? If the damage requires remedial measures, determine the true cause of the damage. Look the plants over carefully, using a hand lens and a flashlight at night if necessary. Don't just guess! If in doubt, ask for help from the local garden store, university, or county extension agent. The shotgun approach of applying a variety of pesticides in the hope that one will work is not an acceptable method of dealing with pest problems.

Suppressing garden pests is not simple or easy. Familiarity with common pests and their controls is the first step toward a healthy, successful garden.

How many kinds of garden enemies are there?

1. **Weeds**—There are 1800 noxious weed species that cause serious economic losses and many more minor troublemakers.
2. **Insects** and **mites**—There are more species of insects and mites than of all the other animal species together, most of them still undescribed. At least 10,000 species constitute noxious pests causing staggering dollar losses.
   Insecticides (insect killers) and acaricides or miticides (mite killers) are used for control.
3. **Diseases**—Fungi, bacteria, and viruses directly cause plant diseases. Of these, fungi are the most important economically and are estimated to produce more than 1500 different plant diseases.Fungicides (fungus killers) bactericides (bacteria killers) are used for control.
4. **Nematodes**—These are tiny round worms rarely seen by the gardener but comprising 15,000 harmful species, 1500 of which cause serious economic losses in crops.
   Nematocides (nematode killers) are used for control.

5. **Miscellaneous**—Snails, slugs, rodents, and other small animals that, except in rare cases, inflict only minor damage.

STARTING CLEAN: STERILIZING AND FUMIGATING SOIL

It is always a good practice to sterilize soil to be used for potted plants or greenhouse benches or for adding to outdoor beds. Packaged soils purchased from garden-supply stores have usually been treated and may be assumed to be relatively free of harmful or undesirable organisms.

If a home greenhouse is maintained, it is wise to sterilize the bench soils periodically to control or eliminate potential pest problems. Whenever bulk soils are purchased to be used either in a greenhouse or for supplementing garden beds, it may seem like a lot of trouble to sterilize the soil before using, but sterilizing could save plants and prevent contamination of adjoining soils.

The most popular, economical, and safe soil sterilant is steam. It is used for most commercial operations and is the treatment most often applied to packaged soils. Used properly it should kill all weeds and soil-borne insects, bacteria, fungi, nematodes, and viruses that are harmful to plants. It will also make heavy soils more granular, improving drainage and aeration.

The most universally recommended conditions for steaming are 180° F. (82° C.) for 30 minutes. Nematodes and insects are killed instantly at 140° F. (60° C.); seeds, fungi, and bacteria die off between 140-160° F. (60-82° C.) after 10 minutes or so. The extra 20° and additional time are insurance that all parts of the soil and its inhabitants are reached. Higher temperatures are avoided to prevent destruction of the organic matter in the soil.

Since "good" organisms, especially nitrogen-fixing bacteria, are eliminated along with the troublemakers, it may be necessary to add nitrogen fertilizers or other supplements until the normal soil bacteria are re-established.

Since steam treatment requires special equipment not generally available in private situations, chemicals are a practical alternative. An ideal chemical for treating soil 1) kills fungi, bacteria, insects, weeds, and nematodes; 2) is inexpensive; 3) is harmless to nearby plants and subsequent plantings in the soil; 4) acts quickly and penetrates deeply into the soil; and 5) is harmless to operator and equipment. Unfortunately such an ideal chemical treatment has not as yet been developed.

Chemicals used to treat soils are of two types, stable compounds which kill when applied as liquids or powders mixed into the soils, and volatile compounds that are applied in liquid form which then release poisonous gases throughout the soil. Temperatures for soil fumigation should be between 50° and 90° F. (10° and 32° C.). Below 50° F. (10° C.) chemical reactions are slowed and biological activity is retarded. Above 90° F. (32° C.) rapid decomposition of the compounds often occurs.

When using chemicals to treat soils, take care both during and after treatment to prevent damage to nearby plants and to plants later planted in the treated soil, since such chemicals may be lethal to desired plants. Soil-sterilizing treatments are best performed outdoors, if possible (never in a greenhouse), and well removed from garden and yard plants, since many of these compounds, especially the volatile ones, give off toxic fumes that are readily carried to nearby plants. Most are also toxic to humans and other animals. Soils treated with volatile fumigants must be tightly covered for a prescribed period, usually 24 hours, and then allowed to air completely before being used.

A sampling of a few commonly used soil sterilants or fumigants is tabled below:

Table 5-1. Common Soil Sterilants

Chemical Name	Trade Names	Effective Against	Volatile	Soil Aeration Time
chloropicrin	Larvacide Picfume	weeds, most fungi, nematodes, soil insects	yes	10-21 days
methyl bromide some formulations are in combination with chloropicrin	Pestmaster Dowfume	weeds, most fungi, bacteria, soil insects	yes	3-12 days
metham or liquid carbamate	Vapam	most weeds, nematodes, soil insects	yes	14-30 days
40% formaldehyde	Formalin	fungi, bacteria	no	14 days
PCNB	Terraclor Dexon-Terraclor	certain fungi, water molds of soil	no	none

Rules For Using Chemical Agents For Weed And Pest Control

1. Always read label before using sprays or dusts. Note warnings and cautions each time before opening the container.
2. Keep chemicals out of the reach of children, pets, and irresponsible persons. Store outside the home away from food and feed.
3. Always store agents in their original containers and keep them tightly closed.
4. Never smoke while spraying or dusting. Avoid inhaling dusts or sprays.
5. When directed on label wear protective clothing and masks. Always wear plastic or rubber gloves when handling insecticides. Traces have been detected on unprotected hands days and even months after contact.
6. Do not spill agents on skin or clothing. If spilled remove contaminated clothing and wash skin thoroughly.
7. Change clothing and wash hands and face after spraying or dusting. Wash clothing before reusing.
8. Cover food and water containers when treating around livestock or pets. Do not contaminate fish ponds or streams.
9. Do not use equipment for applying herbicides for any other purpose, to avoid accidental damage to plants.
10. Dispose of empty containers so they cannot harm humans, other animals, or valuable plants.
11. Observe label directions and cautions to keep residues off edible plants.
12. If symptoms of illness appear during or shortly after spraying or dusting, get medical help immediately.

WEEDS

The Problem

Weeds are looked upon with disdain by most gardeners, yet if asked to define a weed we find it difficult. The cliches stating that "a weed is a plant growing out of place" or "a weed is a plant for which we have not yet found a use" are indicative of the mixed feelings towards weeds.

Certainly most of us find weeds sprouting between our carefully placed garden plants or in our lawns an annoyance at the least. Many of our common field weeds further annoy us when they produce pollen or chemicals that afflict us with allergies or dermatitis. Weeds are esthetically disturbing when they grow up against walls and fences or produce scraggly growth in vacant lots. They obstruct roadway visibility and signs and constitute fire hazards when dry vegetation accumulates, and aquatic species clog waterways and garden pools and ponds. However, these effects pale when compared to the extensive harm caused to crops and other agricultural-related products. It has been estimated that crop losses caused by weeds amount to over $11 billion annually (1979 estimate).

Weeds compete successfully with crop plants to reduce yield. They are more aggressive and deprive the emerging crop seedlings of water, nutrients, and light, and, in some cases, inhibit growth when they secrete toxic substances. They may serve as host plants to insects and diseases, like wild morning glories (*Ipomea* sp.) harboring the destructive sweet potato weevil, or goose grass (*Eleasine indica*) so common in our lawns, helping to spread a barley virus, or species of *Ribes* (gooseberry) serving as alternate hosts of white-pine blister rust.

Weeds affect livestock when they are inedible or produce spines, burrs, fibers, or tendrils that damage the mouths or digestive systems of the animals. Additionally they may also produce poisons that incapacitate or even kill foraging animals.

These problems are rarely of direct concern to home gardeners, so we can better appreciate the fact that the same wild plants that we consider weeds in our garden also cover our countrysides that are scarred by humans or nature, prevent erosion, provide foods and shelter for our delightful wild birds and animals, brighten our fields and roadsides with colorful flowers, and feed bees and butterflies.

Fig. 5-1. Two weed species typical of those that cause problems for humans: (left) Speedwell (*Veronica persica*), a common lawn weed that spreads extensively if not eradicated; (right) Water Hyacinth (*Eichhornia crassipes*), an aquatic weed common in the waterways of our southern states.

Most weeds invade by means of seeds which they produce in such prodigious quantities that it is easily understood why the battle continues even after the application of highly effective controls. Common pennycress (*Thlaspi arvense*), which appears frequently in flower beds, produces 70,000 seeds per plant, burdock (*Arctium lappa*) produces two kinds of seeds (see below) at a rate of 31,000 per plant, and common dock (*Rumex crispus*) may produce 40,000 seeds on a single plant.

Weeds of annuals and biennials are always distributed as seeds, while perennials often spread as bulbs, corms, roots, rhizomes, and stolons as well as by seeds. Some perennials do not bloom in their first year or so but nevertheless manage to distribute themselves by vegetative structures forming ever-widening colonies spreading from the parent plant.

Our gardens offer ideal locations for many weeds, with freshly turned soil and a minimum of plant competition. Many of our cultivated varieties of plants are far removed from their native habitats and/or have been subjected to cultural modifications to the point that they are incapable of competing successfully with the well-adapted weeds.

Weed seeds are dispersed by the natural agents of wind, water, and animals and by the gardener who buys container-grown plants in contaminated soils, plants packets of contaminated seeds, or supplements his garden soils with weed-laden top soil.

Weed seeds can be found most abundantly in the upper 1″-4″ (2½-10 cm.) layer of soil. The seeds are often buried more deeply in the soil by cultivation, by blowing or washing of seeds into cracks and crevices, by burying by earthworms and rodents, and in some cases by the activity of special appendages of the seeds that respond to atmospheric moisture and dryness by twisting and turning until the seed reaches a soil level with a constant moisture content.

Weed seeds are frequently highly adapted to a long dormancy, sometimes of many years' duration. Some weeds, notably the burdock (*Arctium lappa*) and lambsquarters (*Chenopodium* spp.), produce some seeds that germinate readily the first year and a second kind of seed that remains dormant the first year and will germinate after that only if conditions are favorable.

Generally a unit of soil is capable of producing a certain amount of vegetation, and in a garden of cultivated ornamentals invaded by weeds the well-adapted aggressive weeds will usually dominate unless controlled.

Methods of Control

To cope with weeds one could adopt the "English method", as pictured in Fig. 5-2, in which the gardener simply lets the weeds grow and pretends to enjoy them. There are obvious limitations to this method.

We can avoid introducing some weed seeds into our gardens by buying only high-quality packaged flower and vegetable seed from reputable firms. It is less likely to be contaminated with weed seeds and plant debris than is bulk seed. Potted plants and purchased bulk soils should be examined closely for weeds, weed seedlings, or other evidence of potential weed problems. A few weeds are unavoidable, but if excesses are apparent, buy from another nursery.

There are various national, state, and local controls that help prevent dispersal of weed seeds. The Federal Seed Act of 1939 protects growers from purchasing mislabeled or contaminated seed in interstate or foreign commerce. More recently President Ford signed into law the Federal Noxious Weed Act in 1974 which provides point-of-entry inspections of foreign imports, weed surveys, eradication programs within the United States, and quarantines to prevent establishment of noxious weeds. It is to the advantage of gardening and non-gardener citizens alike to support and encourage the application of such laws.

Fig. 5-2. The lovely gardens in Salisbury, England. With such an abundance of flowering plants a few weeds are not likely to be noticed.

In our own yards and gardens we can exert more direct controls. In small areas hand pulling and cultivating may be sufficient. In larger vegetable gardens a long-handled hoe or cultivator will effectively retard weed growth but leaves a somewhat less tidy appearance. If nearby vacant lots or wooded areas constitute a source of weed seeds, it may be possible to mow such areas to reduce seed production and minimize contamination of the garden.

Along with actual removal of weeds from the garden, the practice of mulching can effectively prevent or deter weed-seed germination. Mulch is nonliving material that will exclude light. Hay, straw, raked-up leaves, grass clippings (if no weed killer has been applied), fir bark, wood chips, barley hulls, peanut shells, chopped corn cobs, peat moss, sawdust, newspapers, black plastic sheets and even scraps of indoor-outdoor carpeting can be used as mulch. Organic mulches are additionally useful if they decay and can be worked into the soil later as humus-producing materials, but beware of using fresh manure (see Chap. 2, p. 61). Soil temperatures beneath hay, grass clippings, and plastic sheeting may reach killing levels and eliminate some weed seeds in this way. Care must obviously be taken to prevent damage to the roots of desired plants.

Synthetic Herbicides

When pulling, hoeing, and mulching are not sufficient to control weeds, the fourth method of attack involves the use of chemicals designed to kill unwanted plants. Such herbicides are especially useful in very large gardens or when applied to lawn grasses to eliminate broad-leaved invaders or weed grasses such as crab grass. They also solve the problem of weed-infested patios, driveways, and walkways, especially in the spaces between patio blocks or in gravel or bark-filled areas. Some of the newer herbicides minimize the necessity to weed flower and vegetable gardens if applied at specific times.

There are currently almost 200 different herbicides available, most of them designed for use on extensive croplands or orchards although the home gardener will face a bewildering assortment at any nursery or garden-supply store to meet more limited needs.

Herbicides can be grouped by their chemical structures, by their biological effect, by the plants to which they are applied, or by their method of application. Chemically, herbicides are usually organic compounds, although a few inorganic sustances are also available.

All pesticides, including herbicides, are produced for use not as pure chemicals but as formulations and are sold under a variety of trade names. Only very small amounts of the pesticide chemicals are required for effect, quantities much too small to be easily applied without overdosing. Thus formulations are designed to supply the correct amount in a convenient, economical form, with substances, such as surfactants and wetting agents, added to enhance the toxicity of the pesticide and allow for easy application as well as protection against adverse storage or transit conditions.

For herbicides, such formulations are available as:

1. solutions of water or oil
2. emulsifiable concentrates
3. wettable powders
4. granules and pellets

Of concern to the home gardener are the biological activities of herbicides and the proper application for desired effect. Herbicides must find their way into the cytoplasm of the cells of the plant before they can be effective. To control post-emergent (growing) weeds, some herbicides are applied directly to the soil and are then absorbed by the roots of the plants. Soil-applied herbicides are highly effective and tend to kill weeds at the youngest stages of growth. When herbicides are taken into plants through the roots, the poisons may be active in the roots themselves, or they may be translocated to other sites within the plant where they usually kill slowly by interfering with normal cell functions.

Weed-killers that are absorbed or adsorbed by seeds are called **pre-emergent herbicides**. When they are taken into a seed along with the water needed for germination the seed may be immediately killed. Herbicides that are adsorbed onto the outer surface of the seed coat are later absorbed by the developing seedling and terminate its growth usually before the seedling emerges above the soil.

Herbicide sprays may be applied directly to leaves, stems, and buds. In general, such applications are translocated to the living cells where they affect cell membranes or diffuse into the cytoplasm.

Contact herbicides and systemic herbicides are the two forms of control available. **Contact herbicides** primarily weaken cell membranes to the point that cell contents leak out and the cell dies. This action kills quickly, within a few minutes to a few hours.

Systemic herbicides interfere normal plant functions and metabolism and produce chronic or slow-acting toxicity. Such herbicides are widely translocated within the plant and may affect different cells in different ways. Generally the functions most frequently disrupted are cell division, tissue differentiation and elongation, and other functions involving growth regulating hormones, formation of chlorophyll and cells walls, and enzyme-requiring activities such as photosynthesis and respiration.

Both contact and systemic herbicides may be found in various formulations, some of which are soil-applied and some of which are foliage-applied, although contact herbicides are more likely to be applied above ground while systemics are more readily taken in by roots from the soil.

Even those herbicides applied above ground are often partially sprayed onto the soil as well or washed into the soil from plant surfaces, so that the fate of most herbicides depends on what happens to them in the soil. The selection of an herbicide is often determined by these considerations.

In simplest terms, herbicides may be physically removed from the site by any of several agents, or they may be decomposed and thus deactivated. If applied in recommended amounts, much of the herbicide will be taken into the plants to bring about its toxic effects. That which is not may be temporarily adsorbed onto soil particles, leached from the area of application by rain and ground water, or, if of a volatile nature, vaporized into the atmosphere. The physical removal of the herbicide does not often change its chemical nature, a fact that explains why plants and animals some distance from the site of application may be affected by air, water, or soil contaminated by herbicides.

Herbicides will, however, eventually decompose, some more readily than others. Sunlight brings about deactivating photochemical reactions in some herbicides which are not yet fully understood. The uptake of herbicides by plants and the subsequent cellular activities involving them will result in chemical degradation even while the activities are fatal to the plant. The most important decomposition of herbicides is accomplished by soil microbes either within the site of application or at some point to which the herbicide is eventually removed.

Residual herbicides that remain in the soil after their intended purpose has been served can cause many problems, not the least of which is their damage to desired plants or crops in the soil. Therefore it is important to:

1. Select the herbicide appropriate to your needs,
2. Apply as little as necessary to be effective,
3. Apply the herbicide at the recommended temperature and season,
4. Cultivate the soil and in other ways encourage microbial activity.

Each herbicide has its own unique features that determine its applicability. One herbicide may be more phytotoxic, plant-selective, shorter- or longer-lasting, more easily applied, or slower to leach than another, or have any of a number of other characteristics that would make it preferrable. Depending on the weed problem, the gardener can gain reasonable control by applying herbicides either before the seeds germinate as pre-emergent treatments or after the plants are established.

The accompanying table summarizes information on some selected commonly-used garden herbicides.

Table 5-2. Common Garden Herbicides

Chemical Name	Trade* Names	Information	Persistence	Toxic Effect on a 150-lb. Human
2,4-D	Dacamine Emulsamine	Contact spray for broad-leaved plants, after emergence. If washed off plant, may enter through roots. Use on lawns.	1-4 weeks	Moderately irritating
2,4,5-T**	Esteron 245	Post-emergent spray for broad-leaved plants, especially woody shrubs and trees. Good for lawns and home grounds.	1-4 weeks	Moderately irritating
silvex, ** MCPA	Weedone, etc. Kuron	Compounds related to above and similar in action. Combinations of these compounds are synergistic and require less chemical than when each is used alone.	1-4 weeks	Mildly irritating
amitrole	Weedazol, Amitrol-T	Inhibits chlorophyll formation in woody plants, perennial weeds and grasses. Used as foliar spray.	1 week	Nontoxic
AMS	Ammate X	Used to control woody plants and vines. Good for brambles and poison ivy.	no lasting effect on soil	Mildly irritating
2,3,6-TBA	Benzac Trysben Zobar	Used mostly for broad-leaved weeds, enters, through roots and foliage.	light treatment 6-8 weeks	Mildly irritating
DCU	Crag	Used mainly as pre-emergent treatment for crab grass.	1-4 weeks	Nonirritating
dalapon	Dowpon	Good for orchard-grass control.	2-4 weeks	Mildly irritating
TCA	TCA	Perennial weed-grass control in vegetable gardens.	3-10 weeks	Mildly irritating

Chemical Name	Trade* Names	Information	Persistence	Toxic Effect on a 150-lb. Human
paraquat	Paraquat	Non-selective herbicide and crop desiccant used as post-emergent foliar spray.	Inactivated by soil contact	Moderately irritating
diquat	Diquat	Post-emergent, non-selective compound similar to above but especially useful for aquatic weeds.	Inactivated by soil contact	Moderately irritating
cacodylic acid	Phytar, Ansar 560, Rad-E-Cate	Non-selective, post-emergent general herbicide. Foliar spray for ditch banks, fence rows.	Inactivated by soil contact	Mildly irritating
simazine	Princep	Pre-emergent or early post-emergent foliar spray for non-selective control of vegetation.	3-12 months	Mildly irritating
sodium chlorate	Atlacide	Non-selective rank weed killer. Sterilizes soil. May cause spontaneous combustion in dry vegetation.	Indefinite	Nonirritating
borate	Borax	Non-selective permanent herbicide. Good for driveways, patios.	Indefinite	Mildly irritating
sodium arsenite	Triox, Atlas A	Non-selective weed control and soil sterilant. Also for individual plant treatment.	Indefinite	Causes burns and blisters

*Trade names listed are examples only and do not imply endorsement.

**Recent studies indicate the possibility of more toxic effects on humans than were previously recognized, and these compounds may become banned.

INSECTS AND MITES

The Problem

Insects and their arthropod relatives have inhabited the earth since ancient Paleozoic times and were among the first terrestrial animals. They have plagued every other form of life to some degree, including humans, and will probably still be thriving long after the earth is no longer inhabited by people.

Currently we are fighting an insect horde that destroys crops, devastates forests, infests fruits and vegetables, attacks flower and foliage plants alike, and can turn the relaxing delight of gardening into a frustrating nightmare. The number of species of insects and mites destructive to ornamental plants alone in the United States is estimated to be at least 2500. Those attacking crops and wild plants add considerably more to this total.

Insects have six legs and a body consisting of a head, thorax, and abdomen (Fig. 5-4). There are an estimated 5 million species of which fewer than one million have been named. They range in size from microscopic crawlers to moths with 12" (30 cm.) wingspans.

They are notable for and distinguished

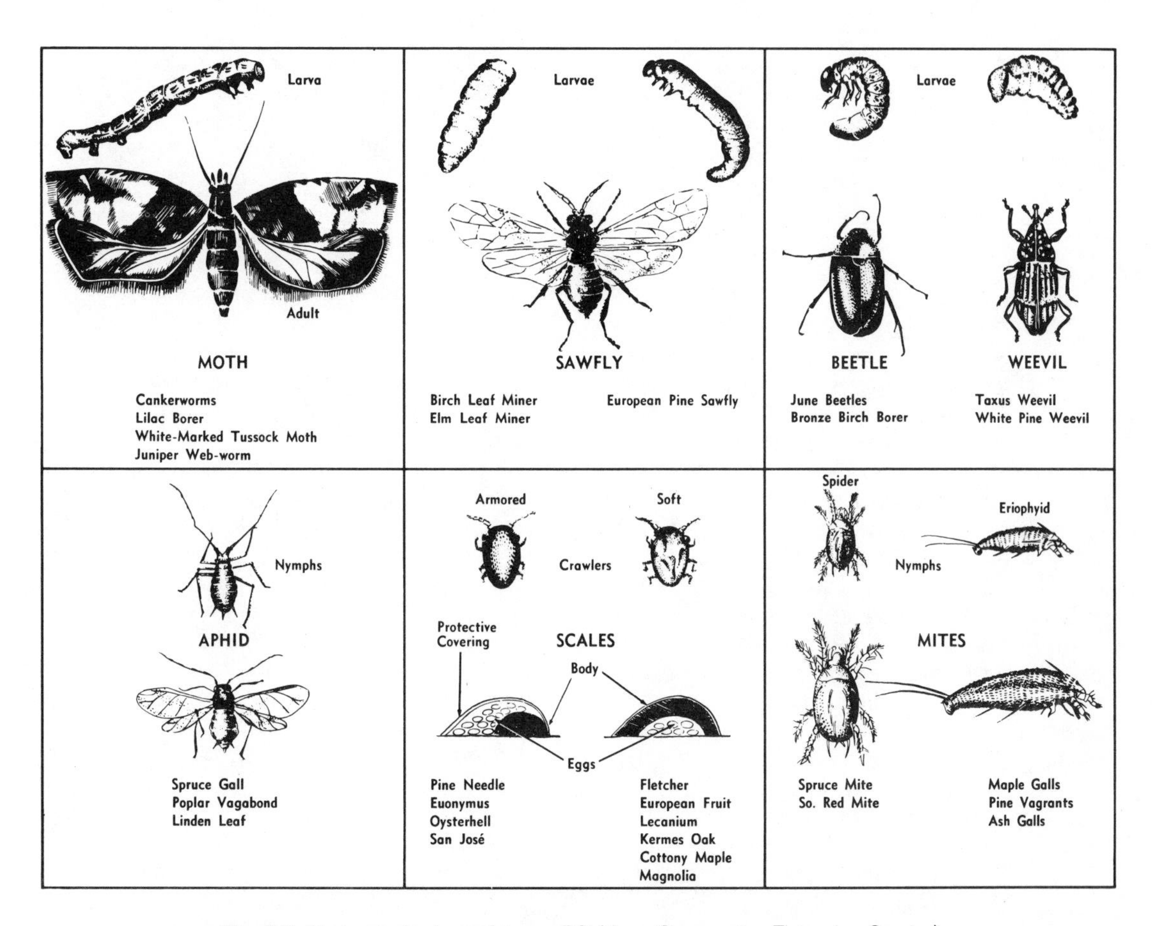

Fig. 5-3. Pests on shrubs and trees. (Michigan Cooperative Extension Service)

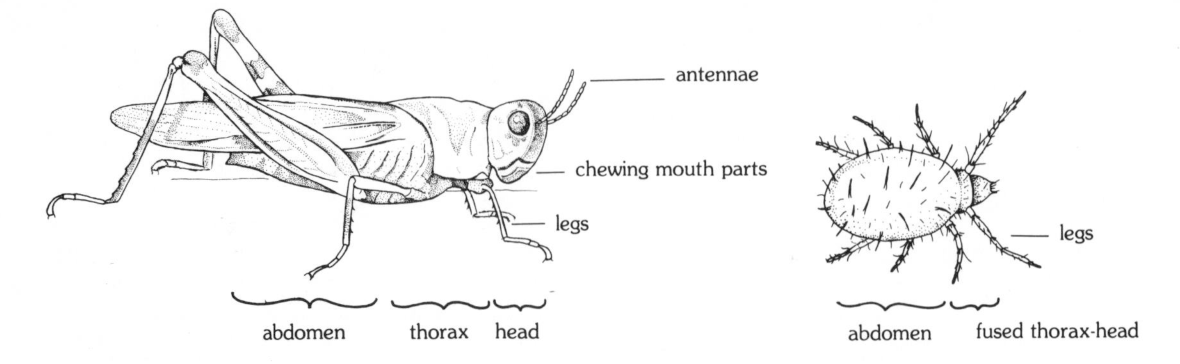

Fig.5-4. The grasshopper, a typical insect, has chewing mouth parts and a voracious appetite for the leaves of either wild or cultivated plants. The mite, closely related to spiders, is a very small wingless pest that sucks the juices from plants.

in part by their patterns of metamorphosis which are illustrated in Fig. 5-5. Some, such as grasshoppers and cockroaches, have a gradual metamorphosis in which the egg hatches into a nymph which sheds its wingless skin several times until the last molt that produces the adult winged form. The nymph is at all times similar to the adult and recognizable as a member of the species.

In incomplete metamorphosis the adult form is usually very different from the larval form. Dragonly nymphs, for example, are voracious inhabitants of a pond, eating mosquito larvae, minnows, and small tadpoles with ease. The last of several molts changes them suddenly into the graceful, though still carnivorous, dragonfly hovering over the cattails, an incredible transformation.

The beetles, butterflies, moths, and flies with their complete metamorphosis seem like something from science fiction. The tiny egg hatches into a larva (variously called a grub, caterpillar, maggot, or worm), feeds on unbelievable quantities of food, molts periodically as it grows, and at the appropriate time transforms into a quiescent pupa. A number of changes occur during this period until at last there emerges the adult form with six legs and two or four wings, often in brilliant colors and elaborate patterns.

Insects are well known for having many working parts, especially mouth parts. They feed either by chewing or sucking. The nectar-sucking bees and the chewing carrion beetles are among the many insects we consider beneficial to our way of life. The grasshoppers that chew up our grain crops and the aphids that suck out the juices in our prize-winning roses are not so favorably viewed.

Garden insects can be found anywhere on or near plants. Transient species flit from plant to plant and are especially difficult to control, since insecticides may simply drive them away before the effect can kill them. Many of the sucking insects establish colonies of several generations on a single plant. Applying insecticides may eliminate some of the feeding insects, but later the unaffected eggs hatch and rebuild the colony in surprisingly short time.

Occasionally plants begin to weaken and die abruptly because of the activity of soil insects, most of which attack roots and other underground structures, destroying root tissues or interfering with the transport of water and nutrients through the vascular tissue. The root weevils and related larval forms within the soil transform into winged but non-flying adults that can often be seen feeding on leaves. Such damage is usually

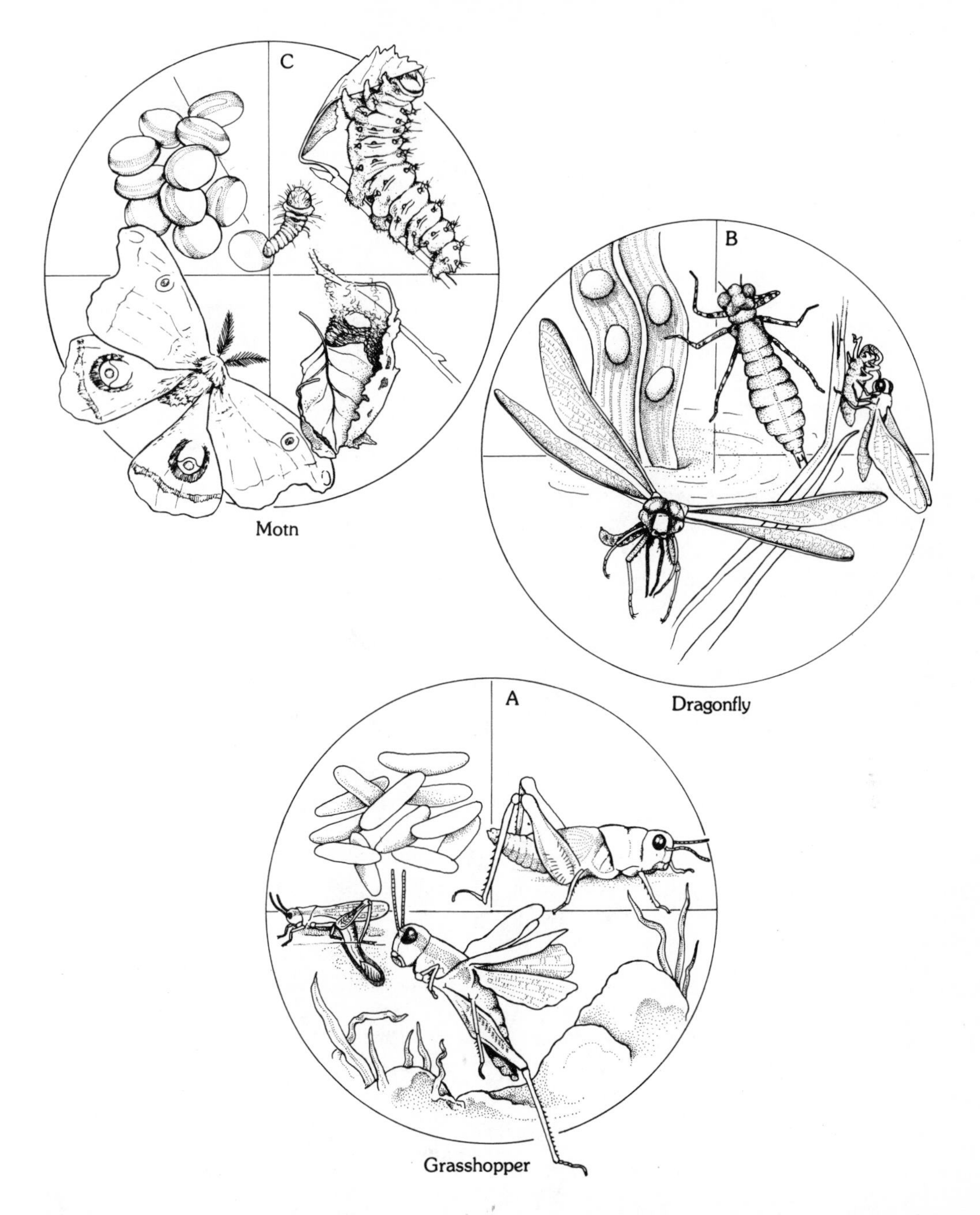

Fig. 5-5. Three patterns of metamorphosis in insects. a) Gradual metamorphosis in which the young and adult forms are similar to each other; b) Incomplete metamorphosis with a larva that transforms directly into an adult; c) Complete metamorphosis, a four-stage life history with a quiescent pupal stage between larva and adult.

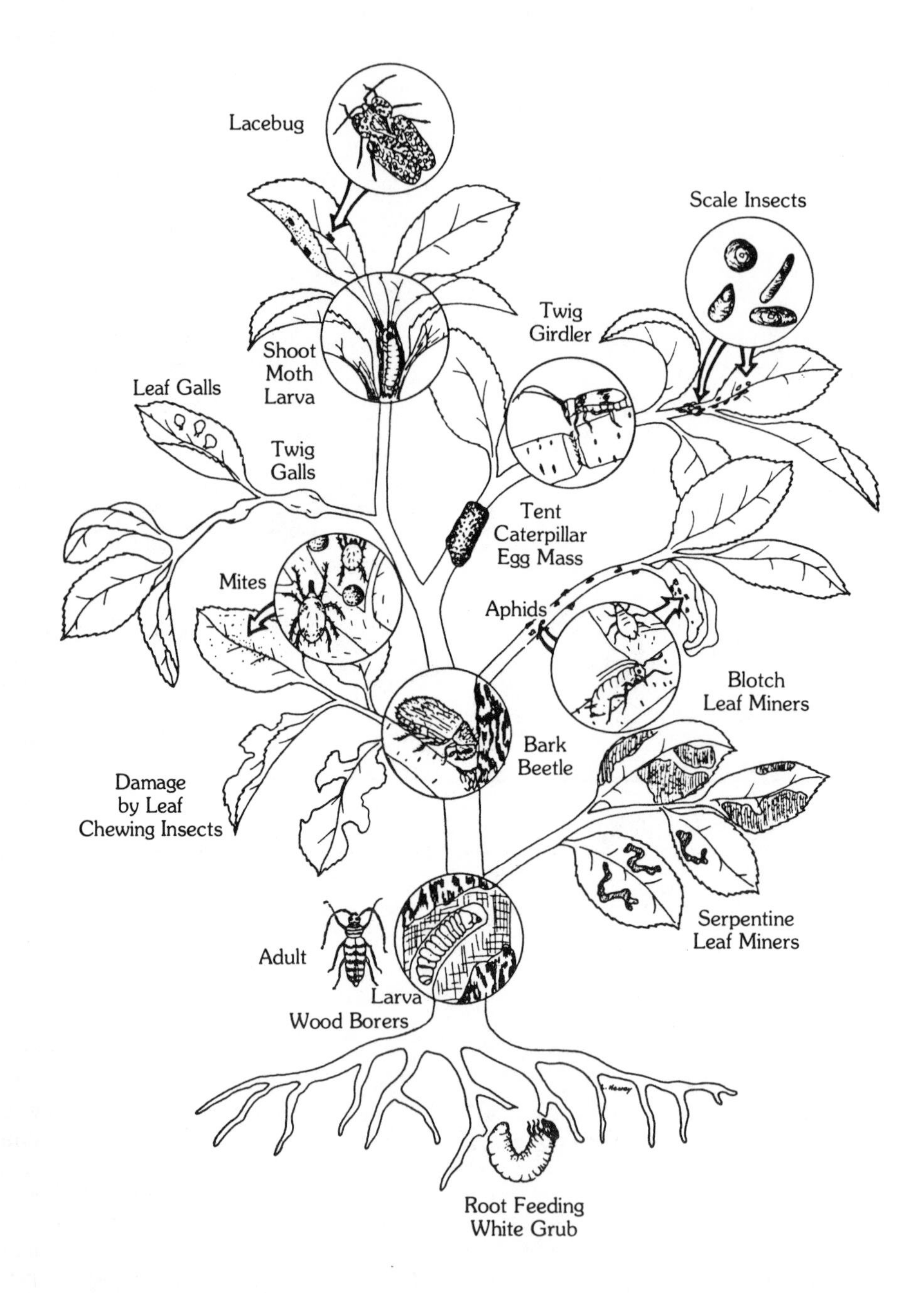

Fig. 5-6 Insects that feed on trees and shrubs. This figure provides search images for damage as well as understanding of relationships between insect and plant. (from Johnson and Lyon, courtesy Cornell University Press)

insignificant, but the adults are a sign that larvae may still be present within the soil causing root damage.

The eggs of boring insects are laid on susceptible plants, often in crevices, wounds, or scars. The larvae of these beetles or moths hatch and bore into the plant, settling in the bark, wood, leaves, or petioles, depending on species. During their development they expel quantities of sawdust or **frass** (a combination of plant material and excrement) to the outside as they cause damage by tunneling within the plant. The tunnels weaken the health and strength of the plant and make it susceptible to wind damage and infestation by bacteria and fungi.

Mites are arthropods with sucking mouth parts that, like insects, are major plant pests and disease carriers. They are anatomically quite similar to insects except that the body parts are fused into one piece with no separation between head and abdomen. They are related to spiders and, like them, have eight legs and no antennae. This similarity probably accounts for the fact that some of the plant mites are called such names as spider mites and red spiders. Mites and insects are compared in Fig. 5-4.

Organic Controls

To minimize insect problems or, in fact, pest problems of any kind, it is important to have vigorous plants. Fertile soil, adequate light, sufficient water, and good aeration will produce strong plants that are reasonably resistant to serious damage from pests and disease and that recover quickly when damage does occur.

Organic gardeners, in particular, use repellent plants in both flower and vegetable gardens. Interplanting or companion planting is simply a matter of using plants believed to repel insects in close proximity to plants needing protection from insect pests. In some cases the plants selected, rather than being repellent, loosen or add nutrients to the soil or offer shade or wind protection for the threatened plants so that their general growth and sturdiness are improved and the plants become more resistant to insect attacks. It is usully difficult to determine specifically what effect one plant has on another or on an insect, and the mechanisms for such effects are unknown, but organic gardeners do claim that certain plants are beneficial to certain others and seem to ward off certain pests and diseases.

For instance, garlic and chives planted with roses are said to prevent blackspot, mildew, and aphids, although exactly how the presence of these plants affect such fungi or their spores or the aphids that often spread them has not been shown. Geraniums are said to deter Japanese beetles and parsley to ward off rose beetles. Marigolds are often planted almost anywhere in the hope that they will drive away soil pests such as nematodes and insect larvae. *Calendula, Salvia,* and *Dahlia* are also believed to rid the soil of these pests.

The annual herbs coriander (*Coriandrum sativum*) and anise (*Pimpinella saxifraga*) and the perennials tansy (*Tanacetum vulgare*) and rue (*Ruta graveolens*) are often planted in the belief that they control aphids, and pyrethrum (*Chrysanthemum coccineum*), the flowers of which are used for some commercial insecticides, and feverfew (*C. parthenium*) are said to be effective as broad-spectrum repellents to many insects. Basil (*Ocimum basilicum*) may be planted as a border or near patios to discourage flies. Some experiments have shown that secretions from the roots of marigolds repel nematodes. However, the others remain in the "reportedly effective" category until verification is substantiated.

Organic gardeners also use "botanical" sprays, made from plant materials, to kill or repel insect pests when they do become a problem. Directions for making and using these and some other effective and recommended substances will be found in the following section.

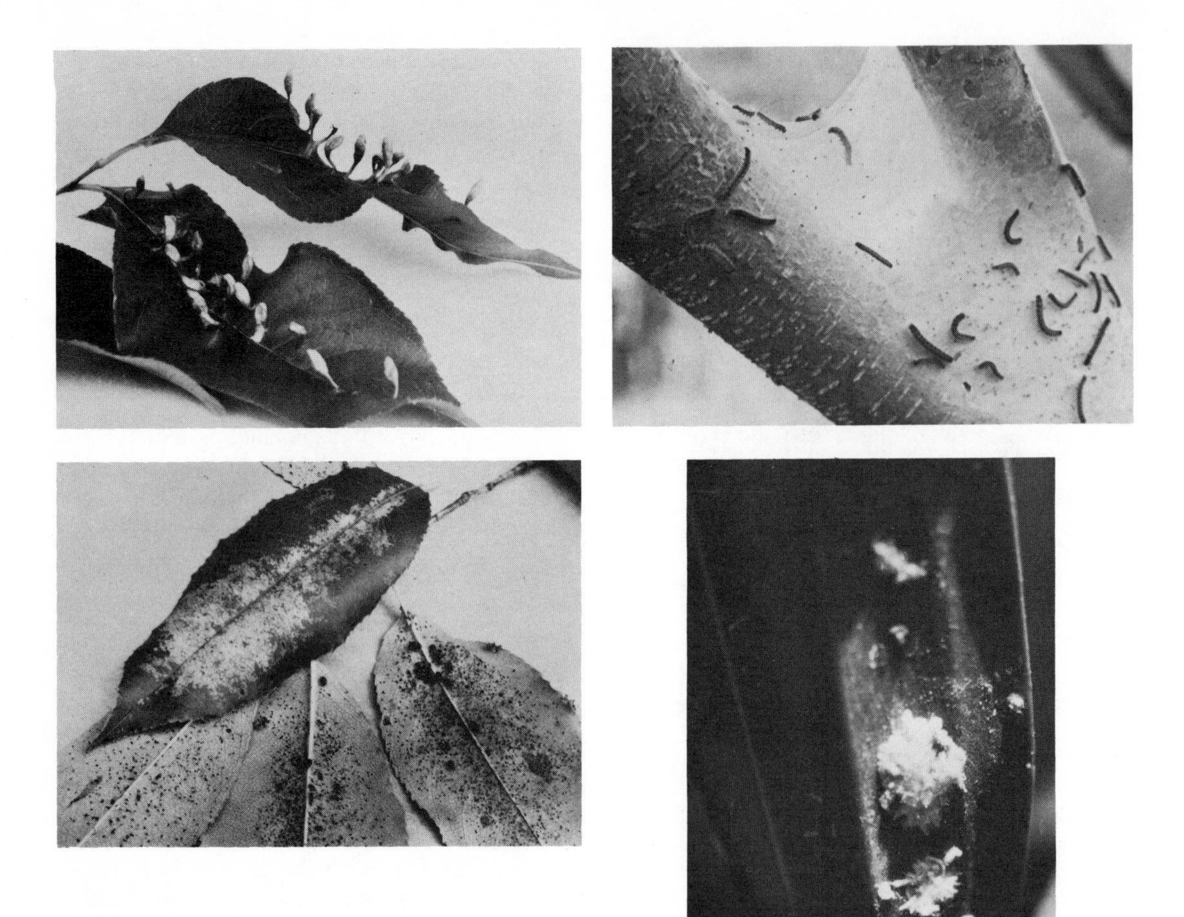

Fig. 5-7. Some examples of plant damage or disfigurement and the pests that cause them. (above, left) Spindle galls on *Prunus*. Gall-causing insects produce a growth regulating substance that stimulates a plant to produce a distinctive form of gall for the insect species. Galls are unsightly but rarely damaging to the plant. (above, right) Tent caterpillars (*Malacosoma americana*). These larvae of a small, undistinguished moth build their webs between branches of trees and forage for newly opened leaves during daylight hours. To remove the web and its occupants hammer a nail partially into a narrow stick at right angles to it and then twist the stick within the web. (lower, left) Lacebug (*Corythucha pruni*) or *Prunus* leaves. These true bugs feed on the underside of the leaf and the damage they cause extends all the way to the upper surface. (lower, right) Mealybugs (*Pseudococcus* sp.) on a greenhouse orchid. The small oval bodies of these insects are covered with white powdery wax in long filaments all around the body. When a number of them are in a group they take on the appearance of tufts of cotton adhering to the leaf surface and in its axils. Mealybugs feed by sucking plant juices, damaging leaf tissues.

BOTANICAL SPRAYS (Recommendations for organic gardeners)

Botanical sprays are a biological method for control of damaging insects. Since these materials are toxic to beneficial insects as well, they should be used only as a last resort and as an alternative to synthetic chemicals. Botanical sprays are organic substances derived from plants, and fall into two categories: 1) those that are toxic to insects and 2) those that discourage their presence. A number of these can be made at home from extracts of plant juices. To prepare home-made sprays:

1. Select plants with strong or disagreeable odors, such as onions, garlic, mint and geranium leaves, chives, turnips, and cauliflower seeds.
2. Add equal amount of water and combine in blender.
3. Strain liquid from mixture.
4. Dilute a small spoonful of the extract with a pint (half liter) of water.
5. Spray onto infested plants.

Suggested spray recipes for specific problems:

1. To control aphids on roses, grind up green shallot onions, add water, and strain. Use for three consecutive days.
2. For caterpillars, especially tomato and cabbage worms, combine one bulb of garlic or onion with 3 large spoonsful of cayenne or red pepper and 1/2 gallon (2 liters) of water. This will also discourage ants if sprayed on their nests or hills.
3. For heavy insect infestations on roses or other perennials, grind together 3 hot peppers, 3 large onions, 1 whole garlic bulb. Cover with water and let stand overnight. Strain and add enough water to make 1 gallon (4 liters) of liquid. Use three times daily for two days. It may be necessary to repeat after a heavy rain. The strained mash can be buried in the soil around the plants for control against crawling insects.
4. To discourage chewing insects, grind up several hot peppers, add water and a spoonful of dishwashing liquid that will act as a wetting agent so that the mixture will cling to plant leaves.
5. For spider mites on roses, combine 1/2 cup (1/8 liter) buttermilk with 4 cups (1 liter) of wheat flour to 5 gallons (19 liters) water. Mix the buttermilk into flour and then add water. The spray will envelop the mites, glue them in place on the plant and kill them when it dries. It is said that the mixture causes the mites to expand rapidly until they burst.
6. For general insect control mix brown sugar, to attract the insects, with tartar emetic. The resulting internal distress is usually fatal.
7. One substance that can be freely used because of its lack of any toxicity is molasses. Diluted in 50 parts of water, it can be sprayed on plants to simply glue the insects in place until they expire.

Synthetic Insecticides

Fly swatters and whisks were early forms of insect control, but it wasn't until 1760 that a nicotine insecticide was derived from tobacco and used to combat the aphids that were so destructive to the plants in European formal gardens of the time. By 1800 pyrethrums extracted from certain daisy-like flowers (*Chrysanthemum coccineum* and *C. cineraefolium*) were being used to kill fleas, an important step in controlling the spread of bubonic plague. By the mid-1800's rotenone compounds, also derived from plants (notably *Lonchocarpus nicori* var. *utilis* and roots of *Derris* sp.), were being used as poisons against leaf-eating caterpillars.

Inorganic compounds were also recognized as potential insect-control agents. Flowers of sulfur (powdered) and sulfur compounds were commonly used, and, starting in the late 1800's, compounds containing arsenic, fluorine, boron, copper, and zinc offered controls over certain insects.

Organochlorine Compounds

The synthesis of organic insecticides began in Germany in 1892 with a dinitrophenol compound which was quickly followed by others of its type. The extraordinarily effective DDT however did not join the arsenal until much later. DDT was actually synthesized in 1874 by a German Ph.D. student, Othmar Zeidlar, as one of several compounds which he produced, for now-obscure reasons, as part of his dissertation. It wasn't until 1939 that its potential as an insecticide was determined by the Swiss entomologist Paul Müller. In 1942 the U.S. military quickly discovered its usefulness in combating insects that were carriers of such dread diseases as typhus, malaria, yellow fever, and plague. Its applications during and following World War II clearly saved millions of human lives and considerable grief and discomfort.

The success of DDT led the chemical industry to explore the possibilities of using other biologically active compounds as pesticides. DDT is an organochlorine compound, and further research on related compounds produced some of our best-known synthetics. Some of these are listed in the insecticide summary on page 149.

Even today there is no clear understanding of the exact mechanisms by which organochlorines kill insects. They are known to act as poisons that affect the central nervous system, judging from the symptoms of tremors and convulsions that are produced. Organochlorines are rapidly absorbed by insects, insoluble in water and therefore not washed away by rain, and not easily destroyed by light or oxygen. In the early days, the unusual persistence and wide range of effectiveness were viewed as their major virtues.

Organophosphorus Compounds

Research during World War II for nerve gases gave impetus to peacetime research on another group of chemicals known as organophosphorus compounds. These chemicals kill insects by inactivating cholinesterase and acetylcholinesterase, enzymes which are important in the transmission of nerve impulses. Inactivation fatally disrupts an insect's nervous system, and since these transmitters are not unique to insects, organophosphorous compounds can kill other animals including humans.

The first commercial insecticide of this type was TEPP, produced in 1944. The nation's most widely used organophosphorus insecticide today is methyl parathion, first marketed in 1954. Others that are generally used today are listed in the insecticide summary (p. 149).

Compared to organochlorines, the organophosphorus insecticides break down more rapidly to innocuous substances, do not accumulate harmfully in organisms, and are active against a narrower range of insects. These properties help to reduce the destruction of desirable predators and insects. Organophosphorus compounds are fairly expensive, however, and since they are degraded rapidly must be applied more often. Because of their toxicity they must be handled with great care.

A number of organophosphorus compounds act as **systemic insecticides** (i.e., those that are transported internally in plant vascular systems). They are able to penetrate the leaves or enter the plant via the root system. The plant therefore does not have to be resprayed as often and the effect is well distributed to all parts of the plant subject to insect attack.

Many of the organochlorines have been, or soon will be, banned for public sale because of their detrimental qualities, while the organophosphorus compounds are all still generally available in garden stores.

Carbamate Compounds

The carbamate chemicals constitute another major group used as insecticides. Isolan, introduced in 1953, was the first of this group and was used against aphids and other sucking insects. Those listed in the summary are widely used today. In general, the carbamates are used for insects resistant to the organophosphorus compounds. They also kill by inhibiting cholinesterase and acetylcholinesterase, and many of them can also act systemically in plants. They are expensive and difficult to produce and are dangerously toxic, in certain formulations, to pollinating insects.

Pyrethroids

A rapidly growing field of research is on pyrethroids, synthetic compounds similar to the natural pyrethrum derived from flowers of certain composites. Allethrin was the first of these and was marketed in 1949. Resmethrin followed in 1967 and NRDC-143 in 1973.

Like natural pyrethrum the pyrethroids are safe for use around the home because of their low toxicity to humans, other mammals, and birds. They are commonly used to control household insects such as flies and roaches, mosquitoes, and insects on ornamental plants. They are generally much safer than the organophosphorus compounds and the carbamates.

Applying Controls

To combat and control insect pests when they do attack our plants, we must first identify the culprits and determine that the supposed pest is actually causing the damage attributed to it. Once the identity has been established, the extent of the damage and the size of the pest population should be assessed.

Remedial measures must be appropriate to the size of the problem. A single houseplant showing early signs of insect damage may only need to be isolated and sprayed with water or have a few leaves picked off to remove the problem. Plants growing outdoors are subject to a wider variety of pests. Routine observations of garden plants often reveal fairly good-sized caterpillars or beetles that are easily picked off before they can inflict serious damage. One does develop a less squeamish attitude toward such removal when a favorite plant is at stake.

If the problem is more extensive, botanical sprays or chemical controls may be needed. **Contact** sprays or powders are those that affect the insect internally, primarily by entering the respiratory or tracheal system which consists of hollow, branched tubes for dispersing respiratory air to body tissues. The extensive nature of such a system is shown in Fig. 5-8. Contact insecticides, in some cases, simply clog the spiracles (tracheal openings) and the insect suffocates. Others are carried further within the insect's body to damage tissues or the nervous system. Contact insecticides kill very rapidly but must be applied directly to the insect.

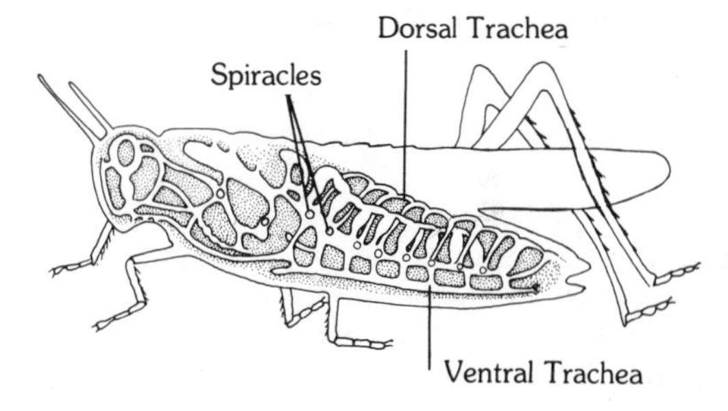

Fig. 5-8. Respiratory system of an insect showing the system of tracheae opening to the outside through spiracles, pore-like openings.

Systemic chemicals are water-soluble and enter the plant's vascular system through the roots or leaves. Some of them act as poisons against chewing insects, while others are fatal to sucking insects that feed by extending a long proboscis into the poisonous vascular fluids. Application of systemic insecticides is by liquid sprays and foams or as granules applied to the soil. Granular formulations consist of the insecticide on an inert carrier. Soil moisture releases the chemical from the carrier and it is then absorbed by roots. Granular applications, unlike sprays, do not drift in the air to nearby plants when applied.

In the past 15 years integrated pest control has been attracting a lot of attention. This requires scientific study of the insect problem and the careful application of a variety of control techniques. Some

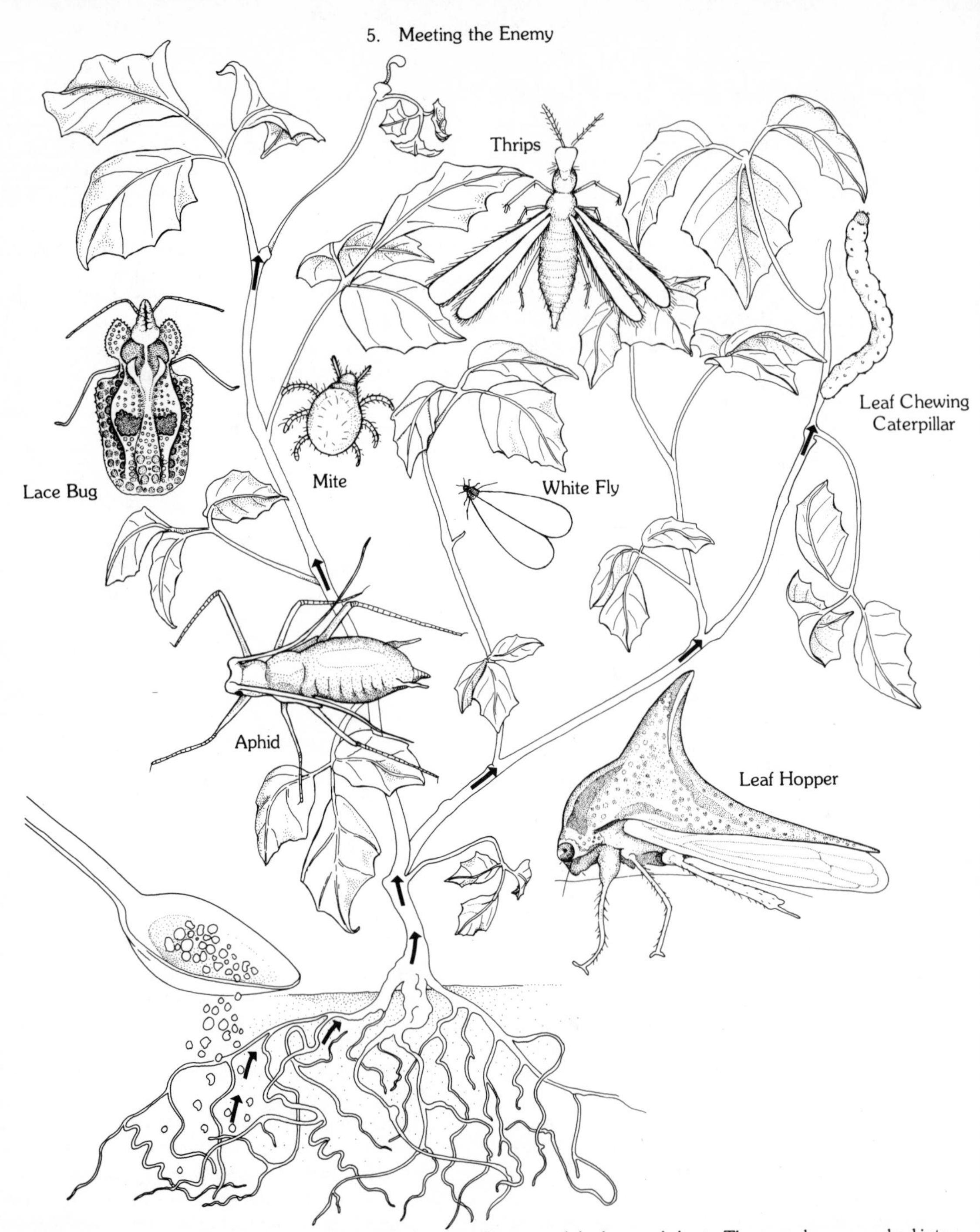

Fig. 5-9. Systemic insecticides are often applied as granules around the bases of plants. The granules are worked into the soil, and water carries the insecticide into all the tissues of the plant, which are then poisonous to insects that feed upon them.

Fig. 5-10. These pictures show the effect of spraying a synthetic growth regulator on a yellow mealworm. On the left is a normal beetle, on the right a normal pupa. The grotesque form in the center is the affected insect with an adult head and thorax and juvenile abdomen. (USDA photo)

observers believe these diversified approaches could reduce the use of conventional insecticides by as much as 40%. If this could be achieved, the benefits would be far-reaching ecologically, financially, and sociologically.

Most of the less conventional methods that replace or augment insecticides are of fairly recent development and include the following:

1. The Sterile-male Technique—Male insects are laboratory-reared and then made sterile by exposure to gamma radiation or sterilizing chemicals. When the sterile males are released in an insect-infested area, the females they mate with produce non-fertile eggs.

2. Insect Growth Regulators—These are extracts of insect juvenile hormone. This hormone, secreted by glands near the brain, prevents the metamorphosis of larva to pupa before the proper time. By artificially applying the hormone to larvae, grotesque forms intermediate between larva and pupa are formed that do not survive (see Fig. 5-10). These substances are effective in small dosages, are not toxic to other organisms, and do not persist in the environment. They are, however, expensive and require precise timing of application.

3. Chitin Inhibitors—The brittle exoskeletons of insects and other arthropods are composed of a polysaccharide called chitin. Larvae that are treated with chitin

inhibitors have very thin, fragile, malformed outer coverings that provide almost no protection for the insect and, in fact, frequently rupture. The insect has little chance for survival. Research is recent but progressing on these compounds. They are nontoxic to other life-forms and effective at low dosages; a mere half ounce applied to an acre of forest provides good control of the gypsy moth. Unfortunately, as with so many other pesticides, beneficial insects and other arthropods will be affected as well.

4. Pheromones—Insects give off and respond to species-specific chemical compounds called **pheromones**. They provide an odor trail that insects follow between a food source and their nests, they are emitted as alarm substances to warn others of danger, and they are used to identify other members of the species or as sex attractants.

Exceedingly small amounts of a sex-attractant pheromone applied to a trap will draw male insects over distance of several miles. Such methods are still in the research stage and not commonly used yet commercially, but pheromone-baited traps are used in estimating the size and distribution of populations of selected insects.

The so-called confusion method requires application of a pheromone over a wide area to permeate the air. Males of the species become so disoriented that they cannot locate females, and matings rarely take place. Gossyplure, a synthetic pheromone, duplicates the scent of the female pink bollworm moth that is destructive on cotton. In 1978 it was approved as the first pheromone-like pesticide for commercial use.

5. Bacterial or Viral Insecticides—Pathogenic or disease-causing bacteria and viruses can be used against certain insects. The first bacterial insecticide, containing *Bacillus popilliae*, was first offered in the early 1940's. Now marketed as Doom or Japidemic, it causes fatal milky-spore disease in Japanese beetles. Another product, marketed as Thuricide, Biotrol XK, and Dipel, contains *Bacillus thuringiensis* and is effective against a wide range of moths and butterflies, beneficial as well as destructive, and is applied only when pest populations are a problem. Such controls are slow-acting but specific in their action and do not harm plants or other animals including human.

Several viral insecticides are currently being field-tested but none is commercially available at this time.

Fungal and protozoan forms of control may join the insecticide arsenal sometime in the future, but now they too are experimental subjects only.

A summary of current and early insecticide and miticide compounds follows.

BIG PROBLEMS FROM SMALL NEMATODES

Nematodes are microscopic roundworms, sometimes called eelworms, that live in soil, water, and plant tissue. Most soil-inhabiting nematodes are free-living, non-harmful, and useful members of the ecosystem because of their role in decomposition. Fig. 5-11 shows a few as they appear under the microscope.

There are, however, 1239 identified of nematodes that are parasitic on plants, attacking all parts of them. In fact, nematodes may be responsible for much of the damage attributed to insects and fungi.

Nematodes cause plants to decline slowly. Above-ground parts may become stunted and chlorotic, with considerable die-back and loss of vigor. Turfgrasses are especially susceptible to nematode damage when nematodes feed on and injure root systems. Vegetables may be hosts to species that form root knots and galls as shown in Fig. 5-12.

Since nematode damage may be so easily mistaken for any of a number of other plant problems, a laboratory analysis of the

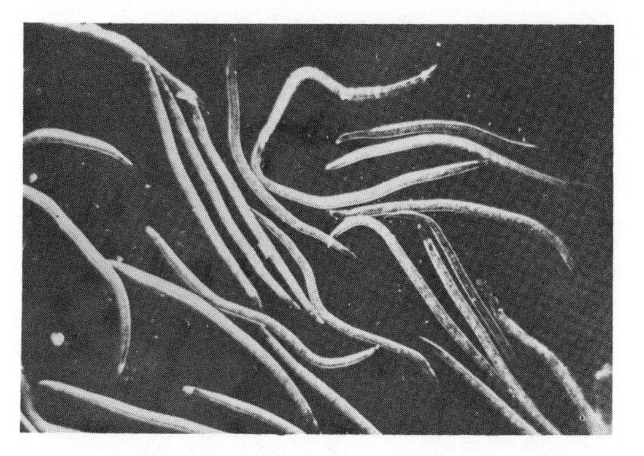

Fig. 5-11 Nematodes or eel worms are so tiny that a microscope must be used to see them clearly. In spite of their small size they cause considerable damage to both economic and ornamental plants. (courtesy Cornell Extension.)

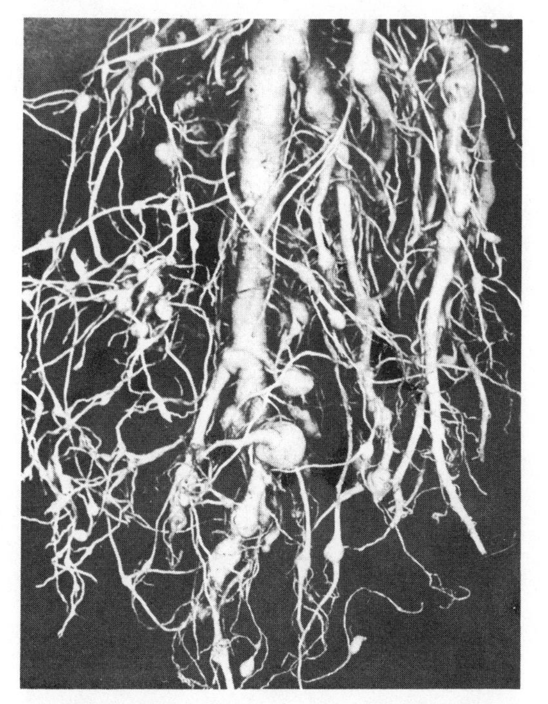

Fig. 5-12. Root-knot nematode damage. Although there are many kinds of nematodes, the root-knot species is the most destructive to vegetables. They also attack strawberries and most flowers. Roots have swellings, galls, or knots in which the nematodes live while they rob the plants of food and obstruct the flow of water upward in the plants. (courtesy Cornell Extension.)

Table 5-3. Summary of Insecticide and Miticide Compounds*

Type of Compound	First Used	Type of Action	Representative Types** (common trade names in parentheses)	Toxicity (humans and other animals)
Nicotine	1760	Contact insecticide	Nicotine sulfate (BlackLeaf 40)	Low toxicity, non-lasting
Pyrethrum	1800	Stomach and contact insecticide and repellent.	Pyrethrins (Ortho Indoor Insect Plant Spray), (d-Con House and Garden Spray)	Low toxicity, non-lasting
Rotenone	1848	Stomach and contact insecticide	(Plant Marvel I-Bomb) Often used in combination with pyrethrins.	Low toxicity, non-lasting

Type of Compound	First Used	Type of Action	Representative Types	Toxicity
Arsenic and certain other inorganics	1865	Various.	Discontinued.	Dangerously toxic
Dinitrophenols	1892	Contact insecticides	Discontinued.	Not determined
Organochlorines	1939	Contact and stomach poisons. Disrupt nervous system.	(DDT, endrin, heptachlor, aldrin, toxaphene, lindane, chlordane, methoxy chlor (Marlate).***	Highly toxic, many now banned. Very persistent, accumulate in tissues.
Organophosphorus compounds	1944	Contact and stomach poisons. Inactivate nerve-impulse enzymes.	TEPP, methyl parathion, parathion, malathion (Isotox), diazinon (Spectracide), disulfoton, chlorophyrifos (Dursban), phorate, fonofos, azinphosmethyl, dimethoate (Cygon), monocrotophos, methidiathion.	Highly toxic, nonaccumulative and and non-persistent.
Carbamates	1953	Contact and stomach poisons. Inhibit nervous-system enzymes.	Isolan, carbaryl (Sevin), carbofuran, methomyl (Lannate), aldicarb, (Temik), propxur (Baygon), bufencarb.	Toxic, non-persistent.
Pyrethroids	1949	Contact and stomach poisons.	Allethrin, resmethrin, NRDC-143 (Most indoor sprays are pyrethroids).	Low-toxicity, non-persistent
Bacterials	1940's	Cause bacterial infection in pests	*Bacillus popilliae* (Doom), (Japidemic), *B. thuringiensis* (Biotrol XK), (Dipel), (Thuricide).	Non-toxic, harmful only to pest

* For extensive lists of pesticides and supplementary information see *The Gardener's Bug Book* by Cynthia Westcott and *The Pesticide Book* by George W. Ware.

** Trade names listed are examples only and do not imply endorsement.

*** Methoxychlor is the only organochlorine readily available for home garden use. All others are either banned or use- restricted.

soil and root tissue is necessary for diagnosis. Affected lawns appear stunted and yellow-spotted, with frequent bare spots in irregular patterns. The warmer climates are more likely to have nematode problems, but outbreaks have recently occurred in the northeast states.

Methods of Control

To prevent nematode damage to greenhouse plants, fumigate the soil with metham (Vapam), methyl bromine (Pestmaster), or nemagon which will eliminate the pests. Lawns treated with a new experimental nematocide called ethoprop showed improved growth and vigor in just four months.

To control nematodes in outdoor gardens, add humus or leaf mold which will provide a richer soil than that favored by nematodes. Such additions are also likely to add certain species of fungi, primarily belonging to the group known as the Fungi Imperfecti, that are predaceous on soil organisms. About 50 species of several genera prey predominantly on nematodes, which they capture by producing sticky knobs (macabrely called "lethal lollipops") or adhesive or constricting loops and rings that hold onto or close around a nematode so that it cannot escape. After capture, the body of the nematode is invaded by strands of the fungus that feeds upon it until the body is completely decomposed.

Soil should also be turned over to expose it to the cleansing qualities of sun and wind. Nematodes cannot survive for long out of a moist location. Rotate or relocate plants that may be susceptible to nematodes, and destroy any plants that are infested. Interplant a few marigolds, which seem to be repellent to nematodes. In severe infestations, dig up and burn infested plants and then leave the ground fallow for a year or more to eliminate most of the remaining pests. Above all, plant healthy stock in well-structured, fertile soils and there should be no nematode problems.

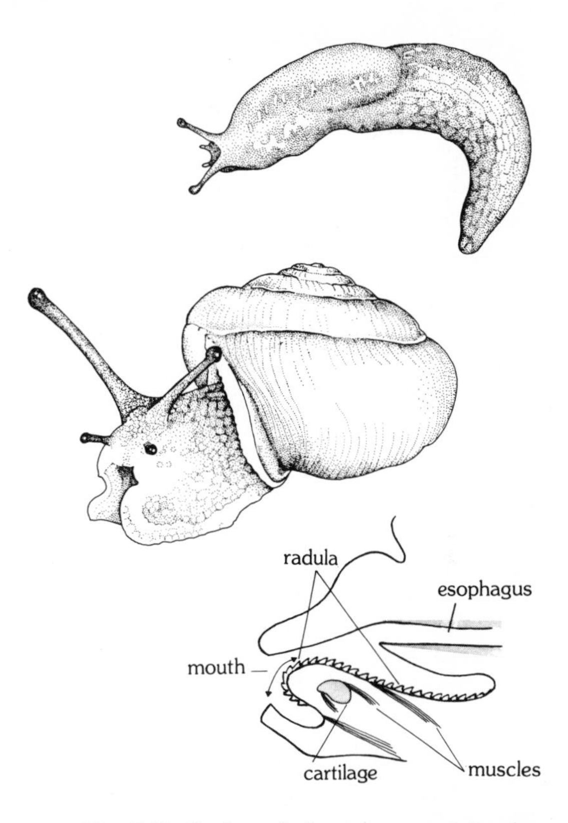

Fig. 5-13. Snails and slugs chew vegetation by means of a ribbon-like radula of horny teeth that moves rapidly back and forth like a file to grind and rasp the food into small enough particles to be digested.

GARDEN SNAILS AND SLUGS

Snails and slugs are mollusks that eat roots, leaves, blossoms, and fruits. Snails have external shells; slugs have no shell (see Fig. 5-13). They are most active at night and prefer cool, moist habitats. Mats of vegetation or mulch, stepping stones or boards, or closely planted vegetation all offer favorable environments for snails and slugs in which to live and lay eggs.

As they move about they leave bits of excrement and a trail of slime that, while unsightly, may be used to locate them beneath protective coverings.

Methods of Control

Among the many natural predators on snails and slugs are snakes, turtles, birds salamanders, and shrews. In some locations, however, snail and slug populations occasionally get out of hand. To reduce their populations, go into your garden at night, when they are most likely to be crawling about, and, as you locate them by flashlight, pick them off by hand.

Tender seedlings, which are a favorite food, sometimes can be protected by tin cans, open at both ends, shoved several inches into the soil around the plants. A foot-square board placed flat on the ground near an infested area may serve to attract the pests as an egg-laying site. Check under the board several times a week and destroy those you find, or, for the gourmet, try them in the following recipe. Garden snails can be an asset as well as a liability, as this method of biological control will prove.

1. Feed captured snails on scrap lettuce leaves or corn meal for a week.
2. Soak living snails in 3 or 4 changes of weak brine, 1/2 hour each.
3. Stew gently over low heat for 3 hours in a little red wine with herbs, carrots, etc.
4. Serve on bread or pasta, or pop back into shells with a bit of garlic butter.

They are excellent. If you have doubts, talk someone else into preparing them the first time you eat them.

Garden snails (*Helix*, like the one in Fig. 5-13) are introduced from Europe, so you will not cause any ecological problems by eating them. However, you should avoid eating native species of snails, any snails not kept captive and fed clean food for a week, or any which die during that time. Not so good with slugs.

Shallow dishes of beer placed here and there in the garden at soil level will attract slugs that fall into the dish and drown.

If infestations become severe, as is more likely to occur in warm climates, chemical controls may be required. Metaldehyde powder sprinkled as a barrier between plant rows will kill the mollusks that try to cross it and prevent the infestation from spreading. The newest product, containing the chemical mesurol, is called Slug-Geta and is said to be far more effective than previous remedies.

In southern California, the latest control is a carnivorous snail, the decollate snail, that has been introduced against the destructive brown garden snail. After several years there seem to be progressively fewer garden snails as the decollate snails become more widely established. Fortunately the decollate snail has no taste for vegetation.

PLANT DISEASES—CURABLE AND INCURABLE

Plant diseases of any kind all require a susceptible host (and weak or damaged plants are most susceptible), a **pathogenic** (disease-causing) organism, a method of distributing the organism either naturally or by human activities, and the proper environment for the development of the disease.

Fungus Diseases

Plant pathogens are fungi, bacteria, or viruses. Of these, fungi are the most common and variable in effect. Diseases caused by fungi encompass those variously called by the names rust, canker, smut, bulb rot, brown rot, root rot, blight, blister, fairy ring, leafspot, black spot, snow mold, clubroot, powdery mildew, downy mildew, damping off, etc. Some common fungus problems are shown in Fig. 5-14.

Fungus diseases are primarily spread by the reproductive spores which are carried by wind, water, insects, or humans. Despite their microscopic size, most of them are surprisingly rugged and capable of withstanding adverse conditions until they find the proper host upon which to germinate. The fungal hyphae (see Chap. 1, p. 40) then penetrate plant tissues where

Fig. 5-14. Fungus diseases are especially unattractive on garden perennials, as these three examples show. (left) Powdery mildew, *Erysiphe cichoracearum*, on *Phlox paniculata* is particularly troublesome during a spell of rainy weather; (center) Black spot, *Diplocarpon rosae*, on roses can be controlled by regular treatments with rose dust; (right) A typical case of "Delphinium blacks". The plant is first attacked by the Cyclamen mite, *Steneotarsonemus pallidus*, which weakens the plant and makes it susceptible to a fungus (possibly a bacterium) that further damages the plant and produces a lampblack-like powder within the buds and distorted new leaves.

they parasitize cells, producing disfigurement and damage to the host plant. No part of a plant is immune to fungi, but a healthy, robust plant will be more resistant to attack then one weakened by stress.

Fungus problems may be minimized by practicing the following methods of control:

1. Plant disease-free plants and seeds only. Don't introduce the disease into your garden, expecting to cure it later.
2. Provide fertile, well-drained, well-aerated garden beds for plants, and fertilize and water regularly to maintain health.
3. Control weeds and insects that may attract or transport fungi or spores to plants or weaken them.
4. Remove and destroy plant debris promptly. Rotting vegetation among plants provides an ideal habitat for many pathogenic fungi.
5. Prune out infected plant tissues wherever possible.
6. Disinfect and scrub ALL garden tools regularly, especially after use on or near infected plants.

Chemical Controls

Soil fumigants and sterilizers are effective as fungicides to kill fungus spores and mycelia or as fungistats to prevent their development. These are especially useful in preventing attack on seeds and emerging seedlings as well as transmission from the soil to plants after they are further de-veloped.

Fungus powders and sprays can be applied to foliage and stems to prevent fungus attacks. It is important for these formulations to provide a film of fungicide between the plant and the attacking spores. Such preparations will not kill fungi already established, and parts already infected are best removed, if possible, and destroyed to prevent spreading.

Some of the newer systemic preparations that are taken into roots and distributed throughout the plant will prevent fungus penetration, and in some cases, kill fungi already parasitizing the plant.

Bacterial Diseases

Many of the diseases caused by bacteria could easily be attributed to fungi since the resulting symptoms on plants are so similar. Thus we have stem and root rots, leaf wilting and spotting, stem cankers, lesions on fruits, etc. Some of the bacteria will produce different symptoms in different parts of a plant or at different stages of the plant's development.

The diseases caused by bacteria are readily transmitted between plants by insects, air currents, or splashing water. Several plants in a group may develop leaf spots such as those illustrated in Fig. 5-15, which disfigure and weaken the leaves. There is rarely anything that can be done to cure a plant once it is infected by bacteria. Antibiotics have been used with limited and questionable success and usually at unwarranted expense. Control of bacterial disease is primarily a matter of prevention aimed at excluding the disease in whatever ways are appropriate for the growing area and the plants.

Sterilized soil and pots and careful watering are a first step. Insecticides against bacteria-carrying insects and thorough disposal of all plant refuse will help. Avoid buying plants showing symptoms, and eliminate those already growing in the garden. Finally, be sure garden tools are regularly scrubbed clean and disinfected. A strong bleach solution or 70% alcohol are good and easily obtained disinfectants.

Fig. 5-15. Plant problems caused by bacteria and viruses are not often seen, and the home gardener is not likely to be familiar with them. The yellowish blotches between the veins of this *Schefflera* leaf are bacterial spots. Excessive humidity encourages such diseases. Infected leaves should be removed.

Viruses

Viral diseases cannot be cured, and the methods of transmission are only slightly understood. The symptoms, while often mistaken as being caused by other diseases or cultural conditions, are in many cases quite distinctive once one becomes acquainted with them.

An overall leaf curl, as distinguished from limited areas of leaf curl caused by insects, will occur quite suddenly on an entire plant. Mottling of leaves and flowers is also usually virus-caused. Stunted bushy growth on evergreens such as junipers and leafy shrubs will, over a period of several years, gradually become apparent on the whole plant. The plant may continue to survive but become less attractive in shape and color as the disease progresses.

The viruses are generally quite specific in the plants they attack, and dissemination cannot be predicted and only rarely explained. Clean garden tools is one of the best defenses against viruses.

6. PLANTS AND LIGHT

he presence, or absence, of light is the most important environmental factor in plant growth and development. The quality, quantity, and periodicity of light, especially in relation to temperature, promote or prevent the many chemical, physiological, hormonal, or other responses that go on in a living plant. Without light, life on Earth would not exist.

RADIANT ENERGY AND THE ELECTROMAGNETIC SPECTRUM

Three hundred years ago, Sir Isaac Newton passed sunlight through a prism and separated it into the rainbow which we know as the color spectrum. These colors represent only a small portion of the radiant energy emitted from nuclear reactions in the sun that convert hydrogen into helium. This energy is in the form of electromagnetic waves that vary in length and frequency and range from the very short cosmic and gamma rays to the infrared and other long waves. The range of these wave lengths can be translated into the electromagnetic spectrum shown in Fig. 6-1. The various wavelengths are expressed in metric units called nanometers which are one billionth of a meter (10^{-9} m). Plants, with which we are concerned, respond physiologically to the portion of the electromagnetic spectrum between 280 and 800 nanometers.

Radiation travels through space as packets or bundles of energy called **quanta** or **photons**. The amount of energy in a photon varies with the wavelength. The longer the wavelength, the lower the energy. Thus photons of, for example, blue light at 420 nanometers posses more energy than those of red light at 690 nanometers. This range from blue to red includes most of the portion of the spectrum visible to humans.

If the intensities of the various wavelengths in sunlight are measured and graphed, a **spectral-energy distribution curve** will be produced like that shown in Fig. 6-2 in which the relative radiant energy of the sunlight is plotted against the wavelengths of the spectrum. The curve will vary somewhat depending on the conditions under which the light is received, e.g., a bright cloudless day will produce higher peaks than a dark cloudy day. The curves shown in the figure represent an average sunny day and a cloudy day.

Light energy must be absorbed in order to be used. **Pigments** are substances capable of absorbing some wavelengths while they relflect or transmit others. If all wavelengths are absorbed, the pigment color is black. If none is absorbed and all colors are reflected, the color is white. The color of an object then is determined by the wavelengths of light which its pigments do not have the power to absorb. The green color of leaves indicates that the pigment chlorophyll absorbs energy primarily from the red and blue portions of the spectrum, to be utilized in the physiological activities

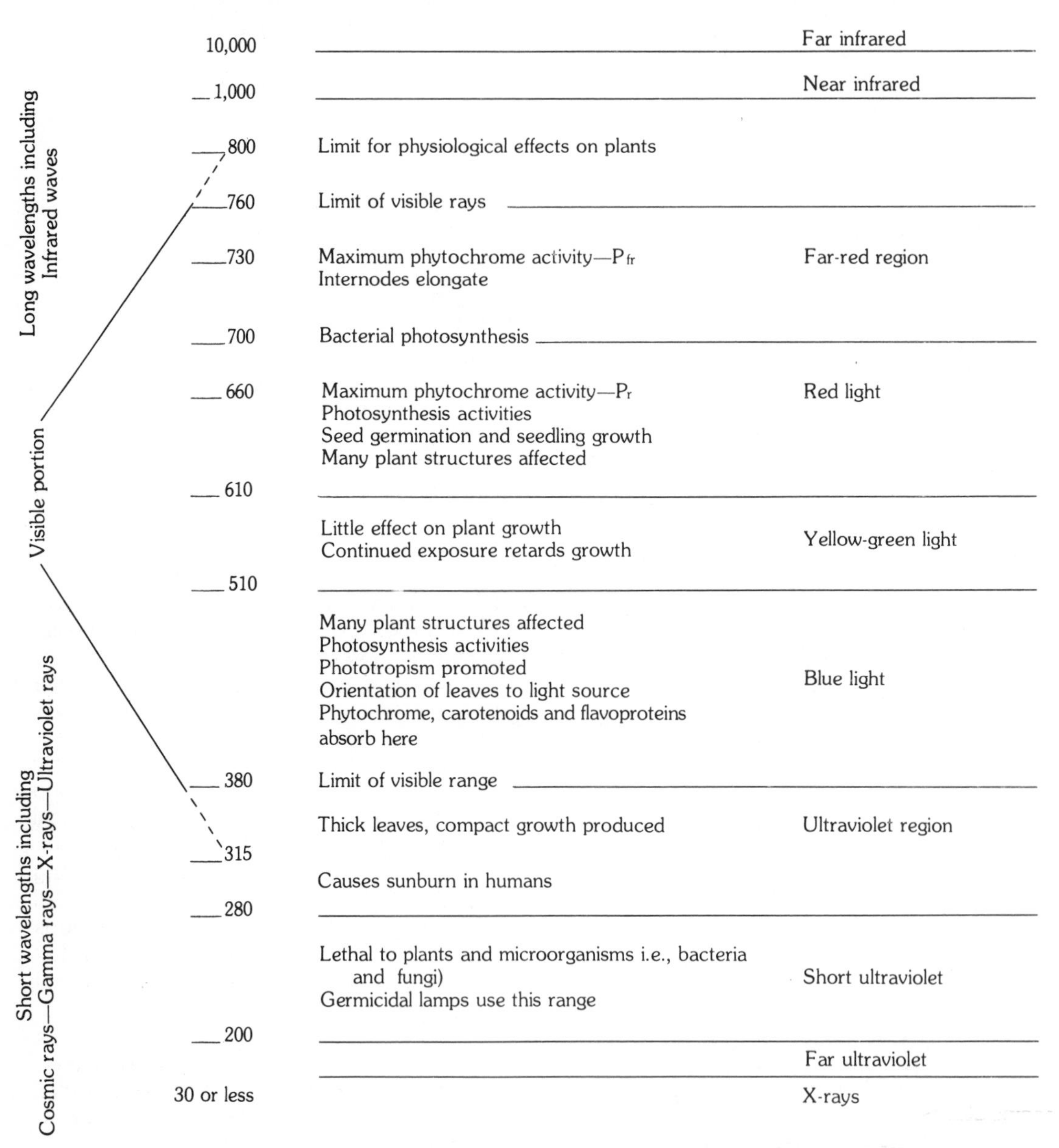

Fig. 6-1. Electromagnetic spectrum.

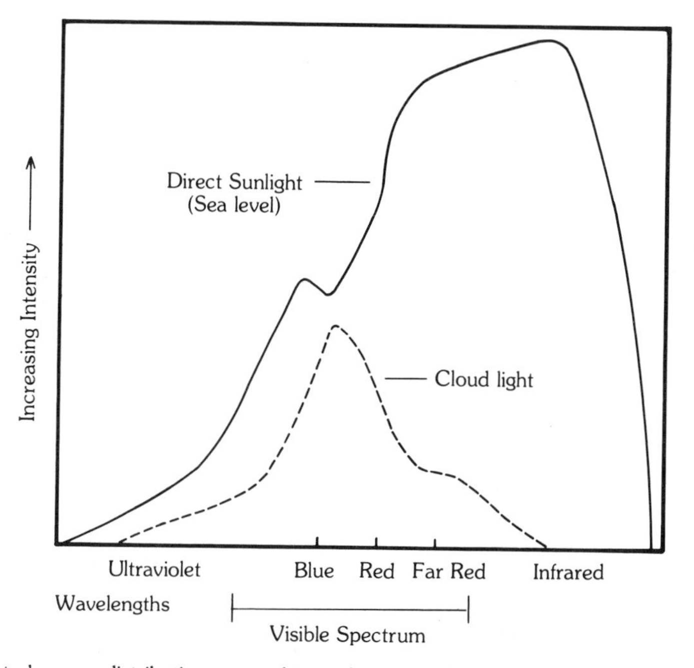

Fig. 6-2. Spectral-energy distribution curves for outdoor light. Direct sunlight (solid line). Cloud light (dashed line). Bright sunlight has a higher intensity of red wavelengths than of blue. On cloudy days the red is less intense and the blue wavelengths may predominate. (Courtesy Mich. S.U. Experiment Station).

of plants such as photosynthesis (see discussion in Chapter 1, p. 19). The green wavelengths are reflected from or transmitted through the leaf (Fig. 6-3), and their energy is unused by plants.

PHOTOPERIODISM—A RESPONSE TO LIGHT

Light plays a role in a number of other plant functions as well as in the essential process of photosynthesis. One of the most important of these is **photoperiodism**, a response to the proportion of light and dark (daylength) in a 24-hour period. Tests with specific plants have shown that temperature and light intensity may play important roles in certain activities. However, especially in regulating flowering, neither is as critical as the lengths of the periods of light and dark. Some plants require specific daylengths in order to bloom, while others will bloom on any daylength but flowering may be promoted or inhibited by long or short days. Changes in daylength are the most reliable indicators of seasons for both plants and animals, and many of their physiological adaptations conform with them. A plant "sensing", for instance, the lengthening days of spring or the shortening days of fall may be stimulated to produce flower buds at the time when other environmental factors are most favorable for its growth and reproduction.

Generally plants can be categorized as short-day, long-day, or day-neutral plants. The terms describe the conditions under which the plants can be brought into bloom. **Short-day plants** grow vegetatively when days are longer and bloom when days are short. These are most likely to be fall, winter, or early-spring bloomers. Chrysanthemums, poinsettias, and kalanchoes are examples of short-day plants (see more

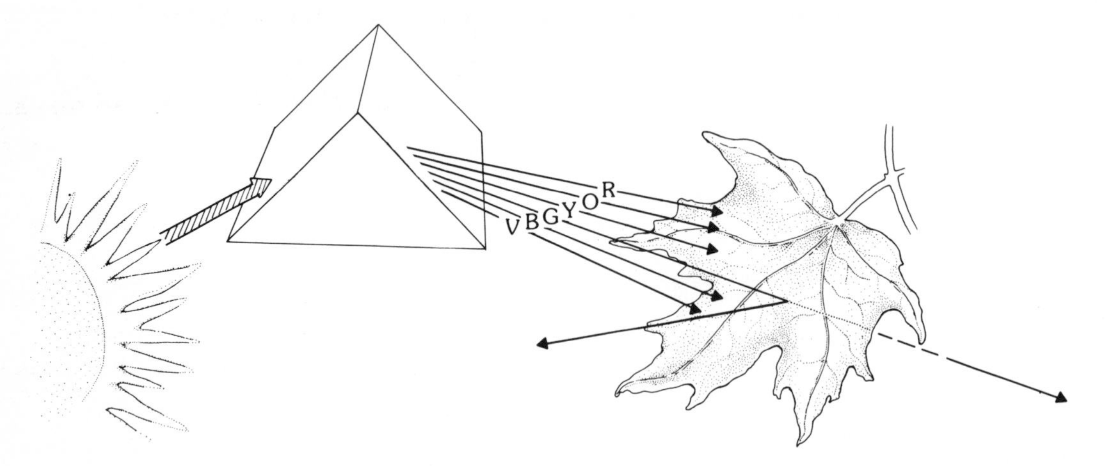

Fig. 6-3. A normal green leaf absorbs light from the blue and red regions of the spectrum but reflects or transmits wavelengths from the green region.

extensive list on p. 159). Delphiniums, gladioli, and petunias are **long-day plants** that grow during the spring and then bloom in mid-summer when days are longest. **Day-neutral plants** are less fussy and bloom when their growth has reached a point sufficient to support flowering, without regard to length of day or night, if other factors are favorable. In general, plants that are native to regions far from the equator will be quite specific in their daylength requirements and are likely to be long-day plants. Tropical and subtropical plants are generally short-day plants. (In addition, plants at high latitudes are likely to require chilling before they produce buds, whereas light intensity is often an important factor in warm-climate plants.)

A number of studies in recent years on the effects of light and dark periods on plants have shown the following:

1. Short-day plants are not affected by interruptions of the light period as long as there is a total period of high-intensity light (sunlight) at least two hours long.
2. Short-day plants fail to flower if the long dark period is interrupted by even a brief flash of low-level light.

Conclusion—Flowering in short-day plants requires long, uninterrupted periods of darkness.

3. Flowering in long-day plants is inhibited by long, uninterrupted dark periods.
4. Long-day plants will bloom under continuous light or·if the dark period is frequently interrupted.

Conclusion—Long-day plant do not require a dark period to promote flowering.

In greenhouses, the gardener can initiate flowering at any time of the year by regulating the light and dark periods, as well as certain other factors, required by specific plants. Thus it becomes possible to have potted chrysanthemums in bloom for Thanksgiving Day or Mother's Day, or Easter lilies (*L. longiflorum*) at Easter time, if temperatures are also strictly regulated, no matter on which Sunday Easter comes that year (see Fig. 6-4).

The spectacular displays of flowers at public conservatories for special holidays and at home-and-garden shows are possible because the plants are all carefully grown according to their requirements and timed to be in full bloom for the occasion. This control often requires increasing natural

Table 6-1. Some Plants Known to be Responsive to Daylength

Long-Day Plants	Short-Day Plants
black henbane	aster
bluegrass	bryophyllum
cabbage	chrysanthemum
candytuft	cocklebur
castor bean	cosmos
clover	cotton
coreopsis	dahlias
delphinium	goldenrod
dill	Japanese morning glory
gladiolus	night-blooming jasmine
henbane	pigweed
hibiscus	poinsettia
hollyhock	ragweed
iris	sorghum
lettuce	soybean
mustard	strawberry
petunia	tobacco
radish	violet
rudbeckia	
spinach	
sugarbeet	
swiss chard	
wheat	

Fig. 6-4. A spectacular display of Easter lilies arranged in the form of a cross for the annual Eastertime exhibit at Garfield Park Conservatory in Chicago. All the flowers shown have been brought into bloom early by controlling light and temperature.

daylength with artificial light. To shorten daylengths, plants are covered with black curtains during part of the daylight hours or grown in windowless rooms with lights timed to turn on and off at preset intervals.

Other studies have shown that daylength can affect the formation of bulbs and tubers in some plants. Irish potatoes and Jerusalem artichokes, for example, are stimulated to produced good-sized tubers under short-day conditions, while bulbous plants, such as onions, usually produce the largest bulbs under long days and short nights. Long periods of daylight also stimulate foliage houseplants to develop more extensive and larger roots.

Many plants which naturally go dormant can be artificially induced to do so or stimulated to produce new growth by exposing them to short ot long days respectively, or they may even be kept evergreen as long as long-day conditions are maintained.

As evidence grew that flowering and certain other activities of vascular plants were influenced by daylength, investigations were extended to other plants and to animals. It is now evident that most, if not all, living things respond to daylength, some in ways that may even be essential to the life of the organism. Vegetative growth and spore production in plants and reproductive activities in both plants and animals are but a few such responses. These organisms are said to be directed by their **biological clocks**. Gardeners, in particular, should take note of the fact that spider mites and garden snails lay their eggs when the days are long, a time that unfortunately (but naturally) corresponds in most areas with the height of the gardening season.

The Role of Phytochrome

The research that followed the discovery of photoperiodism in plants, i.e. the physiological responses to lengths of day and night, contributed many new pieces of information to explain how plants detect and use light energy. Research continues in the field today but among the most important discoveries have been the following:

1. Any plant activity affected by light involves a pigment to absorb the light energy. This energy will alter the pigment molecule to bring about a reaction in the plant cell. Chlorophyll, as a pigment, has long been long recognized for its role in photosynthesis. The challenge now is to explain light-affected plant activities other than those involving chlorophyll.

2. The pigment involved in photoperiodism was found to be a light-sensitive proteinaceous molecule called **phytochrome** which is present in leaves and other parts of plants.

3. Phytochrome exists in two forms. **P_r is a physiologically inactive form capable of absorbing red light.** P_{fr} is an active form that absorbs light from the barely visible far-red portion of the spectrum. An absorption spectrum is shown in Fig. 6-5 to show the relative amounts of different wavelengths that are absorbed by the phytochrome pigments. Each form converts to the other when it absorbs light, the P_r form utilizing primarily red light to convert to P_{fr} and the P_{fr} utilizing primarily far-red light to convert to P_r. P_r is the more stable of the two forms and may last almost indefinitely in the dark. P_{fr}, although physiologically active for a period of time, will revert back to the P_r form in the dark or decompose and be lost as an active factor.

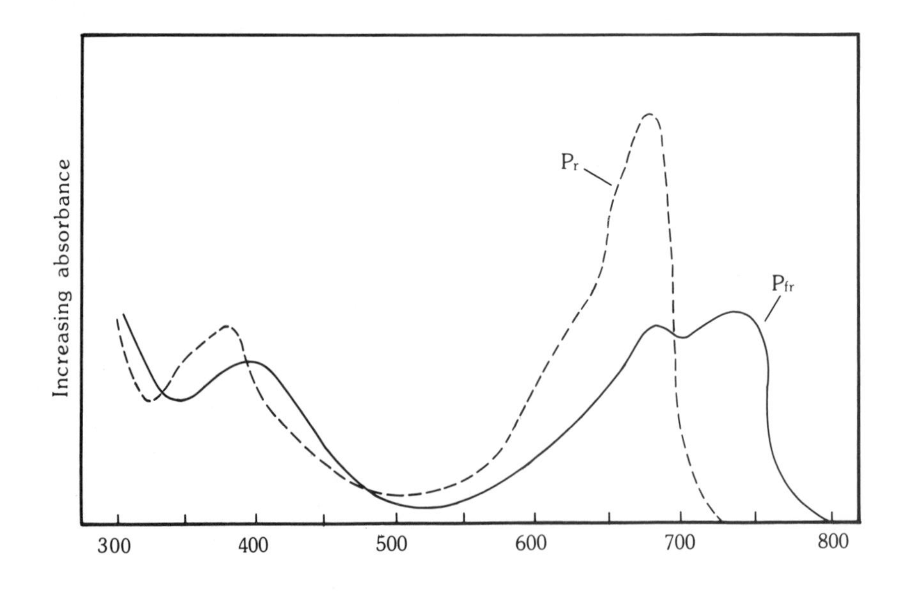

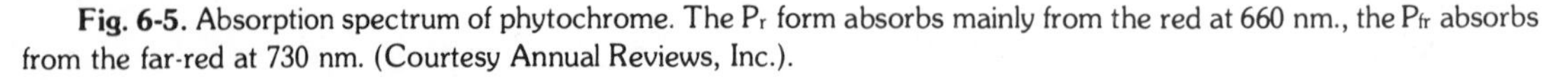
Fig. 6-5. Absorption spectrum of phytochrome. The P_r form absorbs mainly from the red at 660 nm., the P_{fr} absorbs from the far-red at 730 nm. (Courtesy Annual Reviews, Inc.).

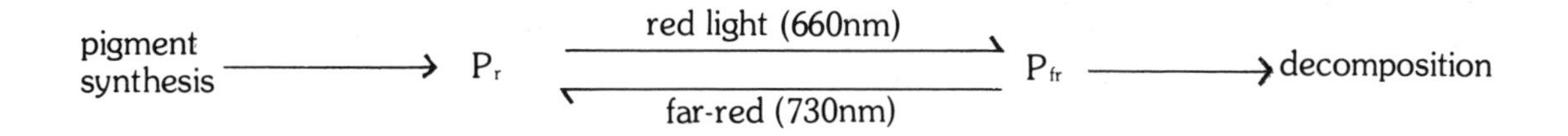

The experiments that led to the understanding of phytochrome activity are now classics of scientific research. They began as an extension of the experiments that established the recognition of general effects of daylength. Growing plants were again exposed to varying conditions of long and short days. This time, however, individual wavelengths of light were used as well as normal daylight and dark. It was found that red and far-red influence flower production more than any other wavelengths.

Interruption of the dark period of short-day plants by either red or white light (which contains red) inhibits flowering, while exposure to far-red or red immediately followed by far-red during the dark period does not. Long-day plants, however, can be induced to flower by short dark periods accompanied by long light periods or by interrupting, even briefly, long dark periods with flashes of red or white light, or by far-red light followed by red light.

(It should be noted that even very low intensities of light (5-10 candles) are sufficient for interrupting the dark period required by short-day plants. It has been many gardeners' misfortune to have plants which they hoped to bring into bloom fail to do so because a light was briefly turned on in the growing room. Even street light or moonlight shining through a window near sensitive plants may inhibit flowering.)

It is apparent that short- and long-day plants respond differently to changing concentrations of P_r and P_{fr} but a definitive explanation is not yet available. It is known, however, that phytochrome is not itself a flowering stimulus but, rather, triggers another substance, possibly the elusive florigen, to carry out this role.

4. Both forms of phytochrome also absorb other wavelengths of light, most notably blue, but to a lesser extent and with fewer, or still unrecognized, physiological effects. Under normal sunlight, the phytochrome system develops a steady state in which the rate of P_r synthesis equals the rate of P_{fr} destruction and the rates of conversion of each to the other are equal. Artificial light, which may differ considerably in wavelengths from sunlight, will produce different growth in plants because of the effect on the P_r P_{fr} balance.

5 Although initial experiments concerning phytochrome dealt mainly with its effect on flowering, it is now known that phytochrome controls many other activi-

Interruptions of dark period	Red	White	Far-red	Red + far-red	Far-red + red
Short-day plants	no flowers	no flowers	flowers	flowers	no flowers
Long-day plants	flowers	flowers	no flowers	no flowers	flowers

ties. It has been found, for example, that many small seeds such as those of lettuce and some common weed annuals will germinate only if exposed to light. The red wavelengths are necessary; thus phytochrome is involved. When seeds are buried deeply in the soil, the stable, inactive P_r form does not bring about germination. When the seeds are exposed to light, the P_r form absorbs red light energy and converts to the active P_{fr} form that induces the physiological activities necessary for germination.

The gardener can benefit from this knowledge by recognizing that certain seeds must be planted close to the surface if germination is to occur. It will also explain why a sudden burst of new weed seedlings often appears in freshly turned-over soils. Seeds that had remained dormant in the darkness within the soil are suddenly stimulated to germinate when brought to the surface and exposed to light. (See also Chapter 3, p. 92 on germination factors.)

Among the other activities affected by phytochrome are stem elongation and general vegetative growth in seedlings, orientation, expansion and movements of leaves, and the development and distribution of chloroplasts, to name just a few. In fact, it can be said that phytochrome, though its action with light, is responsible for regulating the orderly development of many algae, mosses, ferns, and seed plants.

7. Although it is not yet known exactly how phytochrome produces its effects, there have been some plausible suggestions. One is that it influences the manner in which genes, the controllers of all cell activities, are used during plant developoment. It may be associated with major changes in membrane structure and permeability, or with the formation of the chain of chemical reactions that enable a plant to capture and store energy. Finally, evidence indicates that phytochrome may promote or inhibit the production and/or activity of enzymes and hormones, especially those that induce flowering.

GROWING PLANTS UNDER LIGHTS

The first attempts at growing plants under artificial lights (**light gardening**) occured in the 1920's with limited success. Today fresh vegetables are being grown under lights in atomic submarines deep in the sea, and plants of every description are thriving in millions of homes where banks of fluorescent tubes are used instead of a sunny windowsill.

Light gardening has become somewhat of a rage among those who have discovered its advantage. While the greenhouse is still a symbol of a certain degree of gardening skill and interest, light gardening requires no special building, heating, cooling, or water supply. Lights may be set over a few favorite plants on a table, as shown in Fig. 6-6, or a more elaborate set-up may be arranged with a bank of lights suspended from joists in the basement or built in over a table or in a cabinet. Household heat is usually more reliable and stable than that in a greenhouse, which cannot ordinarily in insulated. Artificial lights can be precisely controlled for color, intensity, proximity, and timing, while greenhouses are subject to the natural vagaries of sun, clouds, precipitation effects, etc. Furthermore, while plants in a greenhouse respond to seasonal changes, those grown under lights need never know that summer never ends.

High humidity is probably the only factor easier to achieve in a greenhouse, but more humid conditions can be obtained for light gardens with trays of wet pebbles and/or frequent misting. For large areas a cold-vapor humidifier may be a worthwhile investment.

The Effects of Light Colors and Intensity

Light colors and intensity are important factors that must be considered when planning to grow plants under artificial light. As noted earlier, the blue and red wavelengths

Fig. 6-6. Light fixtures for plants, designed to hold 1, 2, or 4 fluorescent lamps, are readily available in many stores. (Courtesy Sylvania)

of light are readily absorbed by light-sensitive pigments in plants. The yellow and green wavelengths are mostly reflected and scattered and are of little known use to plants. Photosynthesis requires the energy of both red and blue light, and the phytochrome-involving activities utilize the red and far-red wavelengths. Seed germination, flower-bud formation, fruit ripening, tuber and bulb formation, and photoperiodism all depend on the red regions of the spectrum, while the shorter wavelengths at the blue end of the spectrum control petiole and internode lengths, promote normal leaf expansion, and trigger phototropic reactions that bend plants toward a light source. (Phototropism is discussed in Chap. 1, p. 25, and illustrated in Fig. 6-7.)

Whenever light is absent or inadequate, seeds grow into plants that are **etiolated**, that is, have tall, spindly growth with abnormally elongated stem internodes, minute unexpanded leaves, and an absence of chlorophyll so that the plant is pale yellow. A mature plant totally deprived of light survives for a while by utilizing food stored in its various tissues, but new growth is hindered and any that does develop is etiolated. When the food supply runs out, the plant becomes wilted, in spite of moist soil, and soon dies.

If a normal houseplant becomes leggy and pale and shows weak growth, it may be becoming etiolated. The plant needs more intense light and, especially if under artificial light, perhaps a better balance of the blue and red wavelengths (see p. 168 for discussion of color from artificial light). Etiolated growth is unattractive and can't be undone. In many cases the plant should be pinched back and the etiolated growth removed to encorage new, stronger, more attractive growth under better, more intense light. (See Fig. 6-8).

In general, plants of the same species growing under higher light intensity as compared with those under lesser, but not deficient, intensity have thicker broader leaves,

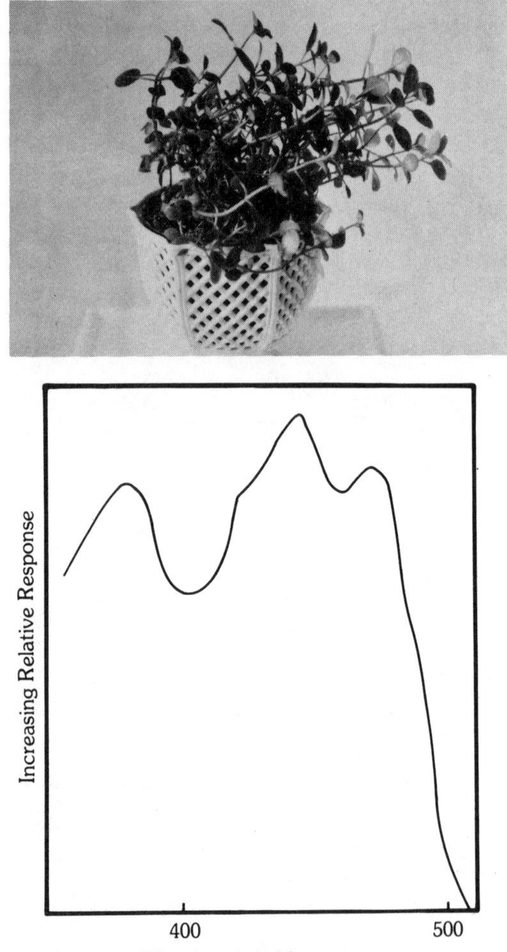

Fig. 6-7. (photo) Phototrophic response in a potted plant. The effective wavelengths for this response are the short waves in the blue region of the spectrum. (graph) Action spectrum for phototropism. (Courtesy AAAS)

shorter internodes, shorter petioles, and better growth because of higher photosynthetic rates. However, the genetic inheritance of a plant limits the range of modification that may occur when leaves are exposed to brighter or dimmer light than that to which the species is adapted. The best results are usually obtained by

Fig. 6-8. The etiolated plant on the left was grown under low light intensity, while the normal plant on the right received adequate light intensity.

providing light intensities approximating those at the upper end of the range occurring in a plant's native habitat.

Measuring Light Intensity

Light intensity is often roughly estimated according to how bright it looks to the human eye. Unfortunately visual sensitivity is greatest for the yellow-green region of the spectrum, and these are the least effective wavelengths for plant responses. The **footcandle**, although nearly obsolete in scientific use, is a unit frequently used in light gardening. By definition, one footcandle **(fc)** is the amount of light *received* from one "standard candle" (as established by the Bureau of Standards) on one square foot of a surface one foot away from the source (see Fig. 6-9). Actually candles have been replaced by other less variable light sources but the term has been retained. For the metric-minded, the **lux** has replaced the

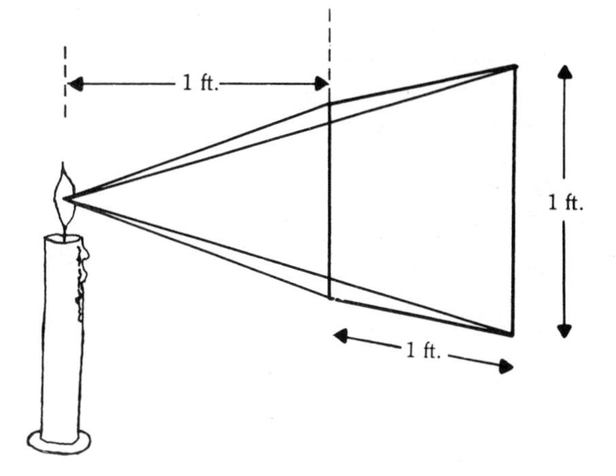

Fig. 6-9. A foot candle is defined as the intensity of light that falls on a square foot of surface one foot away from one "standard candle".

footcandle. One lux is the amount of light received from a standard candle on a square meter of surface at a distance of one meter.

Another unit frequently encountered is the **lumen**, used to express the intensity or brightness of the light being *given off* by the source. One lumen is the total light emitted from one standard candle. All of these units are based on international standards and can be expressed in mathematical equivalents of each other.

$$1 \text{ fc} = 10.76 \text{ lux} = 1 \text{ lumen/ft}^2$$

or

$$1 \text{ lux} = 0.09 \text{ fc} = 1 \text{ lumen/m}^2$$

A present trend favored by lamp manufacturers is to measure light energy in **microwatts per square centimeter.** Watts are expressions of electrical power and give a reading more relevant to an electrically powered light source than footcandles. Although footcandle readings are valuable for ordinary measurements, the microwatt measurements are more precise. Conversion factors have been established for comparing microwatt and footcandle readings. Special instruments are needed to measure light. Simple **light meters (photometers)** such as those used by photographers are sensitive to visual light and give reasonable approximations of the amount of light being received, although they are subject to some distortion by temperature and show maximum sensitivity to the yellow-green region of the spectrum.

You can locate suitable areas in your home for plant growing by measuring the footcandles in various places. The number of foot candles can be determined with an ordinary photographic light meter or with a camera with a built-in light meter. To use these meters, Dr. Henry M. Cathey of the Agricultural Research Service at Beltsville, MD, suggests the following:

> To locate a place that has light equal to 1000fc—
> Set the ASA exposure index to 75
> Set the lens opening to f/8
> Set the exposure time to 1/60 of a second
> For the equivalent of 500fc—
> Use one f stop larger (f/11)
> or—double the ASA index
> or—double the exposure time
> For the equivalent of 2000 fc——
> Use one f stop smaller f/5.6)
> or—half the ASA index
> or—half the exposure time
> Other footcandle readings can be approximated from the above settings.

Alternately, you can determine footcandles by the following method, also using a light meter:

> Set the ASA exposure index to 100
> Set the lens opening to f/4
> Place a sheet of matte-finished paper in the location to be read. Fill the view with the paper.
> Read the exposure time for the approximate number of footcandles at the spot being viewed, e.g., 1/60 = 60 fc, 1/1000 = 1000 fc, etc.

Spectrophotometers and spectroradiometers are instruments that have been developed to determine qualities and quantities of light beyond simple intensity of illumination as detected by human vision. With such instruments, the energy in the individual bands of the spectrum (both visible and near-visible) that are important in the physiological activities of plants can be measured, evaluated, compared, characterized, and in other ways used to further our knowledge of the relationships between plants and light. While such instruments are used primarily in research, much of the information available to home light gardeners from manufacturers of light-gardening equipment comes from the use of spectrometers of various types.

Since intensity and color are two aspects of light that can be controlled with lamps, the indoor gardener should be aware of their effects on plants. Light intensity outdoors ranges from about 5000 fc on a bright sunny day in mid-winter to about 10,000 fc in mid-summer in the middle latitudes of our country. Plants on a window sill in direct sun may experience nearly these levels, but the light intensity decreases greatly as the distance from the window increases. It can be shown, in fact, that the intensity will be reduced by the square of the distance. Thus at 2 feet light intensity is only 1/4 as great as it is at 1 foot and 1/9 as great at 3 feet, 1/16 as great at 4 feet. It is therefore important to consider the distances between plants and light sources. (See Fig. 6-10.)

Light intensity will also be greater during the mid-daytime hours when the sun is high in the sky than it is early or later in the day (see Fig. 6-11). Plants that usually grow beneath taller trees or potted plants behind curtains will both find that their light is greatly reduced and must be tolerant of these levels if they are to survive.

Fifty footcandles is usually considered the survival minimum for most plants. Foliage plants native to dense rainforests are usually successful at this level, although

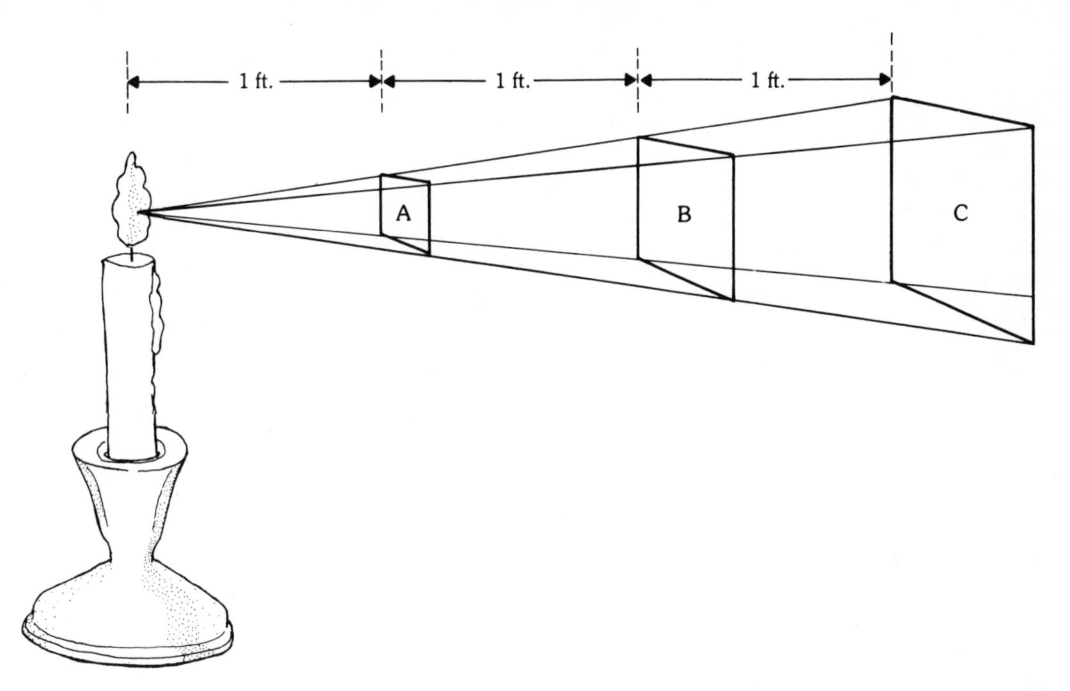

Fig. 6-10. Radiation from a light source is distributed over more area as the distance from the source increases. The inverse-square law applies to light intensity, which decreases by the square of the distance between the light source and the receiving surface. If area A is receiving 36 foot candles of light intensity, then B is receiving 9 fc and C is receiving 4 fc.

they will produce much more luxuriant foliage at levels closer to 250 fc. Plants adapted to the more open areas and mountainsides of subtropical regions will require 250-650 fc, while those of open sunny places may require up to 1400 fc and can usually tolerate much greater intensity. (See list of recommended light intensities for houseplants in Chap. 7, p. 200.)

Artificial lights can provide the necessary light intensity for selected plants. Beneath two standard fluorescent light tubes an intensity of nearly 1000 fc can be achieved at a few inches distance (good for germinting seeds) and an average of about 400 fc at "usual" foliage height as shown in Fig. 6-12. Intensities can be boosted somewhat by using fixtures with reflectors or by placing reflecting surfaces behind, around, or under the plants. As shown in Fig. 6-13, plastered walls and matte-finished white paint reflect with the greatest uniformity,

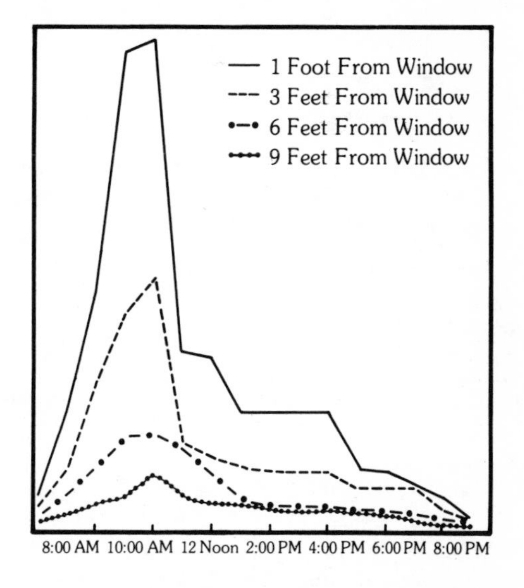

Fig. 6-11. The measurements of light intensity coming through an east window during a bright, clear summer day. The window is 3 feet wide and 5 feet long.

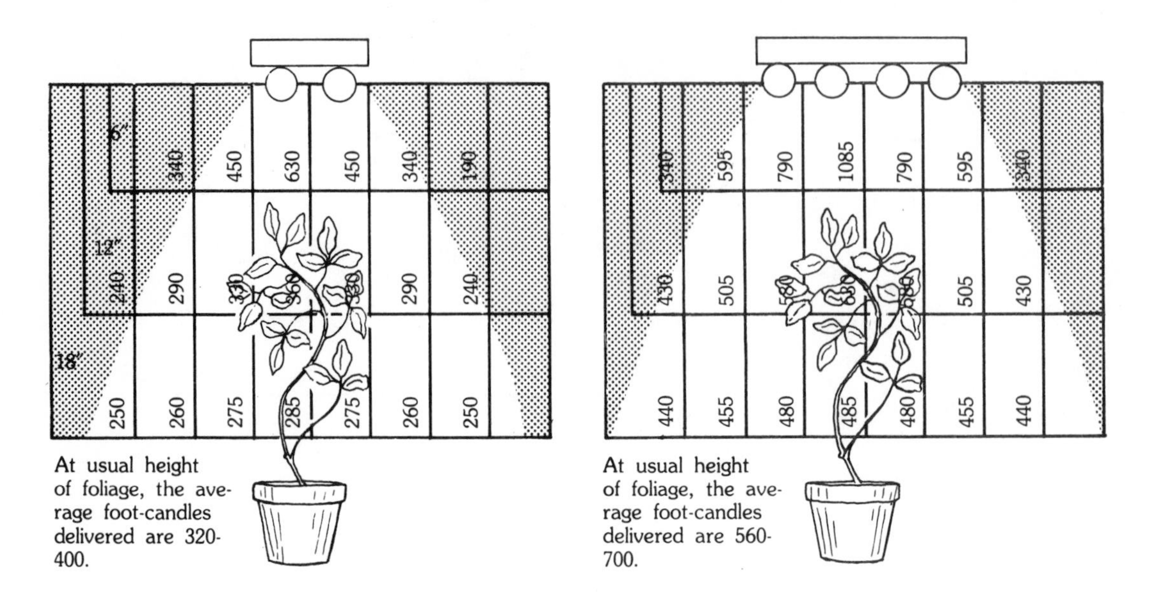

Fig. 6-12. To increase light intensity use additional fluorescent lamps. (courtesy Duro-Lite).

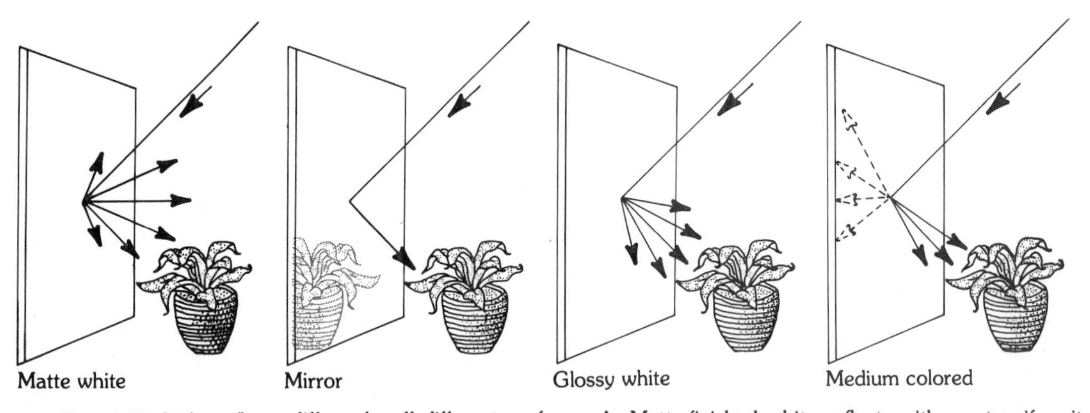

Fig. 6-13. Light reflects differently off different surfaces. A. Matte-finished white reflects with great uniformity, scattering evenly in many directions. B. Mirrors reflect unidirectionally. C. Glossy white reflects less well than matte finishes. D. Colored surfaces absorb much of the light and reflect little of it.

glossy white and mirrors somewhat less. Where low light levels are still a problem, they can often be compensated for by increasing the duration of the light. Timers connected to light banks can control the number of hours of light given to plants with little attention from the gardener except for the appropriate settings.

The colors given off by artificial lights depend on the source of illumination. Incandescent light bulbs give off light from a heated tungsten filament and usually produce light that is primarily yellow to red. If the amount of light generated by an incandescent bulb is plotted on a graph it will produce a spectral-energy distribution

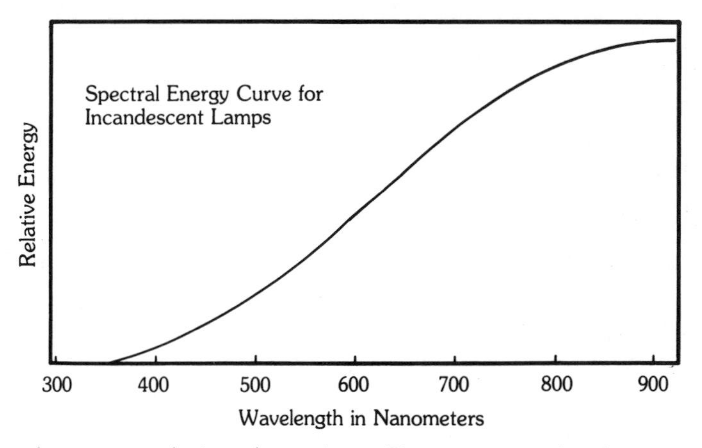

Fig. 6-14. Spectral-energy curve for incandescent lamps. The curve is smooth and increases from the blue to the red range.

curve like that shown in Fig. 6-14, which shows that the curve extends well beyond the regions of red and far-red.

Fluorescence occurs when light energy is absorbed by a substance and reemitted. The reemitted light is necessarily of longer wavelengths because of its lower energy, some energy having been lost in the transformation. In fluorescent lamps an electrical discharge through the tube causes mercury vapor to emit radiation (mainly invisible ultraviolet, as well as some visible wave lengths). The tubes are coated inside with special substances (phosphors) that absorb this radiation and reemit it as visible light. The spectral distribution of light from a fluorescent lamp will depend on the phosphors used for the coatings and on the visible light produced by the mercury-vapor arc, which is very intense but concentrated at particular wavelengths. Spectral-energy distribution curves for some standard fluorescent lamps are shown in Fig. 6-15. (The ultraviolet light cannot be transmitted through glass and does not appear on the graph.) Plant growth and other responses may be plotted against these kinds of curves for some interesting comparisons. One such comparison is shown in Fig. 6-16.

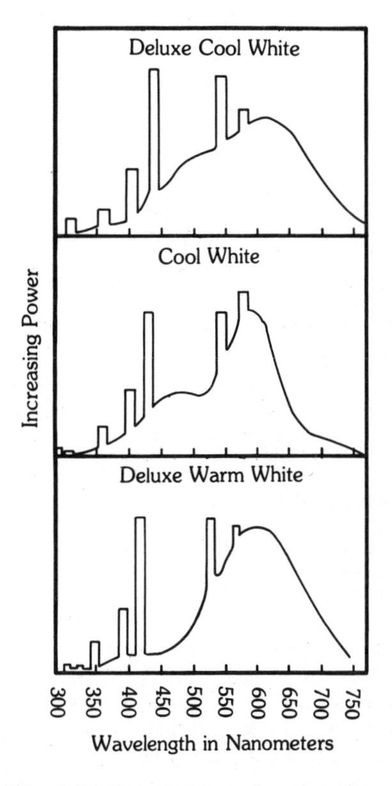

Fig. 6-15. Spectral-energy curves of commonly used fluorescent lamps. The sharp peaks are the result of radiation by the mercury arc of the lamp. (courtesy IES)

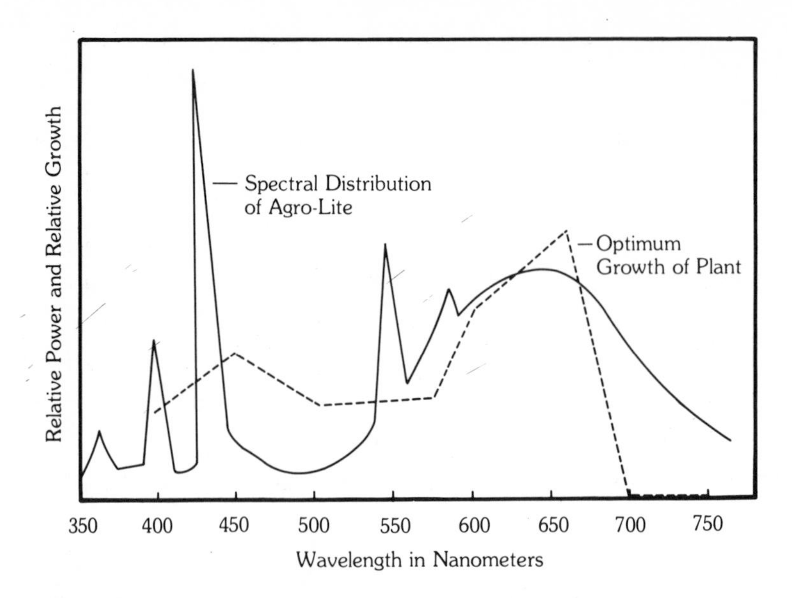

Fig. 6-16. To determine the intensities of wavelengths necessary for optimum plant growth, seedlings of various plants were exposed to increased steps in intensities until they reached a point at which further increase in intensity produced no increase in plant growth (the saturation point). The optimum intensities for each of the colors of the spectrum are plotted on the graph as the dashed line. The distribution of the spectral energy available from a Westinghouse Agro-Lite lamp is represented by the solid line and closely approximates that of optimum plant growth. (courtesy Westinghouse)

Lamps for Growing Plants

Incandescent lamps will provide light for foliage plants that can survive on low levels of intensity. Since heat increases with wattage, it is not possible to supply the intensity needed for seeds or flowering plants without producing damaging heat levels. Lamps with built-in reflectors improve the projection of the beam onto the plants and are often used for decorative effects in the home or yard for enhancement of the plants. In commercial greenhouses and in research studies special lamps are used for increased light levels, but these are not usually suitable for home situations.

Fig. 6-17 shows several kinds of incandescent bulbs. In addition to the standard bulb (A), there are **reflector** (R), **projector** (PAR), and "**reflectorized**" (PS-30) types. The reflectors are commonly used indoors to produce a narrow beam of light, whereas the projector is more ruggedly constructed for outdoor use and several beam diameters are available (see Fig. 6-17). The "reflectorized" lamps (PS-30) are especially designed to produce an economical wide-spread

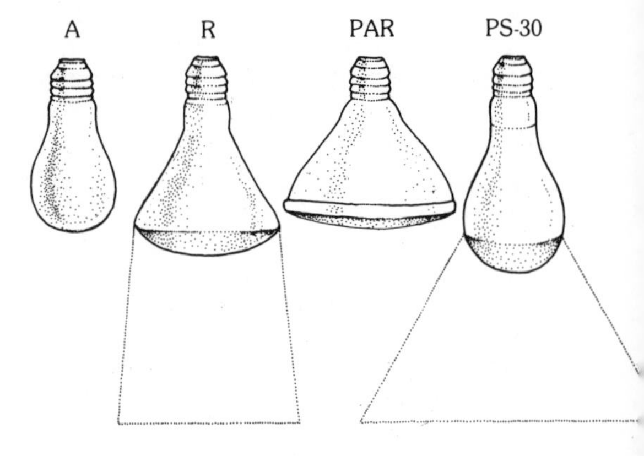

Fig. 6-17. Incandescent lamps. R, PAR, and PS-30 is especially designed for a wide spread beam. (GE)

beam of light for plants in greenhouses under photoperiod control, such as chrysanthemums.

For light gardens indoors, fluorescent lamps are used either alone or with incandescents. A fluorescent tube may be 20 watts and 24″ long, 40 watts and 48″ long, or 72 watts and 96″ long. Most fixtures are designed to hold two tubes or four tubes and may be equipped with a reflector. Growers usually find the 48″ tubes the most convenient to use. A sampling of some of the fluorescent lamps available is shown in Fig. 6-18.

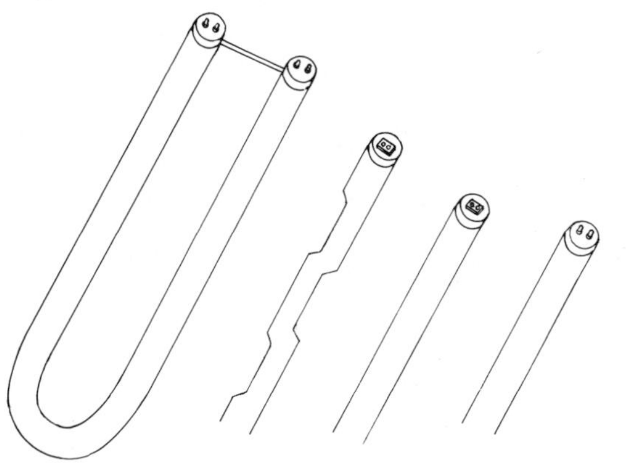

Fig. 6-18. Several types of fluorescent lamps suitable for light gardening. (GE)

The fluorescent tubes available when light gardening was in its infancy produced a cool blue light, good for most foliage growth but lacking the longer wavelengths needed for seeds and flowers. To balance this, incandescents were usually combined with the fluorescents in a light bank to provide the red end of the spectrum for better flowering. This combination proved very effective and is still used by many growers. It is a less expensive source of light than the recently developed special plant-growing lamps.

There is on the market today another type of general-purpose lamp called a warm-white fluorescent that produces some red colors in addition to the blues. The warm whites were developed with the addition of more red-producing phosphors for improved home and special-area lighting. Cool and warm as used here refer to the quality of light produced by these fluorescent tubes and do not imply any relation to temperature.

A well-designed light garden will provide a uniform distribution of a sufficient intensity of light to produce and maintain healthy plants. The quality of the light should supply the right proportion of blue for general growth and red for bloom. The intensity and colors needed will vary somewhat for different plants or stages of growth from germination to flowering. Fluoresecnt tubes, which are three times more efficient than incandescents in converting electrical energy to light, were originally developed for lighting large areas of space at minimum cost, and the phosphors used emitted primarily in the blue region of the spectrum. Both the cool and warm whites were found to be effective for growing plants, and many growers use warm whites alone or in combination with cool whites for general plant lighting.

A light bank for all-purpose growing can be composed of an equal number of 40-watt cool and warm-white fluorescents (the more tubes the higher the intensity) or a combination of 40-watt fluorescent tubes and 25-watt or smaller incandescent bulbs. As a general rule, use 30-40 watts of incandescents to every 100 watts of fluorescents. As you increase intensity, use a higher percentage of incandescent light. Either cool or warm-white fluorescents can be used, although most growers seem to prefer the warm whites because of their higher intensity. A fixture with two fluorescent tubes should be placed over each one-foot width of plant space. If incandescents are also used they can be mounted between the

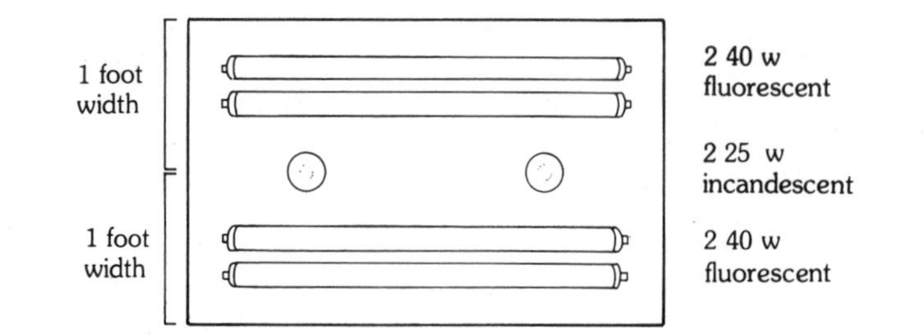

Fig. 6-19. Arrangement of fluorescent and incandescent lamps for a home light garden.

fluorescent fixtures as shown in Fig. 6-19.

There is more energy in a fluorescent tube in the middle than at the ends, and the tubes tend to burn out from the ends toward the center. The intensity from a fluorescent tube is thus greater in the center. To avoid excessively low levels at the ends, the tubes should be replaced about once a year under normal use.

While fluorescent tubes diffuse light over a wide area, incandescent bulbs produce a more concentrated area of light accompanied by higher heat production. To avoid scorching plants it is best to use incandescents of no more than 25 watts.

The combination of fluorescent tubes and incandescent bulbs duplicates sunlight to the satisfaction of most plants and at the most reasonable cost, but as interest in indoor plant growing increased, lamp manufacturers turned to producing special "grow lights" that are designed to supply the most-needed wavelengths. The resulting lamps have been of varying quality and worth, but since research is currently so active, a new more ideal product could arrive at any time.

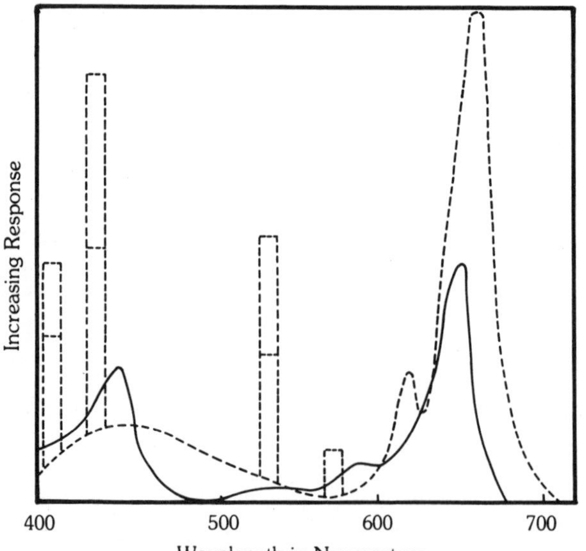

Fig. 6-20. The spectral absorption curve for chlorophyll synthesis (solid line) compared with the spectral energy curve available from Gro-Lux lamps (dashed line). The sharp peaks are a result of radiation by the mercury arc of the lamp. (courtesy Sylvania)

An early entry in the "grow light" market was Sylvania's Gro-Lux, which is still a popular lamp. It produces a spectral curve close to that known to be effective in chlorophyll synthesis. The curves are compared in Fig. 6-20. The later Gro-Lux Wide Spectrum modified the high red and blue emissions and added some far-red for better flowering. Westinghouse's Agro-Lite similarly supplies red wavelengths for the benefit of flowering plants. Test results with gloxinias, for instance, showed that more than twice as many good-sized flowers were produced on plants grown under Agro-Lites as on plants grown in sunlight. It should be noted, of course, that artificial light is constant in quality and intensity and can be applied for longer periods of time than normal daylight hours.

General Electric's Plant Light is offered as an alternative to daylight or cool-white fluorescents for growing plants because it produces a pleasant reddish-purple color that enriches their appearance as it promotes their growth. Duro-Lite's Naturescent matches sunlight very closely, so the plants appear very natural, and the extra wavelengths in addition to the red and blue may possibly provide benefits. Vita-Lite is similar but has an extension into the ultraviolet range (see Fig. 6-21) which can be beneficial in controlling bacterial and fungus problems, since these organisms are adversely affected by ultravioltet radiation. Both lamps are popular with photographers because of the natural appearance of the plants. Cacti and succulents show improved growth when UV is added to their light.

One of the newest lamps on the market is the Wonderlite by Public Service Corporation. This lamp uses a standard screw base and has a built-in reflector but emits light from a special ingredient called "Phospherol". The balance of blue and red is said to be close to that of true daylight, producing excellent plant growth. For commercial establishments and research greenhouses, High-Intensity-Discharge (HID) lamps have been developed that far exceed the efficiency and special energy derived from fluorescent-incandescent systems. They also use screw bases but require spe-

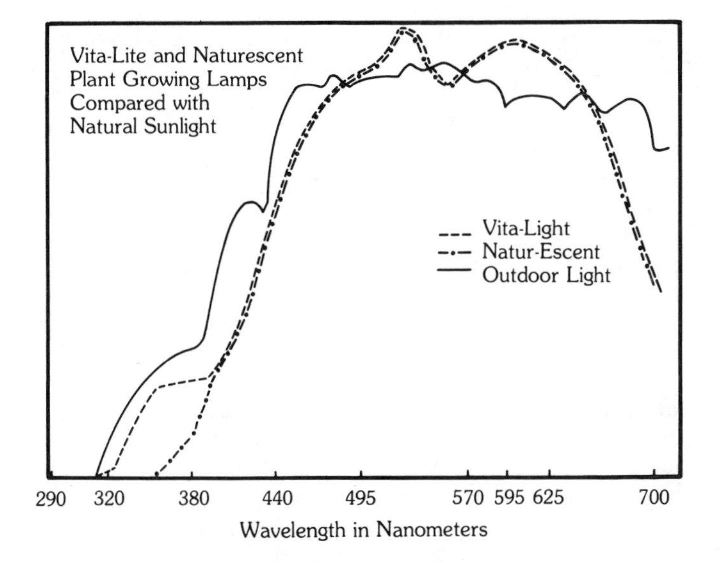

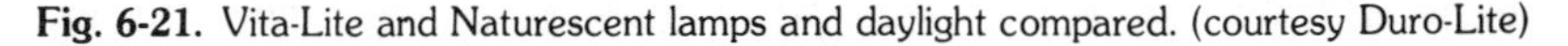
Fig. 6-21. Vita-Lite and Naturescent lamps and daylight compared. (courtesy Duro-Lite)

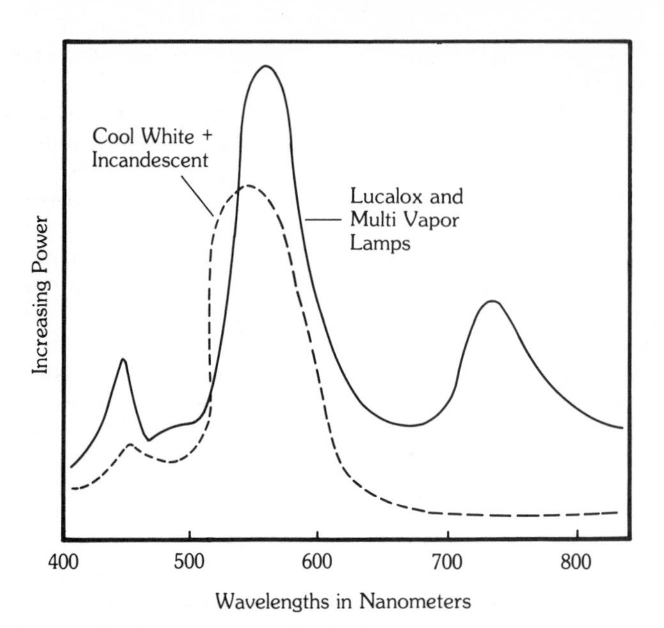

Fig. 6-22. Spectral-energy distribution curves of the high-pressure sodium vapor Lucalox and halide Multi-Vapor lamps compared with the commonly used fluorescent-incandescent combination. (courtesy ASAE)

cial fixtures with ballasts to accomodate them. These lamps are of several kinds such as General Electric's high-pressure sodium Lacalox or the metal halide Multi-Vapor. The spectral-energy distribution curves of these lamps are shown in Fig. 6-22. The mercury-vapor Fluoromeric by Duro-Lite is said to be one of the best in duplicating the natural spectrum. These new lamps are all fairly expensive to buy and install but are long-lasting and use less power to operate. They have proved to be a worth while investment for growers of seedling plants and greenhouse-grown vegetables.

Getting Good Results

Once the growing space has been provided with lights, it will be necessary to provide other good cultural conditions. As light increases so also do the needs for water, nutrients, and other factors. The following suggestions will improve results when growing plants under artificial light.

1. Provide, on an average, 14 to 16 hours of light a day and try to have all plants within 12" of the lights for best intensity.

2. Increase intensity by reflecting light back into the area rather than allowing it to escape into the outer surroundings. Try using metal reflectors, white paint, mirrors, white gravel in the humidifying trays, or loose sheets of aluminum foil hung around the set-up.

3. Increase humity. A relative humidity of 40-50% is ideal. Frequent misting or placing plants on trays of wet gravel may be adequate. If not, a humidifier may be required.

4. Increase water and nutrients. Plants under lights grow more rapidly and thus require more of the substances necessary to produce this growth. A regular fertilizing schedule should be established, depending on the plants being grown. (See Chap. 7 for instructions for growing houseplants.) Many growers add a fraction of the amount of fer-

tilizer needed to every watering. Soilless mixes do not become water logged and can be flushed easily to prevent the buildup of fertilizer salts.

5. Increase air circulation. Plants under lights need more carbon dioxide, and this need will be better met if air is moved through the plants. Still, humid air also increases the potential for mildew and bacterial diseases. An open window or small fan will promote good air movement.

6. Increase temperature. Plants under lights grow better at temperatures 5-10° higher than those normally recommended for the plants. Water plants with warm water, never cold, to avoid shocking plants and cooling the air.

7. Pay close attention to your plants. Keep plants, light fixtures, and surroundings clean. Move plants to the height or distance under the lights most suitable to each one. Try new light placements; perhaps a few vertically mounted lights will increase distribution and/or intensity. One grower recently reported great improvements when he placed a few fluorescents *beneath* his plants to irradiate the under surfaces of the leaves, although others have reported leaf distortions when light is applied to lower plant surfaces.

Growing plants under lights can be an exciting and rewarding activity. Light setups are not difficult to construct, and handsome ready-made stands of many varieties are available on the commercial market. The gardener can almost totally control the growing conditions to suit any variety of plants from flowering gems to edible vegetables at any desired time of the year.

7. INDOOR GARDENING

Houseplants* have been a feature of cold-climate homes for little more than 100 years. The window balconies and courtyards of the ancient homes of Egypt, India, and China, and the 17th- and 18th-century dwellings of the Mediterranean area proudly displayed a wide range of potted foliage and flowering plants. Until the advent of central heating, houseplants in homes subject to cold-winter conditions were nearily an impossibility, and plant lovers were limited to enjoying their plants primarily during the warmer months.

The 1700's and 1800's were years of botanical and horticultural endeavors. Travel between Europe, America, and Asia was well established and there was an urgent need to introduce new productive plants into the New World for food, medicine, and ornamental use, and novelty plants into the formal European gardens. Literature of the time abounds with adventurous and romantic stories of the plant- and seed-collecting expeditions into some of the remote and exotic areas of the world.

When one considers the hazards to which such collections must have been subjected in their transportation it is a wonder any of them ever survived. The swashbuckling sailors of the time were disdainful of their cargoes, and the invention of the Wardian case, a glass-topped wooden box, in 1834 improved remarkably the survival rate of living plants in shipment.

*All houseplant names are according to Graf, A. B., *Exotica*, 9th Edition. E. Rutherford, N. J.: Roehrs Company, Inc., 1976.

Many fine conservatories and greenhouses were established in both Europe and America during this time in which imported plants were grown and studied, and private conservatories were often a feature of wealthy estates. But growing plants in the average home was severely limited.

Fireplaces today are a decorative accessory to our homes. Before the invention of central heating they were a necessity. They produce very warm, desert-dry air when burning and allow room temperature to drop precipitously when they go out. To counteract overly rapid cooling, homes of the 1800's had tall, narrow windows to prevent heat loss while allowing a bit of daylight to penetrate the gloom inside.

The middle-class gentility in their Victorian homes with the tortuously carved furniture and abundant knickknacks were among the first householders to use houseplants. Undoubtedly they were a welcome relief among the heavy wall hangings and the dark horsehair sofas. Obviously any plant chosen had to have a rugged constitution. The *Aspidistra* was aptly named the cast-iron plant and survived well in the turn-of-the-century homes. A number of palms and the reliable rubber tree (*Ficus elastica*) were also used. Figure 7-1 illustrates these plants.

The coal-burning furnaces installed in homes following World War I meant more reasonably even heat throughout the house, but since they required manual stoking to keep them burning, temperatures still dropped as the furnaces cooled. However, this form of central heating, better building materials, and insulation allowed homes to

Fig. 7-1. These plants are representative of those seen in homes at the turn of the century. Left, *Ficus elastica*, the rubber plant, middle, *Chamaedorea*, the parlor palm, and right, *Aspidistra elatior*, the cast-iron plant.

have more and larger windows to admit light. Thus it became possible to introduce plants of a slightly more delicate nature by providing them with a less extreme temperature range and increased light intensity.

Sunporches and bay windows became an envied feature on many of the homes built prior to World War II, and with the introduction of natural gas many of these sunporches were turned into blooming garden rooms.

Homes built since World War II are designed with generous windows and controllable heating systems. In recent years humidifying systems have become increasingly included in building plans; they should improve houseplant growing even more.

Most house plants are grown for their attractive foliage, although all angiosperms can, under ideal conditions, be brought into bloom. Foliage plants are looked upon as an addition to the decor of a room, while flowering plants are usually treated as temporary novelties because of the brevity and environmental sensitivity of the flowers.

Native Habitats of Houseplants

Foliage plants have been selected from a wide range of plant families and native habitats. The large family Araceae is undoubtedly familiar to even the beginning grower since it includes *Aglaonema* (Chinese evergreen), *Dieffenbachia* (dumbcane), and all the other "philodendron-like" plants. Most of the Araceae are tropical or semi-tropical plants appreciative of our warm houses and tolerant of reduced light intensity.

Other plants adaptable to our home environments, in which humidity is so often very low, include the Cactaceae, which originated in the deserts and near-deserts of the Americas, and the Euphorbiaceae, many of which superficially resemble the cacti but originated in Africa and Asia. Most of the succulent Crassulaceae, of Africa, will also survive well.

The mint family Labiatae is well represented as houseplants, and one member illustrates the folly of relying on common names as indicators of anything more than an active imagination on the part of the name-giver. The genus *Plectranthus* is commonly called Swedish ivy but is native to Australia and is not a climber like most true ivies, the common name being derived from the fact that it was supposedly a popular Scandinavian houseplant before it became known elsewhere. The above-named foliage plants are pictured in Figure 7-2.

There are a number of exceptions to the tropical or desert natures of most houseplants. A few species can even tolerate below-freezing temperatures in a dormant state when grown outdoors or be equally successful indoors as year-round foliage plants. Examples are *Saxifraga sarmentosa* (strawberry begonia), *Tolmiea menziesii* (piggyback plant), some of the species of *Hedera* (English ivy), and the hardy species of *Opuntia* (prickly-pear cactus) and *Sedum*. In fact, in startling contrast to the native habitats of most houseplants, *Tolmiea* is a familiar plant in the wilds of coastal Alaska.

Buying a Houseplant

Most easily-grownfoliage plants grow naturally in the lowermost canopy levels of the forests where the light penetration is min-

Fig. 7-2. A few representatives of plant families that have many familiar species as houseplants. (above, left) *Monstera deliciosa*, often called *Philodendron pertusum* when in this juvenile form, family Araceae. (above, center) *Crassula argentea* (*C. arborescens* of the trade), the jade plant, family Crassulaceae. (above, right) *Euphorbia splendens*, the crown-of-thorns, family Euphorbiaceae. (below, left) *Opuntia microdasys*, the bunny-ears cactus, family Cactaceae. (below, right) *Coleus blumei*, a colorful-leaved member of the family Labiatae.

imal. The rooms of our homes also provide only low light levels unless artificially supplemented (see Chapter 6, Plants and Light). So, before buying a plant, estimate the amount of light that will be available to the plant in the spot in which you intend to put it and select a species adaptable to that amount of light (see table at end of chapter). Also keep in mind that variegated leaves have less chlorophyll and need more light intensity than all-green leaves.

Select a plant that ishealthy and appears well cared for, in a pot of adequate size (pg. 187). Keep it segregated from the rest of your plants at home for at least two weeks and watch it closely during this time. Insect eggs should hatch or fungus spores germinate within this time if they are present, and potential catastophes can be dealt with properly.

Don't kill your plants with kindness. Don't overwater, don't repot yet, and don't fertilize. Allow your plant to adjust gradually to its new surroundings. When new growth begins to appear it is a sign that the adaptation has been made. Make one more check for insects and disease and take the final steps you decide are necessary to fit the plant decoratively into your indoor garden scheme.

Houseplant Soils

It is generally a poor practice to use garden soils in houseplant pots. Most of them are too heavy for the comparatively restricted confines of a pot. They often contain herbicides or other gardening chemicals, or diseases or pests which could cause damage to a houseplant (see table of ailments at end of chapter).

If topsoil is used it should be mixed with approximately equal parts of sand, peat, and perlite. Such additions are also required if commercial potting *soil* is used. These packaged soils, while rich and nearly sterile, are so fine and heavy that they hold too much water, and the soil air is reduced below acceptable levels. Roots in these soils soon suffer.

Jiffy-Mix, Redi-Mix, and any of the other commercial preparations based on the original soilless Cornell Plant Mix are excellent media for growing plants with good root systems. These are mixes of pulverized peat moss or ground tree bark, with varying quantities of vermiculite, perlite, and slow-release fertilizers added. These mixes are popular because they are light-weight and nearly sterile and absorb water without the risk of becoming oversaturated, if drainage is provided. They *must*, however, be fertilized regularly and frequently since they contain no minerals to provide essential elements. Check the package for fertilizing instructions. If a fertilizing schedule is not instituted, plants will begin to show signs of ion deficiencies in about two to three months when the mix is depleted of the slow-release fertilizers originally included.

If a purchased plant is maintained in its original potting medium, it is probably growing in one of these commercial mixes. Fertilizing once a week with Hyponex or any other standard houseplant fertilizer will usually supply the plant's needs.

Similarly, epiphytic plants should be regularly fertilized. Most orchids and bromeliads grow naturally in tree tops, snugly clasping the bark with a few stout hold-fast roots. The cup-forming leaves of the bromeliads can be filled with enough water for most of their needs, while the orchids produce pseudobulb growths in their stems for holding water (see Fig. 7-3). Special epiphytic potting mixes are available for these plants. Fir bark or wood chips are the base to which perlite, palm fiber, and slow-release fertilizers are often added. The nutrients are absorbed when the mix is watered periodically.

When you are using newly combined soil mixes or re-using soils from other potted plants it is always a good idea to sterlize the soil. For small quantities, the soil can be spread on a cookie sheet and baked in the oven at 180°F. (82°C) for 30 minutes. If the soil is placed inside a plastic roasting bag the odor is reduced and the possibility of accidentally dumping a trayful of soil into your oven is minimized.

Fig. 7-3. Wild epiphytic plants, such as the Florida bromeliad shown here, are seen in abundance in the trees in the tropical and near-tropical forests of the world. When such plants are cultivated they should be planted in a special epiphytic mix.

If larger quantities of soil are needed, as for planter boxes or large tubs, the formaldehyde method may be more efficient:

Add 2 1/2 tablespoons of 40% formaldehyde (from the druggist) to 1 cup of water.

Pour over and mix thoroughly into one bushel of soil.

Cover with a plastic sheet for 24 hours.

Uncover, stir occasionally, and allow to air for several days until odor is gone.

Commerical soil and potting mixes are pre-sterilized and do not need to be sterilized when used from the package.

A few houseplants require soils that are more acid or alkaline than are the usual standard mixes, or one may inadvertently include substances in a homemade mix which produce soils that are too alkaline or too acid. Whenever the soil pH must be modified for a flower pot or two, simply add crushed egg shells of sea shells or a few linestone chips to make the soil less acid or more alkaline. If the soil is to alkaline or not acid enough, sprinkle a bit of alum on the soil surface or add a few *drops* of vinegar to the water with each watering.

Houseplant Pots

There is a wonderful array of pots available for houseplants today (see Fig. 7-4). The most important feature of any pot is the drain hole. Stones or chips in the bottom of a pot without a hole allow a space for excess water to collect, but when water accumulates above them the soil in the pot will become oversaturated. Oversaturation can occur even in a pot with drainage if the water runs out into a saucer and is allowed to remain there. It should be poured off to prevent water-logged soils.

Clay pots are somewhat expensive and heavy, break and chip easily, and dry out fast. The porousness of the clay, however, makes them ideal for cacti, succulents, and epiphytes such as orchids and bromeliads. These are plants that resent water remaining around their roots very long and need a lot of air in their light sandy or bark-based soils.

Plastic-foam or polystyrene pots are inexpensive, light in weight, and water-resistant but porous, and allow air to pass through freely. The advertisements say they don't shatter, but they do tend to chip away a bit at a time. They are widely used by florists for potted gift plants but are less desirable than clay for large and heavy cacti and succulents since they do not hold up as well.

Fig. 7-4. Several types of flower pots.

Plastic pots come in many sizes, shapes, and colors with and without saucers. They are light in weight and do not allow air or water to pass through. If dropped or handled roughly they can easily crack or split beyond usefulness. They are commonly used for today's houseplants by both the commercial growers and the home gardener. Since they are waterproof they reduce the need for frequent waterings in the low humidity of most homes. Their costs vary with decorative embellishments.

Glazed ceramic pots are decoratively the most attractive. Like clay pots they are heavy and chip easily. Price will depend on size and design. They are available with glaze both inside and out, or some are glazed only on the outside, in which case they absorb some water from the soil. The glaze prevents water seepage to the outside. Some of the designers fail to include a drain hole in the bottom of the pot.

Pots of any kind should always be scrubbed clean before use. Mineral deposits on the edges of used pots are usually carbonates derived from water. The deposits are easily removed, since carbonates react with acids to form more soluble salts, e.g.: $CaCO_3 + 2HC1 \rightarrow$ (soluble) $CaC1_2 + H_2O + CO_2 \uparrow$. If effervescence occurs when a drop of acid is applied to the deposit, carbonates are present. They can be removed by soaking in vinegar or, more quickly, by treating with Sno-Bol, a commercial toilet-bowl cleaner containing dilute hydrochloric acid. Other stains and deposits may also occur, but carbonates from watering are the most common.

Pots as well as soil should be sterilized. Heat-resistant pots such as the clay pots and some plastics can be heated in the oven along with the soil. To sterilize pots of any kind, soak overnight in a tub of water with a cupful of household bleach added.

TAKING CARE OF HOUSEPLANTS

Maintaining a healthy, well-conditioned houseplant is much like maintaining a friendship. Just as one learns to detect certain reactions and behaviorisms and interpret what they mean in a long-time friend, so also may one learn to know certain signs and "behaviors" in one's plants.

Ideally a plant should be turgid, evenly colored, with leaves at close intervals, and possessing an appropriately sized root mass (see page 187 concerning root-bound plants.

The general rule for caring for plants is NOTHING IN EXCESS! Purchased plants have usually been growing in a nearly ideal environment. They often face a shocking change in conditions when brought into their new home. Whether they make it or not depends primarily on whether the purchaser allows them to adapt gradually to their new surroundings.

Watering

Growing attractive, healthy houseplants requires that you know when and how to water plants. Many factors influence the watering of plants. An awareness of these factors will determine your success at recognizing when your plants need watering and how to do it for the best results.

These factors are:

1. The kind of pot the plant is in.
2. The type of soil or potting medium used.
3. The kind of plant and its particular needs.
4. The time of year in relation to the growing season.
5. The size of the plant in relation to the pot.
6. The total leaf area.
7. Environmental factors.

1. As previously described, pots and containers may be made from porous clay, plastic foam, or polystyrene, or of non-porous, water-tight plastics and glazed ceramics. Glass and metal containers are also non-porous. Since the amount of water lost from the soil surface and through the pot affects the rate of drying, it will be apparent that non-porous pots will require less frequent watering than porous ones.

2. The type of potting soil or mix will determine how rapidly water flows through the pot and how much water is retained by absorption and adsorption within the soil or mix. Clay soils, having colloidal-sized particles, drain poorly and retain a considerable amount of water. Sandy soils are quite the opposite. The rate of flow through general potting loam will be highly variable, depending upon the relative proportions of sand, silt, and clay.

Organic ingredients, as well as perlite and vermiculite, if present, will also affect the water-holding properties of the soil. Water adheres to or is adsorbed onto the surfaces of mineral particles but is absorbed into peat moss fibers or into other organic materials. Peat-based potting mixes will allow water to move rapidly through but will absorb and hold water better than soil-based mixes with abundant mineral particles.

Potting soils tend to shrink away from the sides of the pot as they become dry, especially if they are low in sand. Potting mixes will also pack and shrink as the peat moss dries out and shrivels. This condition should be avoided since peat moss is difficult to rewet once it has dried.

3. Different plant species require different amounts of water. It is well known that cacti can go a long time between waterings. It may not be as well known that they will actually be damaged if watered too frequently. Other succulents, especially of the family Crassulaceae (Figs. 7-2), while capable of surviving during lengthy dry spells, will grow rewardingly if given fairly generous amounts of water during their growing season.

At the other extreme is *Cyperus* or the umbrella plant (Fig. 7-5) which grows best when its soil is kept saturated by placing the pot in a saucer filled with water.

Other houseplants fall into a wide range of water requirements between these two end points. Fortunately many of them are very tolerant of their owner's whims and erratic watering schedules. Recommendations for watering specific plants will be found in the chart on page 198.

4. Seasonal growing habits of any plant will determine the amount of water given at different times in its seasonal cycle. Most plants grow more rapidly at certain times of the year and slow down, some to the point of dormancy, at other times. Plants should be given a proper amount of water according to their needs, the rapidly growing plant obviously needing more water than the slow-growing or dormant one (see further details on p. 198).

5. The size of a plant and the size of the pot it is in will also determine water practices. Ideally, of course, a small plant will be in a small pot and a large plant in a suitably

Fig. 7-5. Certain plants have unusual needs. This *Cyperus alternifolius*, a marsh plant, grows best in saturated soils.

larger pot. The roots of the small plant are likely to be young, finely divided, highly absorbent, and actively growing. Water in a small quantity of soil will move quickly through it and drain through the bottom and evaporate from the surface rapidly. This movement will very nicely provide an adequate supply of both water and oxygen essential to young root growth. The surface/volume ratio of soil in a large pot will produce quite different conditions. The larger volume of soil will, of course, hold a greater quantity of water and will drain more slowly, and the surface exposed will be comparatively much less than in the small pot, where evaporation will accordingly be much slower. The soil in the large pot therefore will be wetter longer and will have less air in it. Assuming that a large plant is older than a small plant, its roots will be larger, longer, and more mature. Absorption of water by the plant will be somewhat reduced because the older tissues will have lost their ability to absorb; only the young tips carry on that function.

All of these conditions, then, add up to the fact that the large potted plant will normally require watering less frequently than the smaller one. A plant that is in a pot inappropriate to its size presents exceptions to this and special problems. These are discussed later on page 188.

6. The size of a plant is determined not only by its height and width but also by the number and sizes of its leaves. Plants transpire, that is, lose water, in proportion to their leaf area, and it follows that large, broad leaves will transpire a greater amount of water than will small, narrow ones. The size of the pot and the amount of water given should reflect these differences.

7. The environment in which a plant grows should also influence its watering. Humidity, air temperature, and air movement all affect transpiration rate and, therefore, water needs. Relative humidity refers to the amount of water that air can hold at a given temperature before condensation (rain) occurs. Warm air, with widely dispersed molecules, can accommodate more water vapor than can cooler air with more densely packed air molecules. Since transpiration is the diffusion of water molecules from the leaf into the air (see Chap. 1, pg. 21), we can conclude that the rate of transpiration will be higher under warm, dry conditions than under cool, moist ones. These conditions can be somewhat modified by air movements (or wind), since moving air will carry transpired or evaporated moisture away from the plant and allow continued loss of moisture from the plant into less saturated air.

Plants appreciate humid conditions, however, and will usually grow best if the realtive humidity is at least 30%. At this level, or higher, the plant is not likely to suffer excessive water loss leading to wilting, and the roots will be able to absorb water at a rate equal to or exceeding the rate of loss by transpiration.

For plants then, optimal conditions exist when temperature is moderate, the humidity level is high enough to prevent excessive transpiration but air circulation removes accumulated transpiration moisture so the flow of water and nutrients into the plant is not impeded, and absorption of water by roots is sufficient to prevent any disruption in the flow from root tip to leaf stomata.

To determine when a plant should be watered, observe both the plant and its soil. Soil dryness should be checked slightly below the surface since evaporation will dry the surface before the depths, especially in non-porous pots. Soil color is often a clue to its dryness since moist soils tend to be darker in color than dry ones. And, as noted before, except for sandy ones, soils tend to shrink away from the sides of the pot as they dry. Hefting the pot to check its weight can be a reliable test for water content too, if one is familiar with the weight of the potted plant. As soils dry the weight becomes less, and one can easily become accustomed to judging water content by weight.

An extreme method is suggested for the following situation. On occasion a plant will wilt much more quickly than you might expect or may simply lack the usual turgor you've become accustomed to in its leaves and stems. In these cases there may be too much or too little water available to the plant, but the only way to find out is to remove the whole mass of soil and roots from the pot for close examination.

1. First check the roots to make sure they are disease- and pest-free. If the roots show nodules or mushiness or emit a sour smell they may have serious problems (see Chapter 6 for possible pest problems).

2. If the roots appear healthy, observe the network of roots. As plants progressively fill their pots with roots, they may become potbound (p. 188) and need to be repotted when water can no longer be retained adequately in the pot.

3. Once you have unknowingly let a peat-based mix or a soil containing an abundance of peat moss dry out, the water will drain through rapidly without being absorbed by the dry moss. Bottom soaking of such pots will bring water effectively to the roots and rewet the soil mix.

4. If the condition of the roots seems satifactory and over-dryness is not apparent, check the soil itself. More than one seedling cactus or other small succulent has been overwatered when a large particle of peat moss or aggregate of vermiculite has found its way into a tiny pot in improperly screened sandy mix. The peat moss or vermiculite retains an overabundant quantity of water, much like a sponge, and the tiny rootlets soon suffer and perhaps die, which could kill the whole plant. Ever larger plants may be planted in soils containing such water-holding "sponges" which may damage the plant but not necessarily be fatal. Repot the plant into a well-screened, homogeneous mix.

Plants, like old friends, will signal their needs. Leaves that are limp or flabby, rather than firm and turgid, or whole plants that are wilting indicate a need for water, and one soon learns to detect that "certain appearance" that indicates it is time to water. Plants that are allowed to reach a wilting point before being watered will frequently lose some lower leaves even if the plant itself recovers. It is wise therefore to become familiar with the water requirements of your plants (see table at end of chapter).

There are probably as many methods of watering plants as there are gardeners. Watering on a set schedule is probably the poorest practice because it does not take into account all the environmental variables and the individual needs of the plants. Plants should not be watered a bit at a time either, even if at frequent intervals, because this approach will rarely provide sufficient water. Insufficient water in soils does not penetrate to the deeper levels and will limit root growth to the upper moist level, producing spindly, weak plants.

Most plants are best watered by thoroughly soaking the soil. Then allow it to drain well and do not water again until the soil is dry or nearly so, depending on the requirements of the species. Plants may be watered from the surface or from below by setting the pots in water. Gravity will pull water downward from the surface, and capillary action will move water upward and throughout soils. Plants watered from above are usually saturated when water runs out the drain hole. Plants watered from below are saturated when the soil surface is wet. In both cases allow 15 minutes or so for complete drainage of excess water before returning the plant to its customary location.

Plants should always be watered with warm or tepid water to avoid chilling roots and to speed movement of water molecules both through the soil and into the roots.

Generally plants are not very fussy about the water given to them but high concentrations of dissolved salts should be avoided. Water that is treated in a home water softener may cause problems when the salts used for softening precipitate in the soil in a flower pot. (Salts precipitated from fertilizers can also accumulate in the soils.) These soluble salts create two problems: a deficiency of water in the roots (even though the soil is moist) and a deficiency of nutrient ions (even though they are plentiful in the soil).

You will recall from Chapter 1, page 21 that water normally moves into roots from the soil by osmosis. However, a concentration of salts in the soil around the roots may cause the concentration of water to become greater in the roots than in the soil, and dehydration of the tissues will occur, causing the roots to wilt. The deficiency of water in the roots leads to disruptions in the metabolic activities required to bring nutrient ions across cell membranes into root tissues, and the roots starve. The result of interfering with the intake both of water and of nutrient ions is root dysfunction and eventual death.

In homes with soft-water systems, the water to outside faucets and to toilet tanks usually bypasses the salt treatment and can be safely used for plants. It should also be emphasized that pots with drain holes allow water to flow freely from the pot, carrying excess mineral (or fertilizer) salts from the soil; these salts accumulate in non-draining pots.

The Time to Water

The best time of day to water plants is early in the morning. Transpiration is energized by the sun and thus parallels the daily cycle of radiation received on the earth. In the morning, as the sun rises, leaves warm quickly and their temperature soon rises above the temperature of the surrounding air. At this time the stomata, which were closed during the night, will open, transpiration will begin, and the leaf will cool. The transpiration rate rapidly increases to a mid-afternoon peak and, since the rate of absorption by roots is influenced by the rate of transpiration, water and needed ions will be absorbed at a greater rate from the soil. Transpiration declines late in the afternoon and almost stops at night. Early-morning watering provides water for a plant's critical hours later in the day and prevents wilting and/or scorched leaves (see Figure 7-6). It also serves to prevent diseases in that it allows evaporation from leaf and soil surfaces throughout the day and thus minimizes moist places in which fungi and other disease organisms thrive.

Fig. 7-6. Plants accustomed to indirect light will frequently be scorched, or sunburned, if abruptly moved into direct sun. Scorched tissues are permanently damaged.

Overwatering

Plants are much less obvious about having been overwatered than about being underwatered. Oversaturation usually occurs in pots without drainage holes or when water is allowed to accumulate in the saucer beneath the pot. The results are slow, insidious, and usually fatal.

The roots begin to suffer from lack of oxygen as the excessive water forces the air out of the soil and occupies all of the pores between soil particles. This lack of oxygen leads to metabolic break-down similar to that discussed in relation to salt poisoning (page 185). Root hairs die and decay begins. The decomposition process uses the little remaining soil oxygen and produces excessive carbon dioxide, thereby increasing respiration failure by the roots, and more root tissues die.

Since the shoot of the plant can temporarily draw upon its food reserves, and water is not immediately deficient, it is usually not apparent that the plant is in trouble until it is too late to remedy the situation. Other than attempting to dry out the soil with increased warmth or raking there is little that can be done if the top of the plant has become flaccid or limp. Seldom will a plant lover feel more devastated!

Dormancy

Houseplants are selected for their ability to remain attractive year around in our homes, and very few of them undergo true dormancy. All plants, however, have seasonal periods of active growth alternating with a semi-dormant period when there is little or no new growth. In northern climates, this rest period usually occurs between Thanksgiving Day and Valentine's Day and coincides with our shortest days. During this time, water and fertilizer should be minimized (or eliminated for some plants, e.g. cacti). Cooler temperatures should be provided, if possible, and direct sunlight should be avoided. About mid-February the plants will almost suddenly appear more vital and new growth will appear. Normal watering and fertilizing may be resumed, and the plants can be returned to their customary locations.

Temperature

There are a few things more destructive to a plant than a sharp blast of cold air. Therefore plants should not be placed near a door that is opened to the outside in winter, by a window that is usually opened at night, or in front of an air conditioner. Even temperatures are highly desirable.

If plants must be on or near windows, be sure the sill is not ice-cube cold or stove-top hot. If so, place several layers of newspaper or other insulating material between the pot and the sill.

If the window glass tends to frost at night in the winter, place a piece of cardboard between the plant and the window. In sunny windows the heat and light from the sun can be damaging. Hang a filtering curtain or move the plant back out of danger.

Be especially considerate of juvenile plants since they are much more susceptible to temperature change than are the more mature plants.

Humidity

As pointed out earlier, most plants grow better in a humid atmosphere than in a dry one. Plants should not be placed by fans, heat registers, or air conditioners. If household humidity is very low, plants can be placed on trays of water-soaked pebbles with the pot bottom resting on the pebbles above the level of the water, or they can be misted with a spray bottle. Commercial humidifiers may also be installed.

If a plant shows signs of increasing dehydration due to dry air, a plastic-bag tent may save the day.

Air Circulation

The number-one cause of fungus diseases of plants is lack of air circulation around the plants. Still air allows fungus spores to remain in place long enough to germinate, and transpiration vapor which is not carried away provides them with their needed moisture.

Good air circulation, as discussed earlier, also increases transpiration rate and water uptake by the roots, producing more vigorous growth.

Slow-moving household air transports not only dust particles but droplets of cooking grease and furnace discharges. They settle on leaves where they clog stomata and reduce light penetration by producing a film over the leaf. Plants, especially large foliage ones, benefit immensely from an occasional sponging off with warm soapy water (not detergent) followed by a clean rinse. If shiny leaves are desired, they can be buffed dry with a soft cloth to bring up the natural oils and waxes, or one of the new commercial plant sprays may be applied. Never apply salad oil, milk, or household wax. They can clog stomata and be damaging in many other ways.

All other plants benefit from a period of growing outdoors when the weather permits. Since summer days outside duplicate the native environment more closely than houses do, plants produce new growth that is more vigorous and colorful than that achieved indoors. A few precautions are in order however.

Protect the plants from tearing winds and pounding rains.

Watch for oversaturation of soils from rain that fills pots.

Check for insects and diseases that may find your plants a gourmet treat.

Above all, do not expect plants accustomed to normal indoor conditions to tolerate intense sunshine with its accompanying heat and dryness. Find a shady spot where they'll thrive, not die.

Repotting

A plant announces when it needs to be repotted in many ways. A healthy plant will sooner or later grow so that it appears overly large in proportion to its container and forms a sheet of roots around the inside of the confining pot, causing the plant to become potbound. When such a plant is watered, the water flows rapidly down the root mass and inside wall of the pot and out the drain hole. Little or none of it finds its way into the compressed pores of the inner soil mass which therefore remains nearly dry. With such rapid runoff, the plant will absorb little of the water and nutrients essential to it and will begin to decline in vigor, and little or no new growth will appear even at the peak of the normal growing season.

Fig. 7-7. Root-bound plants fill the pots so tightly with roots that adequate water absorption by the soil cannot occur.

To rescue this underpotted plant knock the whole soil and root mass from the pot and loosen the roots gently. Some of the older excess roots may be trimmed off. Place the plant into a pot just one size larger than its former one, add some soil, and distribute the roots evenly as you continue to fill the pot nearly, but not quite, to the top. (Allow space so water may be poured into the pot easily.) This procedure of repotting into a larger pot is sometimes called **potting on**.

A small plant in a large pot is overpotted and will also fail to grow satisfactorily. The soils without roots extending rather uniformly through them tend to pack down so that they hold proportionately too much water and too little air, and nutrients leach excessively from them. The soil surface will often be dry because of its ample exposure and, as the little plant begins to weaken, we too often assume it needs more water and proceed to make a bad situation worse! The plant should be gently removed, keeping as many roots intact as possible. To determine a pot size suitable for the plant, imagine the size that would fit comfortably over and around the branches if they were squeezed together a bit. This is usually a pretty accurate way to estimate the proper-sized pot, even if the plant is somewhat deficient at present in its root mass.

Repotting, especially of leafy plants, often causes shock. The flow of water within the plant is interrupted, injuries may occur to both root and shoot tissues, and the root hairs responsible for absorption are torn off. Be sure there are no large air pockets in the soil and that the soil mass is uniformly moistened. Provide moderate temperature and indirect light until the plant recovers. A plastic-bag tent placed over the plant for a few days will increase humidity and reduce transpiration loss until the roots have reestablished themselves as efficient absorbing organs again.

FLOWERING HOUSEPLANTS

Houseplants that are grown for their flowers are considerably more challenging than are foliage plants because of their needs for specific temperatures, increased humidity, and more intense lighting, especially of certain wavelengths. There are a few rules of thumb that can be applied to flowering plants to improve success with them.

Getting Your Money's Worth

Since producing flowers is nutritionally demanding on plants, they should be more heavily fertilized than foliage plants. A standard houseplant fertilizer supplying nitrogen, phosphorus, and potassium is sufficient for most plants. Do not, however, exceed the amounts recommended on the package. Too much can produce excess foliage and reduced buds. Never fertilize dormant or semi-dormant or sickly plants of any kind. Plants need to be in reasonably good health to respond productively to the increased nutrients fertilizers provide.

Water, both in the soil and in the atmosphere, can be critically important to flowers since plants producing flowers require extra soil water for the increased cellular metabolism necessary for flower production and for the essential cation exchanges occuring within the soil. Humid air will help maintain turgid above-ground structures and prevent bud malformation or lack of buds that can occur in overly dry air.

Since photosynthesis is the means of obtaining the necessary food molecules for flower production, it follows that increased light intensity will facilitate that process. Direct sunlight should be avoided, though, because of the possibility of excessive heat and drying. Artificial lights are becoming very popular for flowering plants because they can provide high light intensity without damaging heat levels, and red wavelengths may be added for better flowering (see Chap. 6, p. 168.

Warm temperatures facilitate the movement of molecules within both the soil and the plant, increasing the rates of various activities to those sufficient for flower production.

Other than the direct effects of insects or disease on flower buds, there are two other common ways in which buds can be **blasted** (killed). A sharp cold draft or natural-gas fumes from a stove pilot light or other source will destroy a bud within hours. The bud will shrivel and become mushy. There is no remedy, but the gardener should be alert to save the next crop of buds from a similar fate. Blasted buds may also be a warning to be careful about keeping a pilot light lighted and to check for possible gas leaks, since it is the odoriferous methyl mercaptan added to the odorless natural gas for aid in detecting leaks that actually causes damage to plants and not the natural gas.

A Catalogue of Favorite Gift Plants

There are several categories of flowering plants that are of interest to the houseplant enthusiast:

1. Plants which can be maintained under nearly the same year-round conditions as foliage plants but produce blooms throughout most of the year (Fig. 7-8), or produce blooms at a specific time of year without a dormant period in the cycle (Fig. 7-9).
2. Plants which require a dormant period during which their above-ground structures die down completely (Fig. 7-10).
3. Flowering plants usually purchased in bud from commercial growers for immediate but brief personal enjoyment (Fig. 7-11), but which cannot continue to survive in the ordinary home after blooming.

Here are a few examples of each.

1. *Streptocarpus* and African violets (*Saintpaulia*) are close relatives that can be treated similarly. Provided with a rich soil, long hours of light, and regular feedings, they will bloom all through the year with only a slight slow-down from time to time.

Fuchsia, *Gardenia* and *Anthurium* are wonderful accent plants in a collection of houseplants since they too will bloom rather spectacularly during most of the year. Gardenias grow into tub-sized shrubs, anthuriums are window-sill-sized, and fuchsias will drape sensationally from a hanging basket. Strong filtered light, or for gardenias and fuchsias a few hours of sunlight, will promote flowering. Buds that drop off gardenias indicate a need for higher humidity or a dose of iron fertilizer.

Christmas cactus (*Schlumbergera*) is one of several genera of spineless cacti that resemble each other in both flowers and vegetative parts. All require a 3-month period of 16-hour nights with cooler temperatures to produce the bright pink-red blooms at the appropriate season. Commercial greenhouses now offer holiday cacti for Thanksgiving, Christmas, and Easter. New hybrids may eventually take care of the rest of the year. When not in bloom they can be cared for as normal houseplants.

Azaleas are late winter-early spring bloomers as pot plants. If bought in bud, keep them out of direct sun and avoid overheating, to make flowers last longer. They can sometimes be brought into bloom again

Fig. 7-8. Everbloomers. *Streptocarpus* hybrids of purple or white (left) are not as well known as their cousins, the African violets, but have recently been gaining in popularity. The long-lasting floral structure (inflorescence) of *Anthurium* (right) is most familiar with a red spathe but also occurs in white and yellow.

the next year with good care in an outdoor garden during the summer. They require acid soil mix.

Geraniums (*Pelargonium*) are plants for people with purple thumbs. Few plants will take as much abuse as a geranium and still produce flowers. Their needs are simple, and the plants will yield rootable cuttings even while in bloom. They bloom best when days are long, but under lights they can be brought into bloom again at almost any time.

Orchids are often looked upon as the royalty of the plant kingdom and many growers tend to specialize in them. The plants grow best under lights with regular, specific care. Species and hybrids are available to bloom at any desired time of the year, some of them remaining in bloom for as long as three months.

Aphelandra is a discouraging plant. It looks so attractive when purchased with its bright yellow flowers. But unless you can supply its demands to the letter, it will be an ugly green distortion a year later. It is not recommended for the novice.

2. *Amaryllis*, *Gloxinia*, and *Caladium* (Figure 7-10) grow from bulbs and tubers and bloom in later winter to early spring or summer. In fall they should be dried off and stored for the winter in a basement corner until after the first of the year. Some *Amaryllis* are forced into early bloom for Christmas but will bloom again the following year at their later normal time.

Spring-flowering bulbs of tulips and hyacinths are easy to force into bloom for winter enjoyment. Instructions for forcing are in Chapter 4. After the flowers fade, let the plants die back naturally and plant the bulbs in the garden in the fall for outdoor bloom next year.

3. The last group (Figure 7-11) includes plants that are highly desirable for their flowers but are either difficult or impossible to bring into bloom in the house again; however, some can be added to the garden when climate is mild enough.

Cineraria is a plant to enjoy and then throw out, since it is an annual grown from seed and will die after blooming. In comes in spectacular, almost unbelievable colors.

Calceolaria and *Cyclamen* are grown from seed in greenhouses for sale to northern plant lovers. They can be kept growing in the home but the results are poor and rarey worth the effort. In frost-free climates they can be added to the garden.

Poinsettia plants will retain their leaves long after the flowers are gone, and cuttings will produce new plants readily. If you have the space to maintain them and can provide

Fig. 7-9. The bright yellow flowers and "zebra-striped" leaves of *Aphelandra squarrosa* have attracted many a plant lover to try this difficult plant.

Fig. 7-10. Tulips and other hardy bulbs can be forced for early bloom and are welcome mid-winter gifts. The tender tubers of gloxinia, *Sinningia* hybrids, emerge from their dormancy in mid-winter to produce spectacularly colored flowers.

their necessary periods of darkness, they will bloom again but usually not as spectacularly as those produced commercially. In sub-tropical climates poinsettias become large spectacular shrubs in lawn plantings.

Easter lilies brought into bloom in early spring in greenhouses can be planted in outdoor gardens where they will bloom at their natural time in mid-summer in subsequent years. Most of the liles grown as Easter lilies are *Lilium longiflorum*, which is somewhat susceptible to winter cold in more northern areas.

Lastly, two well-loved plants, *Chrysanthemum* and *Hydrangea*, should be mentioned. In mild climates it is sometimes possible for these gift plants to bloom again in an outdoor garden. In the cold climates they are too tender for the sub-freezing winters. In addition, the commercially grown gift *Chrysanthemum* usually requires a 3- to 4-month period of short days, which it gets where it is cultivated in California, but which is not available before the snow flies in the northern states.

FERNS—PLANTS WITHOUT FLOWERS

Ferns belong to an ancient group of plants whose earliest members lived during the Devonian period of nearly 400 million

Fig. 7-11. *Poinsettia* plants forced for Christmas enjoyment and potted *Hydrangea* are lovely while in bloom but not usually worth trying to maintain afterward.

Fig. 7-12. (left photo) Representative ferns showing the locations and arrangements of the spore-containing sori which appear as black dots or regions on the under surfaces of fronds. In the Boston fern, *Nephrolepis*, the sori form a row along either side of the mid-vein of the pinna. (upper, right) Maidenhair fern, *Adiantum*, with sori along the apical edge of frond. The sori of the holly-fern, *Cyrtomium*, are randomly scattered over the lower surface. (center) The fronds (leaves) of all ferns uncoil from a crozier or fiddlehead that emerges from the base of the plant. The crozier of this treefern, *Alsophila cooperi*, is exceptionally large when compared to those of the smaller herbaceous species. (right) The staghorn ferns, such as this *Platycerium vassei*, produce two kinds of fronds (leaves). Sterile, shield-like fronds protectively cover the roots of these epiphytic plants and the fertile green fronds extend upward like forked "staghorns."

years ago. True ferns, similar to those of today, were the dominant plant life in the coal-producing forests of the Carboniferous Age that followed, but ferns as a group have declined considerably since. Many of the ferns existing today adapt easily to the poorly lighted areas of our homes, especially those that live naturally on forest floors.

You will recall from Chapter 1, page 37 that ferns and other spore-producing vascular plants compose a group with life cycles in which a sporophyte (spore-producing) phase alternates with a gametophyte (gamete-producing) phase. The large leafy plant that we readily recognize as a fern is the sporophyte plant. Fern leaves **(fronds)** arise from a rhizome as tightly coiled **croziers** or **fiddleheads** and uncoil until completely expanded. When the plants are mature, sporangia will be produced in clusters called **sori** on the undersides of the fronds in species-specific patterns. (Some sori look like scattered brown polka-dots, other like tiny crescents, still others form a brownish margin along either edge of a leaflet, etc.)

Each sporangium is pouch-shaped with one row of cells having thick walls to form the **annulus**, as shown in Figure 7-13. When the spores are ready to be shed, the drying annulus shrinks back until it tears sharply in an area of weaker cells called **lip cells**. This action catapults the dust-sized spores out of the sporangium so they can be transported on air currents to a new growing site.

In a moist environment the spore will germinate into a tiny, flat, heart-shaped gametophyte. Gametes will be produced in special structures on the undersurface, and, if conditions are moist and warm, fertilization will occur and a tiny new sporophyte plant will form from the zygote. When the sporophyte becomes able to live independently, the gametophyte dies and disintegrates. The entire life cycle of a fern is illustrated in Figure 7-13.

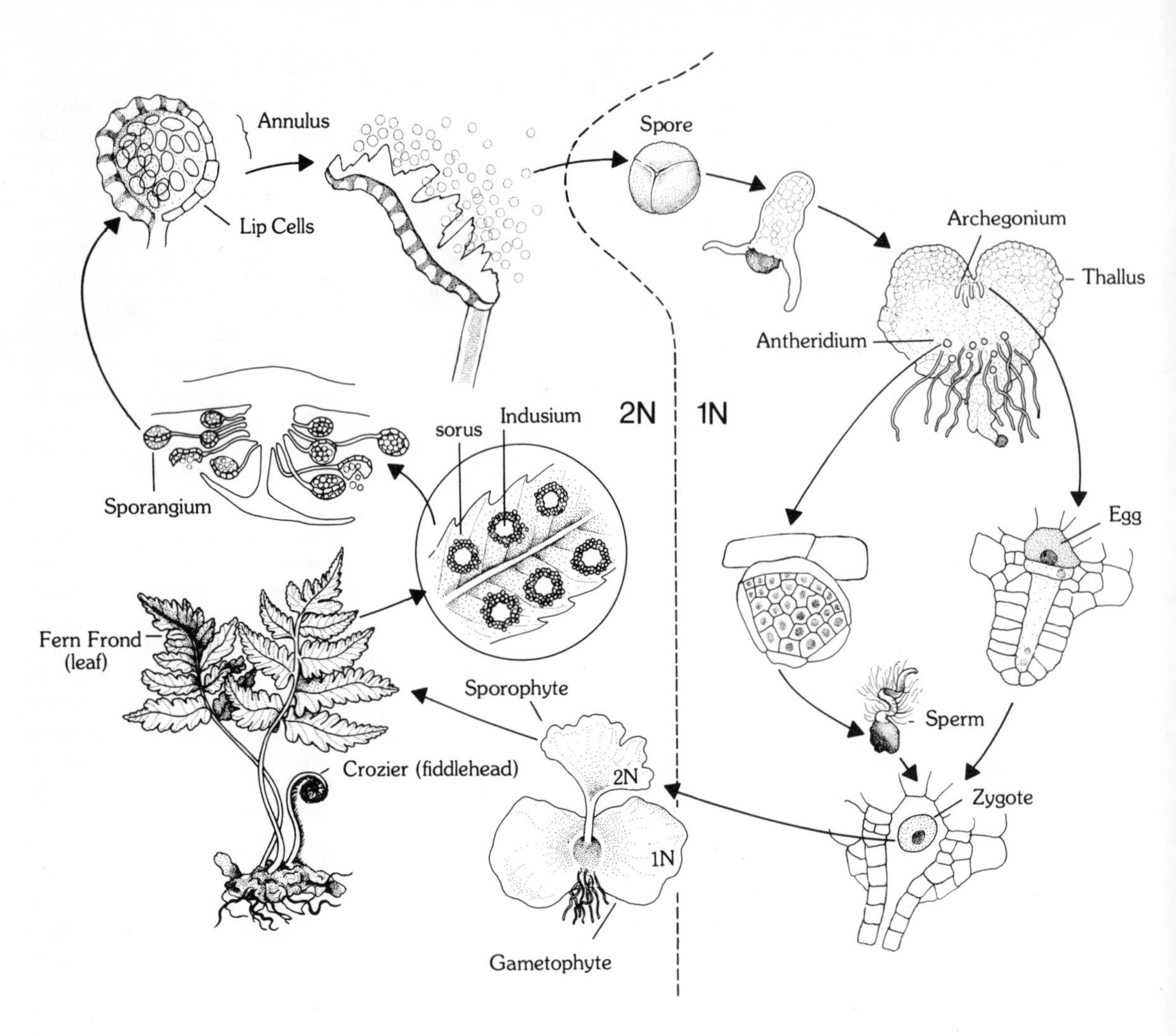

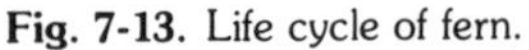

Fig. 7-13. Life cycle of fern.

Ferns From Spores

The first requirement for growing ferns from spores is patience. Some ferns germinate in three weeks, whereas others may take up to several months. As a rule ripe spores, freshly gathered (shake them into envelopes), require the least time for germination. Tiny green specks signal the onset of development of gametophytes. Four weeks to six months after that (depending upon species), the first leaves of the young sporophyte will emerge above the gametophytes like tiny scraps of green lace. It will be several years before the tiny sporphyte plant grows into an identifiable size, and therefore containers growing spores should be labeled as to species at the time they are sown.

Several methods of culturing spores are possible:

1. **Flower-pot culture**— Fill a 4″ clay flower pot with a mixture of equal

parts of sand, peat, and loam. Sterilize the filled pot by baking it in the oven at 180° F. for about 1 hour. Cool pot. Then set pot to soak in non-chlorinated water until both pot and soil are moist. Sift pores very thinly over the soil surface, stand pot in a saucer of water to keep moist, and cover with a glass jar or plastic-bag tent. If invading fungi or algae become a problem they can be controlled with methods described below.

2. **Wet-brick culture**—In a fashion similar to that above, a brick (non-glazed) or an upside-down clay flower pot (without soil) can be moistened and sprinkled with spores for germination.
3. **Agar culture**—Mix together:
   1 g. (1/4 tsp.) Vigoro or other similar plant food.
   12 g. (1 Tbsp.) standard agar (from drugstore).
   1 liter (5 c.) distilled water.
   Pour this medium into Pyrex flask or glass coffee maker. (Avoid metals that may react with fertilizer ions, etc. Glass is harmlessly inert.) Place in an autoclave or pressure cooker for 10 minutes at 15 lbs. pressure. Pour a layer of agar into sterile flasks, petri dishes, or even baby-food jars. When set, sprinkle spores over the surface and cover with a loose lid or cotton plug. Place under lights as recommended below and moisten if culture appears dry.

Bright sunlight can be damaging to germinating spores, and some actually require darkness. Good success can be achieved for most houseplant-type ferns with a 40-watt fluorescent tube above the spores for 14 hours a day.

Fertilization can be facilitated by occasionally flooding the cultures with water to aid in the movement of the sperm to the eggs, producing a greater number of sporophytes.

Damping off, caused by fungi, decimates the gametophytes and topples young sporophytes. It is not uncommon but can usually be controlled. Before sowing spores, first wipe the surface of the set agar with a cotton swab wet with a 10% bleach solution and allow the agar to dry briefly. If mold still appears after spores germinate, remove it with another bleach-soaked cotton swab, or prick out growths with a needle. The germinating spores may also be sprayed with a fungicide. If mold or algae should become rampant in the culture dishes, thinned-out gametophytes can be relocated on freshly prepared medium.

Culture and Maintenance of Sporophytes

As foliage plants, ferns are among the best. Their light requirements are not very demanding. Some species need more than others, but most do well where light is fairly low. They respond noticeably to low humidity with slow growth and brown leaf tips. However, even this browning does not affect their survival, and they will tolerate less-than-ideal humidity for a considerable period of time. Often ferns are grown in bathrooms, where humidity is likely to be higher than in the rest of the house.

A good standard potting medium or soil is adequate for most ferns. A few, such as the staghorn fern *(Platycerium)*, grow best epiphytically and can be attached to a slab of fir bark with sphagnum moss covering the roots.

Contrary to some common opinion, very few species of ferns grow in acid bogs. Most ferns prefer soils that are well drained, frequently rocky, and with a pH that ranges from neutral to slightly alkaline (such as that found in limestone areas). They require very little fertilizing, but, because they are foliage plants, they should be given a bit of extra nitrogen when the leaves begin to pale. A bit of bone meal worked into the soils at repotting time will supply beneficial nitrogen (see Chap. 2, p. 67) for quite a while. To produce a rich green color in ferns, apply a urea-form fertilizer or, more simply, add a teaspoon of household ammonia to a quart of watering water. (Urea-form fertilizers are urea-formaldehyde compounds that release nitrogen slowly and will not cause burning, even under adverse conditions.

Mature ferns are little bothered by insects or disease, but they can be drastically affected by insecticide sprays. Some of them are so sensitive that they can be totally defoliated by the slightest contact. Because of this, it is wise to remove ferns before treating your other houseplants.

Ferns can be most easily propagated vegetatively by either divisions or rhizome cuttings from many species. **Stolons** from the Boston fern *(Nephrolepsis)* and its relatives can be anchored onto a soil surface until they send out new leaves and roots. When the new plants are established they can be potted separately. And lastly, adventitious plantlets that form on the frond edges of the mother ferns, such as *Asplenium*, can be plucked off and rooted in sterilized soil.

HELPFUL HINTS FOR HEALTHY HOUSEPLANTS

1. Packaged potting soils that are rich black in color may not be rich in nutrients. They often have muck added to them so that they will look dark and rich. Such materials are too fine-grained, are leached of most nutrients, and are often acidic. Watch out for cheap and poor-quality soils.

2. White quartz chips on the soil surface will increas air warmth and circulation by reflecting solar heat and thereby reducing the abundance and vitality of fungus spores.

3. Mineral crusts on soil surfaces may be a natural accumulation from local water. These are generally fairly harmless. If they interfere with water percolation or become unsightly simply scrape off the top layer of soil and replace with fresh soil.

4. To avoid overwatering a hole-less pot, place a piece of plastic tubing or a straw vertically into the pot so that it rests just above the bottom of the pot. Add a layer of gravel around the end of the tubing and then fill pot with soil and add plant. Cut off tubing just above soil level. When plant is watered, excess water will drain into gravel layer and can be poured off through plastic tubing.

5. Fertilizer packages usually recommend the maximum amount that may be applied without damaging the plants (while increasing profits). Most plants will do well with a half or third of that amount.

6. When the soil in a large pot or tub is old and compacted and seems depleted of nutrients, you can avoid the difficult job of total repotting by removing about half of the soil from the top. Replace this with fresh soil of good structure and from which substances will soon leach downward, giving a needed boost to the soil below.

7. Whenever a plant is watered, no matter what the soil or moisture requirements, it should always have water poured onto it until it runs out the drainhole. Small amounts encourage surface roots and frequently lead to declining health of a plant.

8. Never fertilize bone-dry soils. Moisten soil a bit first so that salts don't concentrate and burn roots. With fertilizers, less is better than more.

9. Change fertilizers from time to time to get a better balance.

10. If clay pots dry out too fast, double-pot them. Place the clay pot inside another pot, preferably of the non-porous type. Pack the space between them with wet sphagnum moss to keep soil evenly moist.

11. Variegated leaves do not have as many photosynthesizing cells as solid green leaves do and therefore require greater light intensity to produce food. Place them closest to the light source.

12. To clean leaves, use soap, not detergent. Detergents contain wetting

agents that will remove too much of the natural oil and wax of a leaf's surface.

13. The thickness of leaves is a clue to their water needs. Thick leaves of a jade plant need water less frequently than the thin leaves of a *Coleus* plant.

14. Your plant has red or purple leaves instead of green? Don't be fooled! Such plants don't need any less light for photosynthesis than an all-green plant, and may even need more. Turn the leaves over and check the undersides and the veins for green color. Often the lighter green is simply masked by deeper reds and purples.

15. For planter boxes, don't fill them with soil and plants. Put potted plants into the planter and fill the spaces between them with sphagnum moss, bark, or other packable material. If a plant then needs special care it can be easily removed or even replaced without digging up the whole planter. Plants can be rotated for even growth more easily too.

16. Don't plant succulents and cacti in the same bowl. Some succulents need more peat in their soils than cacti can tolerate. Cacti require cooler winter temperatures with little or no water. Most succulents can tolerate warmer house temperatures year around and need a bit more water than cacti even in the winter, if they are not to shrivel (see Chap 8, p. 218).

17. Since the individual plants in a grouping usually have individual requirements, you can take the guess-work out and improve their health by writing each plant's needs (page 198) on a label and placing it in the pot. Then follow your own instructions.

18. If a plant is doing poorly, for any number of reasons, first correct the most obvious problem. Then give it a chance to recover. Give it good light intensity, but not direct sun; keep it moist, but go easy on the water; increase the humidity as much as possible; don't fertilize; provide a gently cool temperature rather than a warm one. Then wait until new growth appears before doing anything further.

19. For some "different" houseplants, take cuttings from *Coleus*, *Petunia*, *Impatiens*, *Browallia*, fibrous *Begonia* or other garden annuals before frost in the fall. They can be rooted and, if given good light intensity, will produce colorful results all winter.

20. There is absolutely no danger in sleeping in a room filled with plants, unless you are allergic to them. The small amount of oxygen taken in and carbon dioxide given off is insignificant. If anything, the increased humidity is probably beneficial to human occupants.

21. Be sure to rotate or turn plants periodically so that all sides are evenly exposed to light for more even growth.

22. There has been an "explosion" in houseplant sales in recent years and prospects are even brighter.

Before 1971 plant sales were between $4-24 million wholesale

In 1971 plant sales were at $38 million wholesale.

1972 plant sales were at $48 million wholesale.

1973 plant sales were at $68 million wholesale.

1974 plant sales were at $111 million wholesale.

1975 plant sales were at $187 million wholesale.

1976 plant sales were at $260 million wholesale.

Where will it go from here?

Table 7-1. Soil and Water Requirements for Houseplants

A. **Epiphytic soils**
Fir-bark or osmunda-fiber media

Plant	Water
Bromeliads in general	1
Columnea	1
Orchids in general	1
Staghorn ferns	1

B. **Sandy soils**
1 part loam, 1 part peat, 2 parts sand/perlite

Plant	Water
Agave	2
Cactaceae in general	2
Crassulaceae in general	2
Echeveria	2
Faucaria	2
Panadanus	2
Sansevieria	2
Sedum	2
Most other succulents	2

C. **Loam soils**
2 parts loam, 1 part peat, 1 part sand/perlite

Plant	Water
Aloe	2
Euphorbiaceae in general	2
Haworthia	2
Kalanchoe	2
Pelargonium	2
Amaryllis	3
Araucaria	3
Beaucarnea	3
Beloperone	3
Campanula	3
Chlorophytum	3
Cissus	3
Clivia	3
Coleus	3
Codiaeum	3
Dieffenbachia	3
Dracena	3
Ficus elastica	3
Gynura	3
Hedera	3
Hibiscus	3
Hoya	3
Ixora	3
Kalanchoe	3
Lantana	3
Neanthe	3
Oleander	3
Oxalis	3
Poinsettia	3
Phoenix	3
Plectranthus	3
Pittosporum	3
Podocarpus	3
Peperomia	3
Schefflera	3
Scindapsis	3
Solanum pseudo-capsicum	3
Syngonium	3
Tolmiea	3
Tradescantia	3
Zebrina	3

1 = Drench and drain. Allow soil to dry between waterings.
2 = Water sparingly. Allow soil to dry thoroughly between waterings. Reduce water in winter.
3 = Drench soils thoroughly. Allow to dry briefly between waterings.
4. = Keep soil in pot evenly moist but allow surface to dry slightly before watering.
5.= Keep constantly moist with no dryness.
6 = Tolerant of or require saturated soils. May be grown in water.

Watering program is dependent on factors of soil, light, temperature, and growing season, and these factors should be considered when applying the information above.

Asparagus	4
Aphelandra	4
Aspidistra	4
Araucaria	4
African violet	4
Aglaonema	4
Abutilon	4
Acalypha	4
Aralia	4
Codiaeum	4
Cordyline	4
Calathea	4
Campanula	4
Chlorophytum	4
Coleus	4
Capsicum	4
Dracaena	4
Ferns in general	4
Ficus pumila	4
Fuchsia	4
Gynura	4
Hedera	4
Hoya	4
Impatiens	4
Iresine	4
Maranta	4
Mimosa	4
Monstera	4
Piper	4
Philodendron	4
Peperomia	4
Plectranthus	4
Pilea	4
Rhoeo	4
Syngonium	4
Schefflera	4
Tolmiea	4

D. **Humusy soils**

1 part loam, 2 parts peat, 1 part sand/perlite

Abutilon	4		
Aralia	4		
Begonia rex	4		
Citrus	4		
Dieffenbachia	4		
Episcia	4		
Ficus	4		
Gynura	4		
Monstera	4		
Oxalis	4		
Pilea	4		
Peperomia	4		
Philodendrons in general	4		
Setcreasea	4		
Tradescantia	4		
Wandering Jew in general	4		
Zebrina	4		
Adiatum		5	
Aphelandra		5	
Begonias (wax)		5	
Caladium		5	
Columnea		5	
Cyclamen		5	
Episcia		5	
Frittonia		5	
Gardenia		5	
Gesneriads in general		5	
Gloxinia		5	
Maranta		5	
Aglaonema			6
Cyperus			6
Spathiphyllum			6
Syngonium			6
Philodendrons in general			6

Table 7-2 Light Requirements for Indoor Plants

Plant	Light
Abutilon	1
Agave	1
Bougainvillea	1
Cactaceae in general	1
Campanula	1
Capsicum	1
Cereus	1
Citrus	1
Gloriosa	1
Hibiscus	1
Ipomoea	1
Ixora	1
Lantana	1
Pelargonium	1
Solanum pseudocapsicum	1
Amaryllis	1 2
Begonia semperflorens	1 2
Begonia (tuberous)	1 2
Beloperone	1 2
Bromeliaceae in general	1 2
Browallia	1 2
Codiaeum	1 2
Coleus	1 2
Crassulaceae in general	1 2
Cyanotis	1 2
Dyckia	1 2
Echeveria	1 2
Euphorbiaceae in general	1 2
Faucaria	1 2
Fuchsia	1 2
Hoya	1 2
Kalanchoe	1 2
Malpighia	1 2
Mimosa pudica	1 2
Neanthe	1 2
Oleander	1 2
Oxalis	1 2
Palmaceae in general	1 2
Pandanus	1 2
Phoenix	1 2
Sedum	1 2
Achimenes	2
African violet	2
Caladium	2
Clivia	2
Coffea arbica	2
Columnea	2
Episcia	2
Euonymus	2
Gloxinia	2
Haemanthus	2
Haworthia	2
Hypoestes	2
Iresine	2
Saxifraga tomentosum	2
Streptocarpus	2
Acalypha	2 3
Aloe	2 3
Araucaria	2 3
Asparagus	2 3
Beaucarnea	2 3
Begonia rex	2 3
Dizygotheca	2 3
Dracaena & Cordyline	2 3
Ferns in general	2 3
Ficus	2 3
Gynura	2 3
Hedera helix	2 3
Impatiens	2 3
Pellonia	2 3
Peperomia	2 3
Plectranthus	2 3
Setcreasea	2 3
Spathiphyllum	2 3
Tradescantia	2 3
Zebrina	2 3
Wandering Jews in general	2 3
Aralia	3
Ardisia	3
Davallia	3
Fatsia	3
Fittonia	3
Maranta	3
Pilea	3
Rhoeo	3
Tolmiea	3
Aglaonema	2 3 4
Chlorophytum	2 3 4
Cissus (Rhoicissus)	2 3 4
Dieffenbachia	2 3 4
Sansevieria	2 3 4
Scindapsis	2 3 4
Syngonium	2 3 4
Aspidistra	3 4
Monstera	3 4
Philodendrons in general	3 4
Podocarpus	3 4
Schefflera	3 4

1 = Direct sunlight
2 = Strong light or filtered sun
1 + 2 = Can tolerate winter sun but protect from summer sun
3 = Good light, no sun
2 + 3 = Increase intensity for flowering or vigor
4 = Lowest light levels
3 + 4 = Can be seriously injured by sunlight or very intense light
2 + 3 + 4 = Best foliage plants, most tolerant of variable light, more vigor with increased light

Table 7-3. Houseplant Ailments

Effect on Plants

PROBLEM		Effect on Plants
Light		
	Too little:	Spindly growth, small leaves, plant may lose leaves. Variegated leaves turn green, green leaves pale.
	Too much:	Leaves pale or bleached often curl. Sun can scorch upper leaf surfaces (see Fig. 7-8), tips can be burned by heat from lamps. Growth retarded, leaves small.
Water (check drainage)		
	Too little:	Plants wilt, leaves curl, roots crowd upper soil level. Leaf tips and edges turn brown, green leaves may drop.
	Too much:	Plant becomes mushy but wilted, roots rot, lower leaves yellow and drop. Corky or blistered spots on lower leaf surfaces appear as edema, especially with high humidity and low light. Water spots (whitish deposits left after water dries) on leaves can damage tissue within leaf.
Fertilizer		
	Too little:	Little or no new growth; plant pale or discolored, spindly, may wilt. Roots dead or dying. Signs of specific nutrient deficiencies may appear.
	Too much:	Failure to flower, brown leaf tips, parchment-like leaf texture, leaves drop off. Loss of healthy color, weak new growth. Roots burned, then rotten. (Never apply fertilizers to dry soil!)
Air		
	Too dry:	Flower buds drop, tips and edges of leaves brown, leaves wilt or drop. Dry air encourages spider mites.
	Too moist:	Edema if combined with overwatering and low light. Excessively humid air encourages fungus and bacterial diseases.
	Too warm:	Wilting and yellowing, brown spots on leaves, edges dry. Reduce humidity and light intensity.
	Too cold:	Leaves wilt and curl, may discolor and/or drop. Stems limp. Brown tips, margins or edges on leaves. Flower buds drop.
Diseases		
	Fungi:	Whole stems, leaves, roots wilt or are mushy, turn limp, brown and die. Corky spots on stems; darkened, irregular leaf spots. Furry or powdery white or gray mold spots on stems or leaves. Seedlings fall over at soil level (damping off).

Control: Increase air circulation, reduce humidity. Avoid wetting foliage, especially at night. Remove badly infected leaves. Treat with fungicide.

Bacteria: Streaks on green stems irregular, angular, yellow, spots between teins or leaves. Whole plant becomes limp, appears distressed and dies without showing specific symptoms.
Control: Increase air circulation, improve general growing conditions. Remove and destroy infected leaves. Destroy dying plants. Sterilize soil (see Chap. 5, pg. 128), pots and tools before using again.

Viruses: Concentric ring spots on leaves, stunted growth, distorted and mottled leaves. Whole plant affected.
Control: Destroy plant and soil. Sterilize pots and tools.

Note: Diseases are often transmitted by insects, mites, and splashing water, so pest control and care in watering are very important. Always isolate individual plants with pest or curable disease problems until cured.

Insects and Related Pests	Appearance of Pest	Effect on Plants
Mealy Bugs:	Pinhead-sized insects forming cottony clusters in leaf axils and along veins. No noticeable movement. Sucking insects.	Plant growth stunted. Leaves pale, mottled, plant weakened.
	CONTROL: Hand-pick or dab with alcohol. Treat with malathion.	
White Fly:	Eggs and larvae stationary on undersides of leaves. Adults flit about when disturbed. Sucking insects.	Plant growth stunted. Leaves become mottled, yellowish, distorted, drop. Whole plant weakened.
	CONTROL: Wash or pick off leaves infested with eggs and larvae. Strips of fly paper will trap some adults. Treat with malathion, pyrethrins, or rotenone.	
Aphids:	Several generations of long-legged, small-sized sucking insects crawl about in masses on undersides of leaves and at stem tips. Light to dark colors.	Plant growth stunted. Leaves yellow, curl and distort, drop when dead.
	CONTROL: Hose off with strong spray. Treat with malathion, pyrethrins, or rotenone.	

Pest	Description	Damage
Scale:	Round, oval or elongated tiny bead-like sucking insects. Movement not readily apparent. Adults with shell-like covering. Light to dark in color.	Stems and leaves mainly infested. Leaves yellow, spotty, drop prematurely. Stems weakened, may die.

CONTROL: Wash plants with soapy water, flick off scale with fingernail. Treat with malathion, pyrethrins, or rotenone.

NOTE: All of the above insects secrete honeydew derived from the sweet phloem sap that they suck from the plants with their long proboscis (sometimes 3 x the length of the insect body). Honeydew trails that are clear, glistening, and sticky are left behind as the insects crawl about or may drop onto leaves and stems below. Sooty mold spores often germinate and grow on the honeydew, producing a charcoal-black coating on the stems and leaves. The mold can screen out light necessary for photosynthesis but does not otherwise harm the plant directly.

Pest	Description	Damage
Spider Mites:	Sucking arachnids. Appear as tiny specks on leaf undersides, build fine webs in leaf axils that can be seen against the light.	Leaves become mottled, poorly colored, distort or twist, finally drop.

CONTROL: Wash plants with soapy water, destroy webs. Treat with malathion, diazinon, or specific miticide.

Pest	Description	Damage
Thrips:	Dense black pepper-like colonies on leaf undersides. Give off unsightly, varnish-like excrement. Insects rasp soft tissues and suck juices.	Leaves become flecked and silvery, later papery, and drop. Flowers often attacked.

CONTROL: Treat with malathion or diazinon. Dust stored corms with malathion powder or flowers of sulfur.

Pest	Description	Damage
Springtails: and Black Gnats:	Small but easily visible insects jumping or flitting over soil surface. Rarely cause damage.	Maggots of gnats may injure roots, and springtails may injure tender parts of plants. Damage usually slight.

CONTROL: Replace upper layer of soil or repot plant into new soil. To treat, soak soil and spray surrounding areas with malathion solution.

NOTE: The first step in the control of almost all pests and diseases is to hose off or wash off plants with mild soap and water. Rinse well. Use botanical sprays or pesticides only if problem is severe and/or plants are very numerous. Some plants may be killed or damaged by chemicals, e.g., ferns, *Hibiscus*, *Cyclamen*, *Poinsettia*, etc. If in doubt, spot-test first by applying a small amount of pesticide to a small area of a leaf.

8. HANGING BASKETS AND TERRARIUMS

Hanging baskets and terrariums are two charming ways of growing plants in which the container plays a special role. The terrarium's greatest hazard is being overwatered, while the hanging basket, in contrast, is most likely to suffer from being underwatered. Each container has its own special requirements, and it is important to understand these needs if plants are to be grown successfully in them.

HANGING BASKETS AND HANGING POTS

Literally any potted plant can be a hanging plant if it is suspended by some means. However not all plants will be attractive if viewed, as most hanging plants are, from below. Appropriate plants are those that extend, at least in some manner, well beyond the confines of the pot and preferably cascade abundantly over the sides.

Plant Choices

The most attractive plants for hanging containers are those that do not have erect, self-supporting stems. Some species of this type will spread over the soil surface or hang gracefully over the edge of the pot or basket. Others may be successfully twined around or tied to a support of some kind. In addition to these, plants that produce plantlets at the ends of long stolons that extend out of the pot are also excellent choices for hanging containers. Suggestions for using these and other plants are discussed in the following sections.

Creepers and Danglers

In the accompanying list of suggested plants on page 213 it will be seen that many of those that creep or naturally seek growing places from which they can dangle are among the most commonly grown houseplants. Generally these are plants whose demands are easily met, have attractive foliage, and grow rather rapidly. They are all easily propagated by stem cuttings so that a potful may be obtained quite readily in a short time.

In dry air these plants tend to lose their older leaves so that the stems become bare with tufts of younger leaves at the growing tips. To cope with this problem cut off the vigorous tips and root them to start new plants, or bend the tips back into the pot, secure them to the soil until rooted, and then cut away the old bare stems. In either case higher humidity should be provided to prevent recurring loss of leaves.

The "creep-and-dangle" plants should be planted one species or variety to a pot and allowed to fill it profusely. Because of their spreading nature it is not a good idea to combine two or more different ones together in a pot since they will soon crowd each other, and the conflict of leaf colors, shapes, and textures may not be attractive. Any creep-and-dangle plant, however, may be added around the periphery of a pot to accompany a plant of an upright-growing habit that otherwise might not be suitable alone in a hanging container.

Fig. 8-1. Hanging plants on planter poles creates an attractive setting in any home.

Climbers

The climbing plants are of a different nature. They will grow most vigorously when allowed to cling to or twist around an upright support. Tendrils and adventitious roots extending from nodes are evidence of their climbing natures. In their native habitat, the tropical philodendrons and *Scindapsis* are found rooted in the jungle floor and twine around giant trees nearly to the tops. As houseplants both varieties are commonly grown supported upon a pole in tubs.

Scindapsis may also be allowed to dangle over the sides of a pot. It will branch readily if the tips are pruned off; they may be used for new cuttings. Most of the philodendrons are too large for the usual hanging basket, but the small and very popular *P. oxycardium* can be exceedingly attractive in a suspended pot, especially if given a framework to twine around in at least a semi-upright manner.

Hoya is another plant that will be at its best if allowed to climb upon a framework. This can be arranged with a small trellis within a hanging basket, and then parts of the plant can be arranged to drape gracefully over the edge so that the beautiful waxy flowers will show to their best advantage.

Nephthytis or *Syngonium*, like *Philodendron*, prefers to twine about an upright support, but the growing ends may be allowed to hang over the pot edge or be woven between the suspension cords to create a luxuriant viney look.

Philodendron, *Scindapsis*, and *Syngonium* all produce some adventitious roots that can be encouraged to grow into the soil for increased support of the plant.

Cissus or grape ivy is a profusely branching plant that arches gracefully with long stems. If planted against a wall, it will climb verticallly holding itself in place with wispy tendrils. Without support it will flow fountain-like out of its pot. To maintain a suitable full form, any branches that begin to form tendrils should be pruned back since these branches tend to become spindly. Pruning will encourage new growth from lateral buds on the remaining stem, and the plant will maintain a luxuriant appearance.

Plant Producers

Both *Chlorophytum* (Fig. 8-3) and *Saxifraga* produce long stolons from which plantlets appear. These may be left on the plants almost indefinitely, producing plants with hanging extensions wafting in space around the pots. *Chlorophytum* also produces small white flowers along with the plantlets on its stolons.

Tolmiea does not produce stolons but rather gives rise to adventitious plants at the bases of its leaves. These may also be left on the plant to produce a plant appearing to explode with plantlets extending well out over the rim of the pot.

Foliage Spray Plants

There are a number of other foliage plants that, while not notable for special hanging characteristics, are nevertheless entirely suitable for hanging containers. In general these include any of the plants that produce long leafy stems from a central root clump. *Asparagus*, many ferns, *Hypocyrta*, and some of the begonias are of this nature. Others such as *Coleus* and *Gynura* produce some varieties that may branch profusely enough from a main stem to produce extended portions that reach well beyond the perimeter of the pot.

Cacti and Succulents

Although we tend to think of succulents as primarily upright plants, these too can be found in forms suitable for hanging. Improved varieties of *Schlumbergera*, *Rhipsalidopsis*, *Zygocactus*, and their hybrids (see Fig. 8-4) have recently brought about renewed horticultural interest in these plants. These are the spectacular "holiday" cacti that fill a pot with jointed green stems

Fig. 8-2. Climbing plants *Scindapsis aureus* (left), *Philodendron oxycardium* (right), and *Hoya carnosa* (bottom).

Fig. 8-3. *Chlorophytum comosum "Vittatum"* (above) with many plantlets produced on extended stolons from the parent plant. *Saxifraga sarmentosa* (below) spreads over the ground by means of long runners that produce plantlets at the tips. In a hanging pot, as shown here, the plantlets dangle gracefully from the parent plant.

terminating in brilliant pink-red flowers.

Other cacti that may be used in hanging pots include night-blooming cereus *(Hylocereus)* which has fleshy clambering 3-angled stems. Its magnificent foot-long white flowers last a single night and close by sunrise. Another night-blooming climber is *Cryptocereus*, with smaller reddish flowers. The rat's-tail cactus *(Aporocactus flagelliformis)* is really much more attractive than its name would imply. The pendulous round, spiny stems dangle profusely over the rim of the pot and produce striking pink flowers even on plants that are very young. The species of *Rhipsalis* are epiphytes found hanging from trees and are often called mistletoe cacti, apparently in reference to the similarity of their habitats.

The family Crassulaceae includes many succulent members that produce densely branching stems that spill attractively over the edge of a pot or that have dangling stems with closely placed pairs of leaves that appear to be threaded on the stems. The common names necklace vine and string-o'-buttons aptly describe these latter kinds. An epiphytic creeper is *Kalanchoe uniflora* that fills a pot quickly because of its ability to root at its stem nodes. Its leaves are round and fleshy. *Sedum morganianum*, another member of the family, produces long tassels of spindle-shaped leaves and is popularly known as burro tail.

Most unexpectedly, the milkweek family includes a genus of climbing succulents. The species of *Ceropegia* have reduced narrow or heart-shaped leaves that have inspired such names as rosary vine, needle vine, and string-of-hearts. Charles Darwin was intrigued by these plants, and they became the subject for studies he pursued in trying to understand the nature of vining plants.

Flowering Plants

Foliage plants are of course a year-around reward, and the plants discussed thus far can be relied upon to maintain and

Fig. 8-4. The "holiday" cacti (Schlumbergera sp.) and the dangling branches of *Sedum morganianum* are succulent plants suitable for a hanging basket.

improve their vegetative appearance with time. There are other plants, though, that may be grown for their flowering branches which will hang gently, displaying colors both brilliant and pastel. Some of them are varieties or species of plants that are better known as upright growers. *Pelargonium peltatum*, *Begonia pendula*, and some of the low-growing azaleas are examples. The others in the list of recommended plants (p. 213) either have pendulous stems which hang naturally or are plants with profuse and extended branches that fill a pot to overflowing. *Impatiens* and *Petunia* are notable for the latter trait.

The recent increase of interest in plants has encouraged growers to try new imaginative plantings, among which have been vegetables in hanging baskets. Miniature tomatoes, cucumbers, small pepper plants, some herbs, and even loose-leaved forms of lettuce can be attractive as hanging plants right up to the time of harvest.

Fig. 8-5. *Aeschynanthus javahicus* is one of the popular "lipstick plants" that produce clusters of bright red flowers in the spring and summer.

PLANTING AND CARING FOR HANGING PLANTS

Basically there are two kinds of hanging containers. Any normal clay or plastic flowerpot can be suspended by wire, cord, chain, or macramé supports. As with all pots it is important that the pot have a drain hole. To prevent damage from dripping water, a saucer can be included with the pot, another potted plant can be aligned below to catch the drips, or absorbent material or newspapers can be temporarily placed below the pot and removed when the dripping stops.

Hanging ordinary pots from poles, brackets, and window frames is an excellent way to utilize all of an indoor growing area at minimum cost. There is, after all, about the same amount of light coming through a window near its top as there is at its sill, and the imaginative gardener will take full advantage of this "free" light.

Commercially available hanging baskets are open frameworks of wire, wood slats, or plastic mesh and are best suited to outdoor situations. To hold soil, the basket should be lined with sheet moss or an inch (2+ cm.) -thick layer of fibrous sphagnum moss. These are available at local garden stores. Both of these materials require a prolonged soaking before using, preferably overnight. Drain and gently press out excess water to minimize dripping. Line the inside of the basket with the moss so that a solid wall is formed up to the top edge. (See Fig. 8-6). Fill to within an inch of the top with slightly moistened soil for the plants to be grown.

For plants other than cacti and succulents, the soilless mixes are excellent for hanging pots and baskets since they hold water well and may not need to be watered as frequently as normal soil mixes. Soil mixes, however, are also quite acceptable. Cacti and succulents, requiring soil dryness between waterings, should not be grown in a sphagnum- or sheet-moss-lined basket, since these materials dry out with the soil

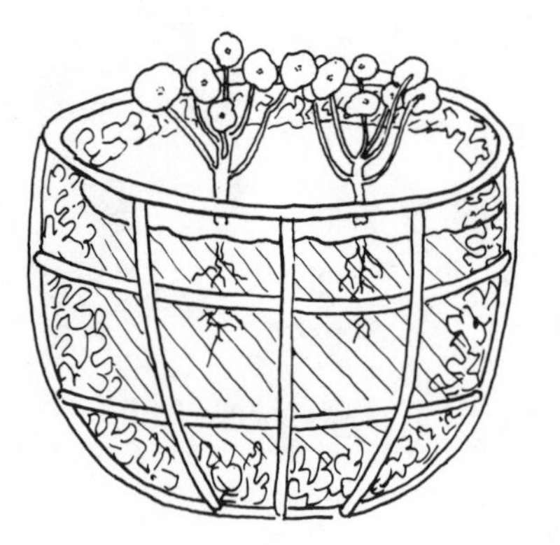

Fig. 8-6. Planting of hanging basket. See text.

and soon lose their attractiveness. (Regular pots, preferably clay, are more appropriate for plants in sandy soils.)

Position the plants in the soil so that they can most easily grow outward over the edge of the basket. To settle the soil firmly against the lining material and to reduce the possibility of air pockets, set the whole basket in a tub of water to soak thoroughly until the soil surface is wet. The water in the tub must not be allowed to flow over the surface of the soil since this may dislodge the plants or wash soil away. Drain until the dripping stops and hang the basket in place. Subsequent watering may be done by pouring the water onto the soil in the basket. Small spreading plants or seeds can be poked into the outer surface of either type of moss lining material so that their roots will contact the soil within. Among the many possibilities are *Helxine* (baby's tears), *Episcia*, *Ficus pumila* (creeping fig), *Pilea depressa* (creeping Pilea), small herbs such as *Thyme*, and flowering sweet *Alyssum*. When these grow they will add even more interest to the basket.

Hanging pots and baskets will require more attention than most potted plants. Since warm air rises, the temperature is usually a bit higher and humidity a bit lower at the hanging height than it is at normal table or floor levels. This means that soils will dry out faster, leaves will transpire greater quantities of water, and plant growth will be greater. Such plants should be watered more frequently and often require extra fertilizer. Baskets that are lined with sheet moss (not fibrous sphagnum moss) may become a lively green again, as they were in nature, if they can be kept moist enough.

Adequate humidity is essential for some of the plants if they are not to lose their older leaves or have their growth retarded. Frequent misting, grouping plants closely together, and avoiding direct sunlight will increase the humidity. Also try to avoid placing plants directly over or near heat registers or next to frosty windows during cold weather. Hanging planters always look best if the plants within are full and abundant with leaves and/or flowers.

Finally, be sure all supporting structures are strong enough to hold your hanging pot or basket when it is thoroughly wet. A planted hanging pot can be very heavy, and if its supports should break it could be expensive to repair the damage and replace the pot and perhaps the plant. If support problems do arise, there are some very attractive pedestals (see Fig. 8-7) available upon which a cascading plant can be safely placed.

Fig. 8-7. Plant pedestal.

Fig. 8-8. The famous Hanging Gardens of Babylon, built in 605 B.C., were one of the seven wonders of the ancient world.

HANGING GARDENS OF BABYLON—HISTORICAL PRECEDENT?

When we have successfully produced a few attractive hanging containers it is not hard to imagine that we have recreated our own Hanging Gardens of Babylon in miniature. Interestingly enough, though, the legendary gardens were actually gardened terraces (see Fig. 8-8) rather than hanging canopies as so often suggested. As one of the Seven Wonders of the World, the gardens were built by Nebuchadnezzar about 605 B.C. for one of his wives who was homesick for her verdant Persian homeland. The Hanging Gardens were an elaborate combination of a terraced hillside and a seven-story brick temple tower or ziggurat representing a symbolic link between heaven and earth. They covered about 3-4 acres with elaborate arches and fountains and soil-filled columns. The plants were primarily blossoming and shade-giving trees such as larch, cypress, cedar, acacia, mimosa, aspen, birch, chestnut, poplar, and palm, all native to northern Mesopotamia. The cooling shade of the trees and the dancing waters of the fountains must have created an impression of loveliness on all who saw them, and despite the total destruction of the gardens later by invaders and earthquakes their fame continues even to today.

PLANTS FOR HANGING CONTAINERS

Creep or Dangle

Plectranthus sp. — Swedish ivy
Senecio mikanioides — German ivy
Pellionia sp.
Tradescantia sp. — Wandering Jew
Zebrina sp. — Wandering Jew
Ficus pumila — Creeping Fig
Helxine soleirolii — Baby's Tears
Hedera helix — English ivy
Setcreasea purpurea — Purple heart
Peperomia fosteri, P. prostrata, P. scandens — all small peperomias
Pilea nummulariifolia — Artillery plant
Hoya minata, H. compacta — Wax plant
Cymbalaria muralis — Kenilworth ivy
Asparagus asparagoides — Baby smilax
Gibasis geniculata — Tahitian bridal veil
Fittonia verschaffeltii — Nerve plant
Muehlenbeckia complexa — Wire vine

Climbers

Hoya argentea, H. carosa — Wax plant
Philodendron oxycardium — Heartleaf philodendron
Scindapsis aureus — Pothos or devil's ivy
Cissus rhombifolia — Grape ivy
Syngonium podophyllum —Nephthytis or five-fingers
Gynura "sarmentosa" — Purple passion vine

Plantlet Producing

Chlorophytum sp. — Spider lily
Saxifraga sarmentosa — Strawberry geranium
Tolmiea mensiesii — Piggy back plant
Episcia dianthaflora, E. cupreata — Flame violets

Sprays

Asparagus densiflorus cv. *sprengeri* — Asparagus 'fern'
Fern in general
Coleus rehneltianius — Trailing queen
Gynura aurantiaca — Velvet plant
Hypocyrta nummularia — Goldfish plant
Begonias in general

Succulents

Sedum morganianum — Burros tail
Ceropegia woodii — String of Hearts
Sedum sieboldii — October plant
Crassula perfossa — Necklace vine
Cyanotis kewensis — Teddy bear plant
Rhipsalis sp. — Mistletoe cactus
Any of the 'holiday' cacti — Christmas, Thanksgiving, Easter

Flowering Plants

Begonia pendula, B. solananthera
Campanula isophylla — Star of Bethlehem
Fuchsia sp.
Impatiens — many varieties
Thunbergia grandiflora, etc. — Clock vine
Pelargonium peltatum — Hanging geranium
Columnea sp.
Plumbago auriculata
Aeschynanthus sp. — Lipstick vine
Petunia sp.
Nasturtium
Alyssum

TERRARIUMS

Growing plants in glass containers originated with the Wardian cases of the last century. Dr. Nathaniel B. Ward, a London botanist, introduced the idea of glass-topped boxes as a means of growing plants that could not otherwise be forced to grow in the smog of Victorian England. Wardian cases (see Fig. 8-9) of many sizes and styles became popular accompaniments of the parlor palms and rubber trees of the

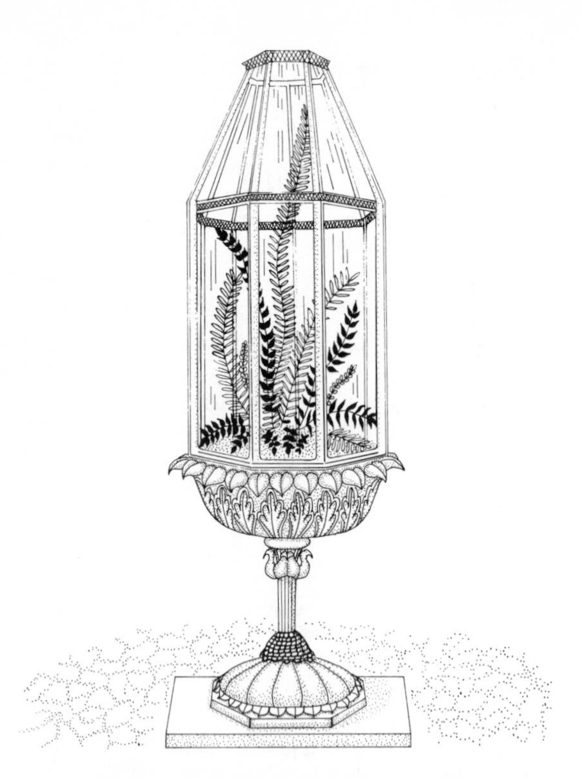

Fig. 8-9. A typical Wardian case developed by Dr. Nathanial B. Ward in the late 1800's.

late 19th and early 20th century homes.

Today we use the term terrarium to designate any kind of glass or plastic container enclosing growing plants. The terrarium may be simply an attractive way of displaying a favorite houseplant, or it may be a more elaborate structure utilizing a number of plants tolerant of the terrarium environment and of each other so that a scene in miniature is created.

The basic elements of a terrarium are a container, plants, gravel and soil, and water. Properly combining these elements will produce an eye-pleasing garden that will require a minimum of care and last for a reasonable length of time.

Containers

To select an appropriate container, one must decide where it is to be placed and what size it should be. Most garden supply houses sell attractive terrariums, sometimes with built-in lights. However the possibilities for other types of containers are endless, and the only real equipment is that the glass or plastic should be clear or only lightly tinted. Darker colors do not permit satisfactory growth. To spark the imagination, here are a few suggestions.

- Bottles — wine, milk, olive, etc.
- Cider jugs
- Jars — jelly, candy, cookie, pickle, etc.
- Bowls — dessert dishes, mixing bowls, serving bowls, etc.
- Covered glass dishes — Pyrex baking ware
- Glass coffee makers
- Brandy snifters
- Decanters
- Glass domes — used for memorabilia, cheese, etc.
- Flasks, beakers, and other chemical glassware
- Aquariums and fish bowls — a perfect way to make use of a cracked or leaky one.

Most containers are used upright, but there is no reason why they can't be placed on their sides if they provide enough growing height in that position and are cradled if they tend to roll.

The terrarium may be covered or uncovered depending on the size of the opening and the kind of plants being grown. For plants that thrive in humid air, such as ferns, *Peperomia*, baby's tears, etc. (see list on page 220), the container should be almost completely closed. A jug with a small opening will probably not need to be capped, because water loss will be insignificant. A wide-mouthed jar or bowl, however, will suffer rapid dehydration and should be covered with its lid or a pane of glass or plastic wrap. If a glass dome is used, the circumference of the dome should fit closely around or within a straight-sided base or bowl in which the plants are growing. For desert or succulent plants the air must not be very humid, so an uncovered wide-mouthed container, such as a fish bowl, is ideal. In all cases it is a good idea to check the odor of the air in the terrarium periodicaly for indications of rot or mold, so that you may detect their presence before they cause serious damage.

Planning the Terrarium

There are four usual types of terrariums: tropical, woodland, bog, and desert. The kinds of plants appropriate to each type will determine the soil to be used, the amount of water needed, and the most suitable container.

Always use small plants initially. They will grow to fill the space available and then

Fig. 8-10. Foliage plants in a terrarium. (A) *Selaginella*, (B) *Sansevieria*, (C) fern, (D) *Asparagus*, (E) *Philodendron*.

begin to crowd each other. At this point they will need to be pruned back, if possible, or taken out and replaced.

Arrange plants so that each has enough soil and space and will receive an adequate amount of light when the terrarium is in its place. Don't allow any plant to deprive others of their needs.

Foliage plants are the most easily maintained in a terrarium, but flowering plants may also be used. *Episcia, Columnea, Kalanchoe, Oxalis* and others that produce small flowers more or less continuously throughout the year are preferred over those that produce overly large flowers approximately once each year. These more elaborate productions usually look out of proportion in relation to the plants and container and can create a disease-attracting decayable mass that may be difficult to remove when the flowers fade.

Terrarium soils should be sterilized to minimize weeds, insects, and diseases. The container itself should also be thoroughly cleaned with a soap and bleach solution. Rinse thoroughly and allow to dry in the sun for further disinfecting.

Do not use soilless potting mixes. Since a terrarium usually does not have any means of draining its excess water, the soilless mixes tend to be overly wet for many plants. These mixes also necessitate regular feeding which, besides being an unnecessary bother, can produce accumulations of fertilizer salts and subsequent damage to plant.

A tropical terrarium should contain plants that like moist soil, high humidity, indirect light, and warm temperatures. *Aglaonema, Philodendron, Plectranthus, Peperomia,* and *Pilea* are a few tropical genera that will do well together. A number of other species normally used as foliage houseplants may also be used.

A standard potting mix of one part sand, one part loam, and one part peat moss is suitable for most tropical and subtropical terrariums. Fertilizers should not be used, since terrarium plants need to be kept small. If the soil is endowed with the elements usually expected, the plants should grow well enough without fertilizers. If deficiencies become obvious, add only enough to alleviate the problem.

A woodland terrarium consists of plants native to our temperate deciduous forests. These are the small plants that we tend to overlook while taking in the majesty of our forest trees on a woodland walk. They survive nicely at normal room temperatures when enclosed in the moist atmosphere of a covered terrarium, although cool tempertures will increase the life span of most plants. *Selaginella* and *Lycopodium* are ground-hugging club mosses; ferns and mosses and liverworts are other spore-producing plants that may be included in the terrarium. Partridge berry *(Mitchella repens)*, rattlesnake plantain *(Goodyera pubescens)*, pipsissewa *(Chimaphila umbellata* and *C. maculata)*, and wintergreen *(Pyrola elliptica)* are ideal additions that are difficult to find and are **on the protected plant lists** of many states. These plants, however, may be ordered from biological supply houses or nurseries that grow terrarium plants and have special permits to collect and sell protected species for purposes of propagation. The U.S. Fish and Wildlife Service carefully controls collecting activities to avoid endangering rare species, and professional collectors respect the rules since they also protect their livelihood. (Wild plants suitable for terrariums may be ordered from most wildflower nurseries or from the following dealer who specializes in plants and supplies for terrariums: Arthur E. Allgrove, 279 Woburn St., N. Wilmington, Mass. 01888.)

Try to duplicate the woodland soils to which the plants are so well adapted. Their tiny roots reside in the upper humusy layer of the forest floor, so a mixture of leaf mold and forest humus, with a small amount of loam added, will be best. Keep the woodland terrarium out of direct sunlight but well lighted.

A bog garden is an ideal setting for the fascinating insectivorous plants. The soggy natural soils in which these plants grow are deficient in certain elements, especially nitrogen. The plants capture live insects and other small invertebrates and secrete digestive enzymes to obtain the nitrogen available from the animal tissues. A medium of sphagnum moss with a bit of sand and humus will mimic bog soils to the satisfaction of most of these plants. Never use fertilizer, and don't supply the terrarium with insects. The plants will do quite nicely without their grisly dinners as long as there is a bit of humus in the soil.

The carnivorous or insectivorous plants (see Fig. 8-11) can be found in bogs

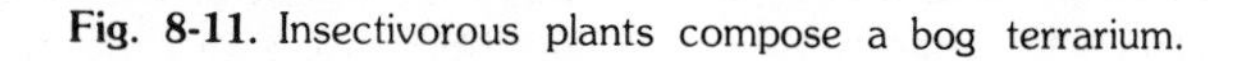

Fig. 8-11. Insectivorous plants compose a bog terrarium.

throughout the world. Bogs are open, sunny, unshaded habitats, so the plants will need at least four hours of sun each day. The Venus Fly Trap *(Dionaea muscipula)* is native to the bogs and coastal pinelands of North and South Carolina, and the cobra plant *(Darlingtonia)* is found only in the coastal ranges of nortern California and Oregon at an elevation of about 5000 feet (1500 m.) Species of the other insectivorous plants grow in peaty or marly bogs from Florida to New Foundland. Pitcher plant *(Sarracenia)*, sundew *(Drosera)*, and butterwort *(Pinguicula)* may frequently be found in the wild, but **state laws prohibit removing them**. Many garden catalogues, however, offer some if not all of these plants which are especially grown for commercial purposes.

The desert terrarium will be occupied by plants adapted to sunny areas with low humidity and sandy soils. The container should be wide-mouthed and never covered, to allow for free circulation of air.

There is a great temptation to plant a desert terrarium with any combination of the plants known to prefer such habitats. Cacti are leafless, usually spiny natives of the American deserts. A succulent may be a plant of any of a number of families that is vegetatively adapted to a habitat low in available water, not necessarily a desert. Their needs will therefore differ from those of the spiny cacti, and these generally make poor companions if planted in the same terrarium.

The requirements for growing spiny cacti differ from those for succulents in the following important ways:

Spiny Cacti

Direct sun at least 6 hours/day.
Sandy soil mix.
Preferred winter temperature—45° F. (7° C.).
Winter water requirements—None from October through February.
Summer water requirements—Allow soil to become thoroughly dry before rewatering. (Blooming cacti should be kept in soil that is well watered but well drained.)

Succulents

Most direct sun, but a few (the aloes) require strong indirect light.
Sandy soil mix with additional humus.
Preferred winter temperature—50-60° F. (10-16° C.)
Winter water requirements—Every 3-4 weeks for most species.
Summer water requirements—Can tolerate dry soils, but most species appreciate more frequent watering than cacti can tolerate.

If you wish to include both cacti and succulents in the same terrarium, some of the plants can be left in their pots and covered to the rim so that they may be removed during the winter months for separate maintenance.

Planting and Maintaining the Terrarium

Begin by placing a layer of gravel on the bottom of the terrarium for drainage (see Fig. 8-12). A layer of nylon net or section of nylon stocking over the gravel will

Fig. 8-12. A rocky terrarium with mosses and baby's tears creating interesting combinations of natural greens and grays. The layers of gravel and soil are clearly visible.

prevent the soil from working downward into the gravel. A thin layer of peat moss will be acid enough to discourage the roots of many plants from growing into the gravel and they will retreat back into the soil area. Add soil, appropriate to the plants to be used, to about 1/3 the height of the container. The soil should be slightly moist so that holes can be dug in which the plants can be planted.

Position the plants and tamp the soil to secure them in place. Attach a short piece of rubber hose to the end of a funnel and water the terrarium by placing the hose against the walls of the container and allowing the water to run gently down. If there is any soil on leaves or stems, hold the hose so that the water will wash it off. DO NOT OVERWATER! To avoid overwatering, add only a bit of water at a time. Allow it to disperse through the soil, and then, if necessary, add a bit more until the soil is properly moistened. If condensation on container walls or guttation (see glossary) on the plants seems to be excessive, the soil may be too moist, and the terrarium should be opened to allow it to dry out a bit.

If overwatering accidentally occurs, uncover the terrarium, rake the soil surface to enhance evaporation, and provide warm temperatures to increase the movement of water molecules into the air. In severe cases turn the container on it side, allow the excess water to collect on the low side, and pour or spoon it out. If this is not feasible, try blotting the surface with facial tissues or paper towels. Or set an electric hair-dryer nearby to warm and dry the air, increasing evaporation from the soil. *Don't* aim blower at plants, however, or you will burn them.

If the container has a small opening it will be necessary to contrive or purchase special tools for planting. One of the handiest is a mechanic's pick-up tool from any hardware store. This is a long flexible tubular wire, available in several lengths, with a claw at one end that is opened and closed by finger controls at the other. The claws are usually too sharp to use on delicate plant tissue without protection but strong enough to grip razor blades (for pruning), brushes (for cleaning), or any other sticks, tampers, spoons or forks, etc. Other useful tools are extra-long laboratory forceps, bamboo sticks or dowels in various diameters, and hot-dog roasting forks. Other tools can be rigged by wiring the needed piece to a bamboo stick. Sets of terrarium tools may also be purchased in several price ranges.

When placing plants through small openings, roll the leaves carefully within each other and drop into a pre-dug hole into the soil below. Firm the soil gently with a bamboo stick. The soil around plants in terrariums need not be quite so protectively covered as for normal potted plants because of the high humidity and lack of drying air currents.

Molds and mildews are the most serious disease problem a terrarium may face. The warm, moist atmosphere is very conducive to these fungus growths. Cleanliness is the best deterrent. Sterilized soil, a clean container, and clean plants and tools will minimize problems. Prompt removal of spent flowers and dead leaves will help. If fungus does appear, remove badly infected leaves, provide circulating air to decrease humidity, and dust plant with fungicide powder. The powder will be unsightly but can be brushed away in a few days when the disease seems under control.

One last step in producing a terrarium may be the addition of items other than plants. There is a wide variety of rocks, figurines, ceramic animals, and other accessories available for terrariums. These may or may not enhance the appearance. Many of them are just too "cutesy" and are easily overdone. In any case, select accessories that are appropriate to the theme of your terrarium. Mushrooms and snails, for instance, would be out of place in a desert terrarium but perfectly acceptable in a woodland terrarium.

PLANTS FOR A TROPICAL TERRARIUM

Aglaonema sp. — Chinese evergreen
Philodendron oxycardium — Heartleaf philodendron
Plectranthus sp. — Swedish ivy
Peperomia caperata, P. prostrata — Emerald ripple, etc.
Pilea depressa, P. microphylla — Artillery plant, etc.
Chamaedora elegans — Parlor palm
Pellionia pulchra — Satin Pellionia
Cissus striata — Miniature grape ivy
Ficus pumila, F. sarmentosa — Creeping fig, etc.
Fittonia sp. — Nerve plant
Hedera helix — English ivy
Hypoestes sanguinolenta — Pink polka dot
Iresine herbstii — Blood leaf
**Episcia dianthiflora* — Lace flower vine
**Columnea crassifolia*
**Oxalis* sp. — Wood sorrel
**Exacum affine* *Plants which flower.

PLANTS FOR A WOODLAND TERRARIUM

Selaginella and *Lycopodium* — Clubmosses
Most mosses and liverworts
Cladonia cristatella — British soldier lichen
Cladonia rangiferina —Reindeer moss (Lichen)
Pellaea rotundifolia — Cliffbrake fern
Empetrum nigrum — Crowberry
Polygala paucifolia — Gaywings
Mitchella repens — Partridge berry
Goodyera pubescens — Rattlesnake plantain
Chimaphila umbellata, C. maculata — Pipsissewa
Pyrola elliptica — Wintergreen

PLANTS FOR A DESERT TERRARIUM

Senecio herreianus — Green marble vine
Hatiora salicornioides — Drunkard's dream
Almost any of the Crassulaceae, e.g., *Crassula, Cotyledon, Echeveria, Kalanchoe, Aeonium, Sedum.*
Haworthia
Aloe variegata — Partridge breast aloe
Aloe vera — Medicine plant
Almost any of the Cactaceae

PLANTS FOR A BOG TERRARIUM

Dionaea muscipula — Venus flytrap
Sarracenia purpurea — Pitcher plant
S. flava — Trumpet plant
S. psittacina — Parrot pitcher plant
Drosera intermedia — Sundew
D. filiformis — Dewthread
Pinguicula vulgaris — Butterwort
Darlingtonia californica — Cobra plant

9. OUTDOOR GARDENING

utdoor gardening can be one of the most satisfying of avocations or a most frustrating, discouraging, and demoralizing one. However it is really quite easy to make it the former and avoid the latter if the prospective gardener keeps three simple rules in mind:

1. Start with good-textured, well-draining soil.
2. Provide lots of sunshine and water.
3. Don't bite off more than you can chew. (Translation: Don't plant more garden space than you can adequately care for.)

In order to follow these requirements it is important to locate the best garden areas, plan them appropriately, analyze their soils (modifying them if necessary), and be prepared to provide water. After these important steps are taken then you can shop through the garden catalogues choosing from the delightful gems you will find there.

GETTING STARTED: LOCATING AND PLANNING THE GARDEN

Outdoor gardens present problems seldom, if ever, encountered when gardening indoors. You have to start with the soils already on the property. Sun and shade areas will vary as the seasons change, and wind, rain, hail, frost, and other climatic factors will have to be dealt with. If a garden site is already established, you may be reluctant to abandon it to relocate for more sun or better soil, preferring instead to try remedial measures. If you are establishing a new landscape it is very important to consider the location of your garden carefully before it is turned over so that it will be most favorably situated for the kinds of plants to be grown. Fruit trees and vegetable gardens need sunshine nearly all day, so over-hanging trees or shady sides of buildings should be avoided. Flowering annuals and perennials too are primarily sun lovers but can tolerate periods of shade, and a few do indeed actually prefer the shady spots. Sun-loving plants are best grown where they will receive at least 4 to 6 hours of sun, and this period should be continuous, because, for plants, two or three hours in the morning and again in the late afternoon do not add up to the same prolonged intensity of light (see Chap. 6), and the plants may do poorly.

A level, well-draining piece of ground will make the best garden, although a slight slope can be tolerated if the soil holds water well and rains will not cause erosion of the bare soil between plants. Where the ground is very steep, raised planting beds or terraces should be used to produce level areas for gardening (additional suggestions for vegetable gardens are on page 237).

Finally, in locating gardens, place them so that tools and water are conveniently close by. Pathways and gates wide enough to accommodate wheelbarrows, rototillers, and any other necessary equipment should be provided. Most average garages can store the necessary tools and equipment for gardening along with the family car, but where a large vegetable garden, for instance, is planted some distance from the house, a nearby tool shed may be more convenient. "Little Red Barn" (Fig. 9-1) buildings of wood or metal are commercially available at reasonable prices.

Fig. 9-1. This little red barn is a convenient place to store gardening equipment.

A good-quality garden hose of ample length with a good nozzle is one of the gardener's best friends. A bit of self-psychology here should quickly convince you that you will be more willing to water your garden if you don't have to struggle with leaky, short, stiff hoses. And it is an established fact that the best gardens are well-watered ones.

When you have selected suitable spots for vegetable and/or flower gardens, you should make an honest estimate of the time, energy, and money you are willing to spend on your gardens. It's so easy to plan big! Whether you are putting in new gardens or reclaiming old ones, start small and enlarge later if you see that you can handle it. Recruiting gardening help is nearly an impossibility these days, aside from the fact that you deprive yourself of some very satisfying activities.

Make a scale drawing of the gardens so that you can estimate the numbers of plants that will be needed (flower garden plans are in Fig. 9-2, vegetable gardens in Fig. 9-12). Your drawings will be referred to many times while you establish your gardens and will prevent time- and energy-consuming errors.

One of the joys all gardeners share is the arrival of the first flower and seed catalogue on a cold winter day when gardens and thoughts of gardening have been dormant since the first snow or heavy winter rain. Now you can vicariously enjoy your gardens while you peruse the colorful pages, selecting the plants and seeds to be planted in the spring. Catalogues are invaluable in planning gardens since they provide necessary information about sun requirements, general cultural needs, blooming times, heights, colors, and other information of use to the gardener about the selected plants. Seed companies do, after all, want your plants to be successful so that you will continue to buy from them in succeeding years.

Always keep a record of the order you place, complete with any plant-variety names. Often, plants that are ordered turn out to be either spectacularly successful or dismal failures, so you want to be sure either to order that variety again, in the case of the successes, or avoid it next year, in the case of the failures. Invariably, unless you have written down the name and can retrieve it from wherever you mindlessly put the list last spring, you will find that the new catalogues seem totally different and identifying the plant is quite impossible.

It is of course possible to label the plants in the garden with plastic or wooden labels or the seed package itself, but don't count on any of these surviving the weather, animals, children, and the garden hoe through an entire summer.

Most skilled gardeners keep a gardening record or diary in which they note names of plants, where and under what conditions they grow best, and any special treatments needed. It's also a good idea to record mistakes and failures so you can avoid repeating them. It does take a bit of extra effort to record your gardening activities, but it can be a great help in planning

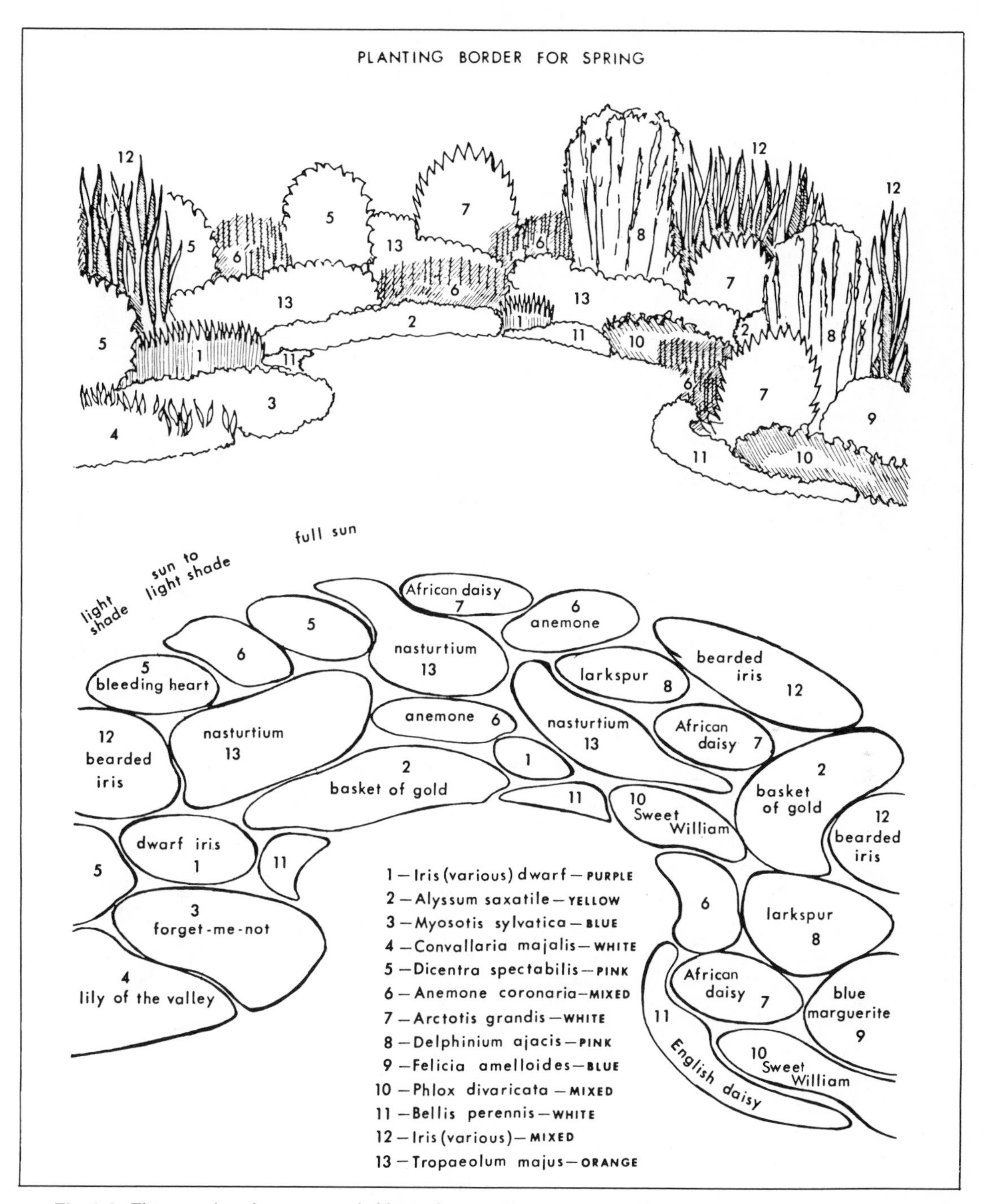

Fig. 9-2. Three garden plans to provide bloom during spring, summer, and fall in the flower border. (from Kramer, **Gardening and Home Landscaping**, courtesy Harper & Row)

PLANTING BORDER FOR SUMMER
bachelor buttons 9
1 hollyhock
painted tongue 10
baby's breath 3
1 hollyhock
4 African marigold
baby's breath 3
2 peony
painted tongue 10
2 peony
7 zinnia
2 peony
7 zinnia
painted tongue 10
African marigold 4
3 baby's breath
French marigold 6
sweet alyssum 8
10 painted tongue
2 peony
9
6 French marigold
sweet alyssum 8
moss rose 5
8 sweet alyssum
moss rose 5
1—Althaea rosea—PINK
2—Paeonia (various)—ROSE
3—Gypsophila paniculata-WHITE
4—Tagetes erecta—ORANGE
5—Portulaca grandiflora—MIXED
6—Tagetes patula—YELLOW
7—Zinnia elegans—MIXED
8—Lobularia maritima—WHITE
9—Centaurea cyanus—BLUE
10—Salpiglossis sinuata—MIXED

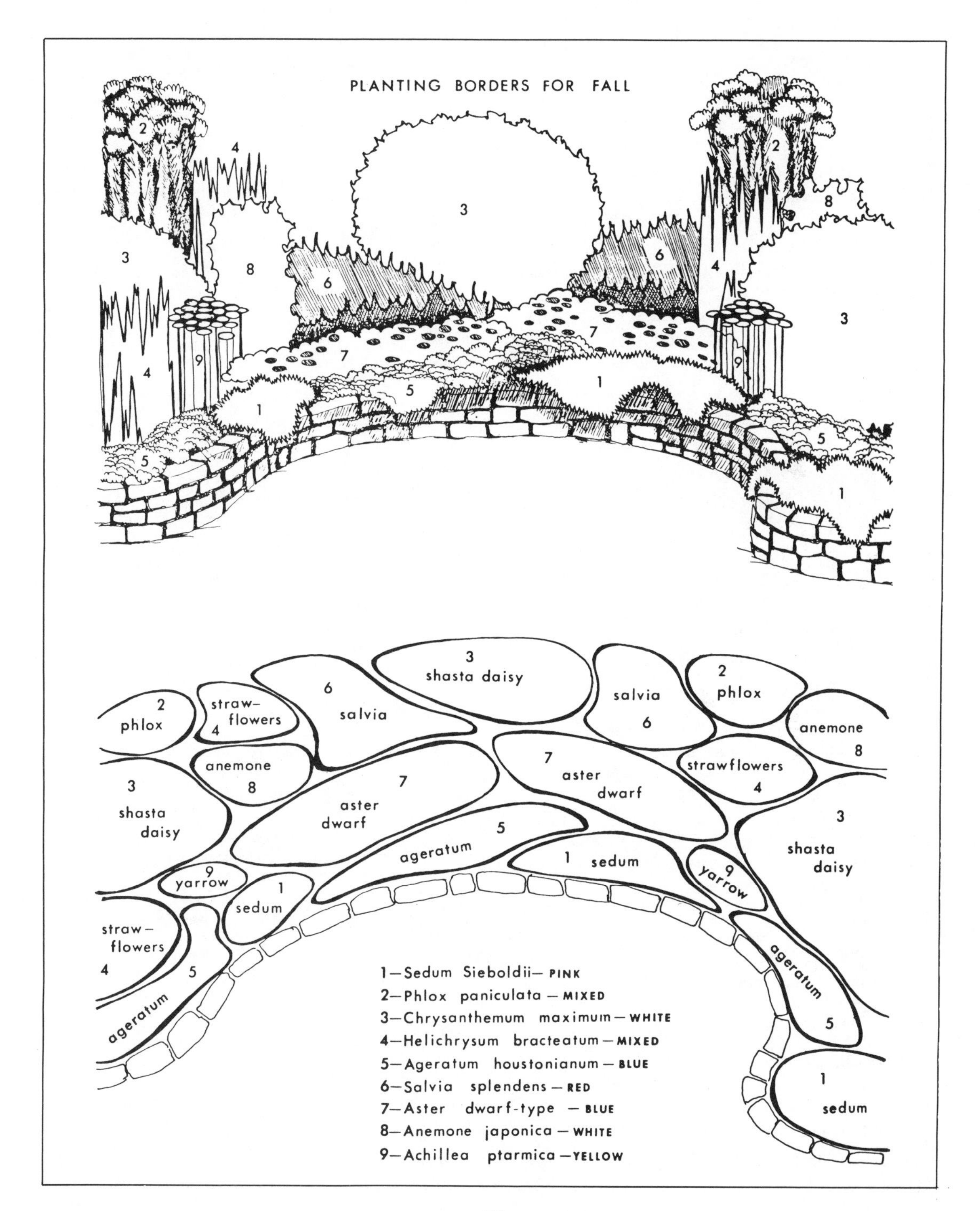
PLANTING BORDERS FOR FALL
3
shasta daisy
2
phlox
straw-
flowers
4
6
salvia
salvia
6
2
phlox
anemone
8
anemone
8
7
aster
dwarf
strawflowers
4
3
shasta
daisy
7
aster
dwarf
3
shasta
daisy
5
ageratum
1 sedum
9
yarrow
9
yarrow
1
sedum
straw-
flowers
4
5
ageratum
ageratum
5
1
sedum
1—Sedum Sieboldii— PINK
2—Phlox paniculata — MIXED
3—Chrysanthemum maximum — WHITE
4—Helichrysum bracteatum — MIXED
5—Ageratum houstonianum — BLUE
6—Salvia splendens — RED
7—Aster dwarf-type — BLUE
8—Anemone japonica — WHITE
9—Achillea ptarmica — YELLOW

next year's garden, and it's great fun to read of your early struggles years later when you are a more accomplished gardener.

Select flowering plants of appropriate heights and compatible colors. Flower beds, which are usually viewed from all sides, may have taller plants in the center and low ones around the periphery. Flower borders against walls or fences are viewed from one side only, and plant heights generally graduate from tall plants in back to shorter ones in front (Fig. 9-4). Flower gardens that present a succession of bloom are usually the most satisfying, and blooming times should be carefully noted in the catalogues.

Vegetable plants are best planted in well-defined rows for easy maintenance. Ideas for vegetable garden plans are available from state extension-service bulletins, 4-H clubs, and sometimes local garden stores, as well as many gardening magazines and books (also see Fig. 9-12). Most novice gardeners underestimate the amount of produce which relatively few plants will yield. Good judgment concerning family tastes and needs means less wasted effort as well as fewer uneaten crops.

Fig. 9-3. Plant Growth Regions of the Unites States. (Shaded areas are mountainous.) Regions having approximately similar growing conditions for the same elevation are shown by numbers within the heavy border lines. There are also certain relationships between the growth regions shown on the map for typical native plants. For example:

Plants from regions 12 and 13 are usually successful in regions 24, 25, 27, and 28. The reverse is rarely true.

Plants from regions 27 and 28 are usually successful in regions 1, 2, 24, 25, and 29.

Plants from regions 1, 2, and 4 will usually grow in regions 2, 29 and 30.

Plants from regions 5 usually will succeed in regions 31 and 32, and the reverse also seems true.

Plants from regions 20 usually do well in region 29.

Nine times out of ten, such moves as these will be successful, always provided the plant is grown in a situation similar to its natural habitat. (USDA)

Fig. 9-4. A well-planned garden will have tall plants placed behind shorter plants and will have pleasing color combinations.

If garden space is limited, or if you don't need or wish to dig a separate vegetable garden, some vegetable plants can be interspersed in a flower garden or even planted in containers or hanging baskets. Some of the catalogues feature selected vegetables for such special situations.

PREPARING THE SOIL

After the plans are completed, the garden plots must be prepared. Whether you are turning over new soil or reworking an old garden, the soil must be turned into a coarse-textured, water-retaining medium. It would be appropriate at this time to send a soil sample to your state agriculture station for analysis and recommendations for proper treatment. Instructions for taking soil test samples are in the conclusion of this chapter on page 243.

Even without such help the novice as well as the experienced gardener can quickly learn to judge soil. Check the texture and porousness. Is it in good tilth (i.e., can it be worked readily)? Does a spade cut easily into it? Pick up a handful of soil. Is it hard-packed clay, powdery-dry silt, or overly sandy soil? A good soil will be dark in color, crumble easily, and feel coarse because of decaying vegetative remains. The presence of absence of earthworms should also be noted, since earthworms are most abundant in soils that are water-retentive and contain a good supply of organic matter.

Next check the depth of the workable soil. In a new development the yard surfaces are often covered with a scant inch or two of poor-quality topsoil, just enough to sprout a bit of grass seed so that the house looks attractive to prospective buyers.

In established yards the soil may be badly depleted because of poor past gardening practices or neglect, even though the depth of the soil may be satisfactory. This is often apparent by the condition of the plants already growing. Stunted growth, poor color, small leaf size, or browning tips could be indicators of low soil nutrients.

Nearby tree and shrub roots can deprive gardens of moisture and nutrients, so your garden should be located some distance from these plants. Sometimes, however, the only possible or suitable garden spot is south of a large tree. If this problem exists, dig a trench about two spades deep between the garden and the tree. This will sever the roots, damage which most healthy trees can tolerate if not overly extensive. Into the trench place a verticle barrier of a long-lasting impenetrable material such as sheet metal or heavy roofing paper. Refill the trench and your garden should be secure from tree roots for several years or more.

If you are anxious to start your gardening in a new yard, you may find it worth the effort and expense to remove some of the poor subsoil and fill the garden to a depth of 8"-24" (20-60 cm.) with good loam, which is readily available from landscape contractors, landscape and gardening-supply dealers, and nurseries. In the case of depleted, poorly structured soils, peat moss, straw or, that old favorite of the organic gardeners, well-rotted manure (from a riding stable or farm) should be added by digging down two spade depths into the soil and working in the organic material.

Once the soil is of a satisfactory structure and depth, additional nutrients can be added with compost and fertilizers. However, even the best fertilizers are wasted or poorly utilized by plants if the soil is not friable and moist so that their roots can spread generously to absorb the nutrients.

The first fertilizers used by humans were those copied from nature—organic debris. Wild plants produce their own compose when leaves, flowers, and twigs fall to the ground around them. The natural process of decay produces nutrient-rich soil that continues to support the plant for all the years of its life. Today, however, we interfere with this natural process in cultivating plants. Vegetables and fruits are harvested, removing from the garden site the nutrients that went into producing them. Flowers are picked for bouquets, or dead flowers are removed because they are unsightly. At the end of the growing season, plant remains are cleaned off the gardens so that fungi and insects can be controlled. And it doesn't stop there! As the leaves of trees and shrubs turn color and fall, they are meticulously raked up from every inch of the yard and stuffed into huge green (how paradoxical!) plastic bags to be hauled off to the dump or incinerator.

Hopefully your yard will have a hidden corner where some of this organic material can be composted into new soil materials. Some composting methods are suggested on page 244. The addition of composted material will keep your soil friable and rich for healthy plants. In fact you can save yourself a considerable amount of work by not raking up all the leaves from around every tree, shrub, and hedge. Your yard may not be as tidy as your neighbor's, but your plants will love you for it.

Organic-gardening claims to the contrary, our genetically improved flower and vegetable plants today require supplements of inorganic fertilizers to reach their full potential as shown in the pictures of the catalogues. Composted organic debris will supply many of the needed ions but usually not enough to meet demands of our large and vigorous plants.

Apply a complete fertilizer such as 5-20-20 early in the spring before turning over the soil for planting. Where snows are heavy, some gardeners spread fertilizers on the melting snows so that the nutrients seep slowly into the thawing soil. Follow the label for application rates, which are usually about 20 lbs./1000 sq. ft. Excess fertilizer will damage roots. Too little is better than too much when using fertilizers.

Turn the soil over with a spade or rototiller so that the fertilizer is well worked into the soil. There is one precaution here. Be sure the soil is in condition to be worked. Squeeze a handful of soil, then open your hand. If the soil forms a sticky mass, it is too wet and will form clods if worked now. If, however, the ball crumbles apart easily, the soil is dry enough to be turned over. After the soil is turned over, the clumps should be raked out and the surface smoothed with the back of the rake. Your garden is now ready to be planted.

Supplemental fertilizers are usually applied later in the growing season to particular plants, especially vegetables and high-demand flowering plants such as roses. Dry fertilizers can be scratched into the surface of the soil and then watered down into the soil. Some concentrated fertilizers, both solid and liquid, are formulated to be mixed with water and poured onto the soil around the plants. Read and follow label directions carefully or you may find yourself with a collection of sick or dead plants.

As long as the soil is maintained in good tilth and drains well there should be no problems from the applications of fertilizers. If, however, the soil becomes compacted through lack of cultivation, or is not kept aerated with organic materials, or has perhaps a low spot where water can collect, fertilizer residues can build up, producing a more acid pH in the soil. For this reason, plants at one end of a garden will sometimes behave differently from those at the other end, and an adjustment of pH may be necessary.

Much has been written about the benefits of mulches, and it is true that a good organic mulch such as buckwheat hulls, grass clippings, or chopped cornstalks, applied to gardens in the fall, will reduce frost heaving which can damage roots and bulbs. In the spring the mulch can be turned into the soil along with an application of fertilizer, improving soil conditions with a minimum of effort.

Most gardening instructions, including some of those above, may be applied equally well to either flower or vegetable gardens, although each has some obvious special needs. Annuals and biennials, whether flower or vegetable, are short-term performers and need similar treatment in either type of garden. Herbaceous perennials however are quite different. They are often the mainstay of a flower bed but you will find few or none of them in most vegetable gardens. We'll look at some of the similarities and then the differences in plants and care for both vegetable and flower gardens in the sections that follow.

PLANTS FOR THE GARDEN

Plants from Seeds

Herbaceous flower and vegetable plants may be annual (lasting one growing season), biennial (lasting two seasons), or perennial (lasting several to many seasons). However these terms in the seed catalog may not be applicable to some plants as they grow in your climatic zone. Many perennial plants are grown only as annuals in the north if they are not winter hardy. Of the flowering plants *Impatiens*, fibrous begonias, and snapdragons (*Antirrhinum*) are three of the most popular. Tender plants with bulb-like structures, such as *Gladiolus* and *Dahlia*, must be lifted and stored each fall and replanted again the next spring.

Conversely, some annuals and biennuals may be extended into an extra year or so where tempertures stay well above freezing all year. To add to the confusion, a number of kinds of plants are available in both annual or perennial species or varieties such as *Alyssum*, hollyhock (*Althea*), forget-me-knot (*Anchysa*), *Campanula*, and candytuft (*Iberis*). Vegetable plants are generally annuals (e.g. radishes, peas, tomatoes) or biennials that are harvested in a vegetative form their first year (e.g. carrots, cabbage, celery).

Seeds are the least expensive means of getting new plants. The best seeds for strong, vigorous plants are available from the long-established seed companies that have the facilities and know-how to grow plants especially for seed production, gathering the seed at the best time and storing it under conditions that will insure maximum germination. Most of these growers compete in the All-America Trials by submitting seeds of their best plants to be grown in each of 55 trial gardens in Mexico, Canada, and the United States. Many varieties are entered but only a few receive the coveted medals. You can expect to pay a premium price for prize-winning seeds, and the supply is often limited.

When two plants with special genetic endowments are selected for seed produc-

Fig. 9-5. All-America Selection plants are exhibited in many gardens throughout the U.S. Frequently the gardens are associated with state agricultural schools and colleges.

tion, crossing them produces seeds which grow into F_1 hybrid plants. Such hybrids are usually more vigorous and desirable than their not-so-specially-bred relatives. They may or may not be All-America winners, but they can be expected to produce excellent results. However, don't plan on buying a few seeds this year and then saving seeds from the plants for next year's garden. Such plants are often sterile. As in mules, there may be genetic incompatibility so that fertilization between two hybrids cannot occur. Even if such is not the case and seeds are produced, the results at best will produce no more than a few plants like the hybrid, the others reverting back to the less desirable parent types.

Plant AA x Plant BB = Hybrid AB but Hybrid AB x Hybrid AB = 1/4 Plant AA, 1/2 Hybrid AB, 1/4 Plant BB.

Seeds may of course be saved from your garden, but even if they are genetically acceptable there may be other problems. Seed viability is often difficult to maintain unless exact temperatures and humidity levels are adhered to, and in our centrally heated dry homes these levels can be difficult to achieve (see Chap. 3, pg. 96).

There are several ways of planting seeds. Hardy seeds of either annuals or parennials can be sown directly in the garden in the fall or in early spring. Tiny seeds can be covered with cheesecloth or burlap to prevent their being blown away. Some fall-sown seeds may germinate at the time they are planted, in which case the little seedlings must be hardy enough to survive the winter. Seeds sown in early spring must be able to withstand cold temperatures and rain while they germinate. The resulting little plants are fairly sturdy, though, and the busy gardener may find it convenient to get these beds ready and the seeds planted when other gardening chores are not so demanding.

A word of caution here about selecting planting times for seeds and transplants. Often only trial and error will determine which seeds and plants can be planted in the fall and which cannot. (Results will of course be recorded in your garden notebook for future reference.) Frequently the planting instructions for flowering plants found in books and magazines follow those of the English gardeners who grew garden flowers for generations in England and introduced many of them into our country over the past 200 years. The problem is that the climate of England is considerably milder than is much of North America's despite its more northerly latitude. The Gulf Stream is responsible for moderating Britain's temperatures, and the surrounding waters maintain its humidity. Many seeds and transplants which can be fall-planted in England and survive the winters are not at all capable of withstanding the rigors of our northern winters and must be spring-planted if they are to survive. (Gardeners in southeastern states, excluding Florida, and northern Pacific coast regions experience English growing conditions and would do well to heed such advice.)

Tender annuals can be started indoors in a greenhouse or hotbed or under lights. Even a brightly-lighted window sill can be used, but the plants may be a bit spindly. In any case, indoor planting will provide good-sized plants earlier than those grown outdoors from seed. Directions for planting seeds indoors are given in Chapter 3.

If the seeds of tender plants are to be planted outdoors, you must wait until all danger of frost is past or, using the Old Farmer's Almanac admonition, when oak leaves are the size of squirrels' ears. Your local weather bureau can give you a more reliable date for your area. Keep in mind, however, that if your garden is in a low spot, cold air may hover over it even after the higher levels of your yard have warmed up.

It will do you no good to try to plant tender annual seeds early. Even if the seeds are not killed by the cold, they will usually not germinate at low temperatures, and if

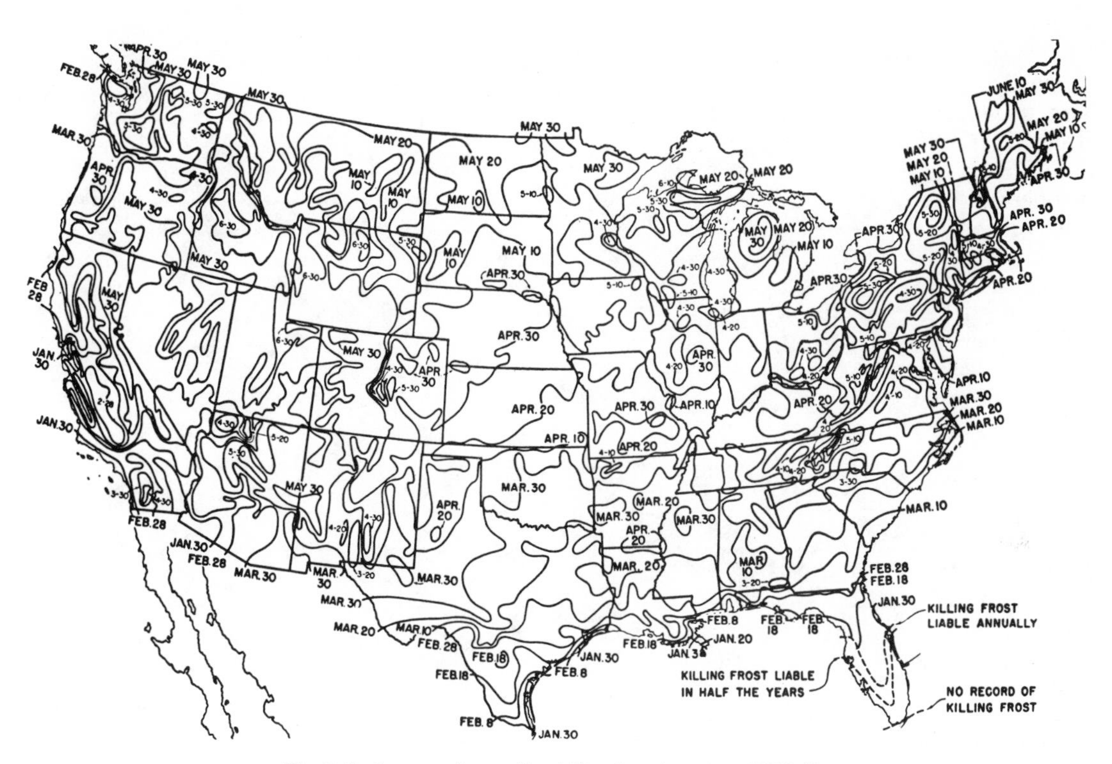

Fig. 9-6. Average dates of last killing frost in spring. (USDA).

any should, the cold soil will damage or kill the emerging seedlings. If you can't wait for the warmer temperatures you'd better plant flats indoors.

When planting seeds of any kind, indoors or out, follow the package directions explicitly. Seeds planted for flowers can be broadcast in the garden for a random or irregular distribution, or planted in rows (as are most vegetables) for borders or edgings. In either case keep the seeds well watered until germination occurs and the seedlings are established, because the emerging seedlings have little resistance to drying out. Ruthlessly thin out your plants as soon as the first sets of true leaves appear on the seedlings to eliminate any competition for those plants remaining. It is of utmost importance to space plants far enough appart for good growth. You may cringe while you commit "planticide" on your radish or marigold seedlings, but the results will be improved may times over what you will get if you don't thin properly. You will see rather dramatic proof of this if you thin one half of a row of carrots and not thin the other half of the row. The unthinned carrots will probably not even be worth harvesting. Remove the extra seedlings very gently, and if the roots are badly entangled it is often prudent to cut off at ground level those plants to be eliminated rather than pulling them out and risking damage to the nearby seedlings you intend to keep. A pair of small, sharp-pointed embroidery scissors will do the job. Space plants according to directions on package. Pelleted seed and seed tapes are immensely helpful in spacing plants adequately and reducing the need for thinning.

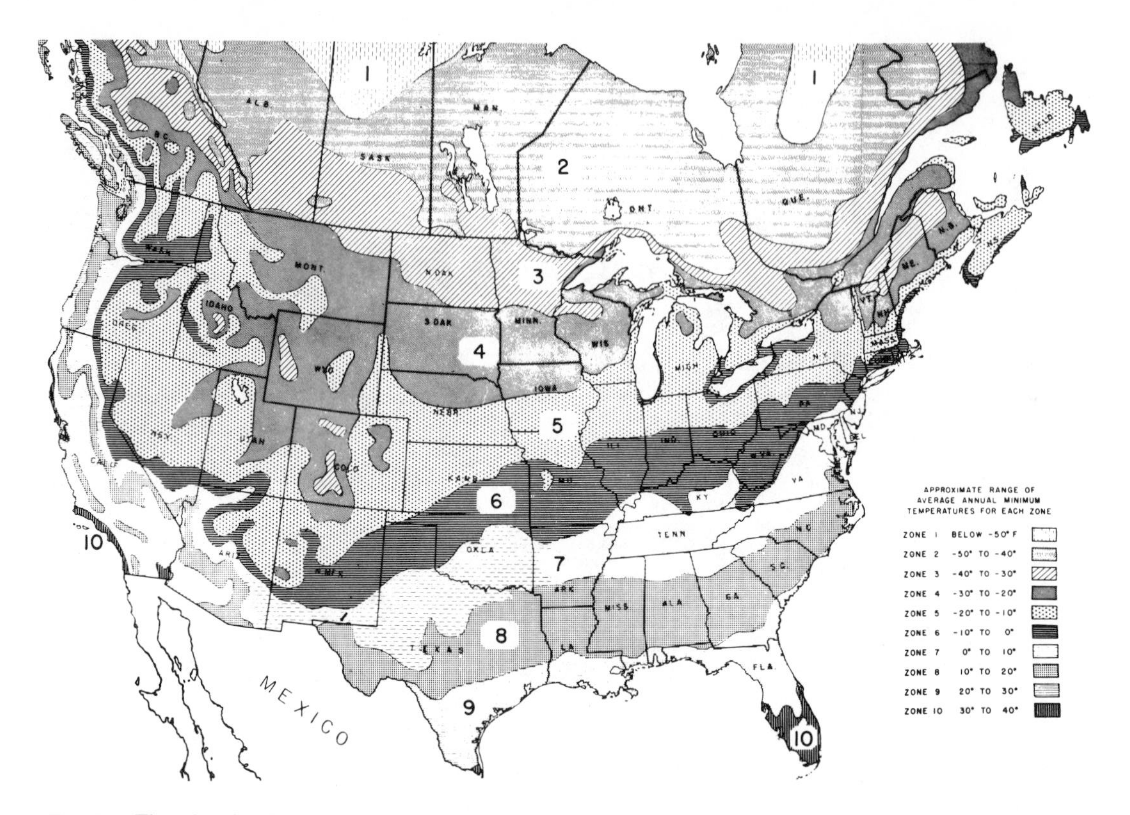

Fig. 9-7. This plant hardiness map is produced by the Department of Agriculture to show areas of winter hardiness for certain types of plants. Commercial plant catalogues and gardening books usually indicate the zones suitable for specific plants. For example, flowering dogwood grows well in Zones 5-9 but would not be expected to withstand the colder winters above Zone 5 or the climate conditions of Zone 10.

If a garden bed is not available to receive fall-planted seeds, a cold frame (see page 239) could be the answer to starting your plants. Other seeds could also be added in the early spring.

Transplanting Seedlings

Transplanting seedlings, whether grown at home or purchased at a garden store, is a way to put plants precisely where you want them at a time your garden is (hopefully) ready for them. A few words concerning transplanting procedure will simplify the project.

1. Harden off your plants in their flats by exposing them gradually to outdoor conditions for increasing periods over one or two weeks before transplanting. Plants growing in flats will also be transplanted more successfully if the soil in the tray is cut into blocks around each seedling about a week before transplanting. This will allow them to recover from some of the transplanting shock before being removed totally from the flat.
2. Plant only healthy seedlings. Weak or deseased plants rarely, if ever, develop into satisfactory mature plants.
3. Do your transplanting on a cool, cloudy day or in the evening if possible. The heat and intensity of a mid-day sun

can be fatal to seedling plants.

4. Remove only one or a few seedlings from the flat or their pots at a time so that their roots do not dry out. Disturb the roots and soil as little as possible.
5. Make sure the hole is large enough to receive the roots without damage. The soil must be damp but not soggy or sticky.
6. Transplant the seedlings to the same depth as or a bit deeper than they were, leaving a bit of a depression in the soil around the stem so that water will flow toward it.
7. Water well with a very fine spray so that the plants are not damaged. Many gardeners use a mild fertilizer solution for this first watering to help the plants adjust more rapidly to their new location.
8. When young flowering plants are at least several inches tall and have several sets of leaves, nip out the growing tip of each plant with thumb and forefinger to increase branching and produce bushier plants.

FLOWER GARDENS

Gardens are usually prepared as flower gardens or for vegetables, and each kind has its own special requirements as discussed in the following sections.

Growing Perennials

Perennials may be grown from seeds, but because of the time required to produce a blooming-sized plant, most gardeners plant container-grown plants or dormant stock that is bare-rooted (sold without soil) in their flower gardens. With a few exceptions most perennials can be planted in the spring, the bare-rooted ones early, as soon as the soil can be worked, the leafy container plants a bit later, after all danger of frost is past. In the western-coastal areas and the south, where winters are wet and mild and summers are dry it is essential to plant as early as possible to establish plants before the rains cease. In the central, northern, and eastern-coastal areas, planting is generally done after freezing temperatures no longer occur. Check a garden catalogue for planting times and instructions for specific plants.

Perennial plants should be planted in a hole wide and deep enough that you arrange the roots loosely within it without crowding. On bare-root stock look for the juncture of stem and root to determine planting depth; plant so this juncture is at or slightly below soil surface. Container plants can usually simply be set into a hole the size of the container. However, if the roots of a container-grown plant are badly crowded and rootbound, it may be necessary to loosen them somewhat and provide a larger planting hole. A handful of fertilizer mixed into the soil around the bottom of the hole will insure good nourishment.

Be sure there are no air pockets in the soil after the plant is in place. Press soil down firmly and water thoroughly. Keep the plants especially well watered for a few days after planting to help them establish themselves.

Most perennials need a bit of cultivating and periodic doses of fertilizer if they remain in the same spot over the years. Many of them are very heavy feeders and may need more than one application during the growing season. (This is also true for perennial vegetable plants.)

The novice gardener often makes the mistake of thinking that a perennial plant needs only to be planted and watered to give years of continuous bloom. Nothing could be farther from the truth! A perennial plant requires perennial care. Aside from water and fertilizer, most perennials will also need to be lifted and divided every few years, although there are a few notable exceptions such as bleeding heart and peonies. Usually older parts of the plant are discarded and only the young, bloom-producing parts are replanted. If the plants can be

Fig. 9-8. How much rain and when: Total precipitation (in inches), by month. (USDC)

moved to a new site at this time the soil is less likely to become depleted.

Perennials usually produce one strong crop of blossoms at a specific time during the growing season. Thus we have spring-blooming daffodils followed by the bearded iris. The phlox and delphiniums show their colors in mid-summer, and the chrysanthemums conclude in the fall. Repeat bloomers such as roses, gaillardias, geraniums, and Iceland poppies are among the relatively few exceptions.

For a stronger, longer lasting plant, faded flowers should be removed promptly before seeds form. Seed and fruit formation require a great deal of energy from the plant that is better spent in producing roots and leaves that will lead to better flowers next season.

Fig. 9-9. When a perennial plant gets too large for its place in the garden, it should be dug from the soil, divided into smaller portions, and replanted in suitable locations.

Maintaining a Flower Garden

A gardening chore commonly objected to is that of staking plants that tend to be tall and gangly. However it must be done if all your other efforts with them are not to be in vain. A plant snaking along the ground struggling to lift a cluster of flowers upward is a pathetic sight. Even worse is a tall spike of flowers, such as *Delphinium*, broken over by its own weight.

Green-stained bamboo stakes in assorted lengths and thicknesses are the least expensive stakes from the garden store. They are strong enough to support most plants, but they do split and deteriorate after a few seasons of use and need to be repurchased regularly. Various kinds of metal rods and rings last a bit longer but tend to rust and get best out of shape. For some reason they are easy to bend out of shape but almost impossible to bend back again. (Or maybe it only seems that way.) However they don't rot off under ground as bamboo or wooden stakes often do. Plastic and fiberglass stakes are more recently on the market, but their initial expense is fairly high and their length of service variable and, at present, uncertain. Some of the bushier plants can be adequately staked by pushing a sturdy cluster of dead woody shrub or tree branches into the soil next to the emerging shoots of your garden plants which will then grow upward within the branches to be supported by them.

Most plants have to be tied to their stakes at one or several points. The paper- or plastic covered wire twist-ems available at all garden stores are commonly used. However, soft string, raffia, or strips of nylon stockings can also be used. It is best to wrap a tie firmly around the stake where you wish it to stay so that it doesn't slip down, becoming useless.

Whether you have seedlings or full-grown plants in your garden, a good layer of mulch will keep the soil moist, reduce weeds, and moderate soil temperatures. Organic mulches will also enrich the soil if they are turned under when the soil is worked the following season. Peat moss, however, should be used with caution. It looks very attractive scattered over a garden surface but will dry out and draw moisture away from plants. It is then exter-

Fig. 9-10. Many garden plants need support when their flowers are in bloom. (left) Leafless woody branches are easily concealed in bushy plants. (middle) Green bamboo or metal stakes will keep top-heavy plants from breaking over. (right) For viny or sprawling plants a section of wire fencing can be used.

mely difficult to rewet.

A fine protection against insects and disease is to start with healthy plants and keep your garden clean. Damping-off, a fungus disease that topples young plants over at ground level, can be prevented by treating seeds with benomyl solution. In fact many commercial seeds are pretreated before packaging.

Attacks on mature plants by either fungi or insects should be dealt with promptly but only at an intensity appropriate to control the problem. Fungicides should be appplied as preventives on susceptible plants before the fungi appear. Insecticides should be used only after the pests become abundant enough to become objectionable by causing noticeable damage.

Since many of our loveliest flowering plants are far removed from their lands of origin, they often grow only with difficulty in their new homes. It is therefore important to eliminate the competing weed plants which, given a chance, will grow rampantly. Even if weeds are removed, the desired plants will still be struggling against each other to a certain extent, so it is important to provide a rich fertile soil and to replenish it with fertilizers at appropriate times. Scraggly growth, pale leaf color, and skimpy blooms are a plant's ways of calling for help. Compost adds a wide variety of useful ions to the soil, and inorganic fertilizers of known content will add specifically needed ions to remedy recognized problems.

Fertilizers high in phosphates and potassium (potash) will produce strong roots, and the size and abundance of flowers will be increased. There are special fertilizers formulated especially for flower gardens, and these should be used in preference to other less specific fertilizers. Fertilizers high in nitrogen stimulate leaf growth and, while fine to use on lawns, will produce heavy foliage and reduced bloom if applied to the flower garden.

Above all keep your flower garden well watered. A plant with ample water will be turgid and its leaves will be lifted to receive abundant sunlight. Its roots will readily take in those needed ions in the soil solution, and insect pests will be discouraged. How often you water will depend on the soil and subsoil. Good humusy garden soil will hold water well but will require more frequent watering if your garden is in a naturally sandy locale than it will in a clay-based or gravely or rocky area. Also, a sunny, breezy garden spot will dry out more quickly than a shady area along a hedge or wall. If in doubt dig up a trowelful of soil and check on the moisture well below the surface. If water is needed, sprinkle the garden until the soil is thoroughly wet to a depth of at least several inches. If only the surface is dampened, roots will crowd upward, where they will not only be subject to damage from drying out and cultivation but will also be deprived of many of the nutrients available only at the deeper soil levels.

Seasonal Care

Care of the flower garden will vary with the season. In the spring the gardens should be picked clean of weeds, preferably by hand, if not overly extensive. The job will be easier for the rest of the year if this can be done thoroughly. Last year's mulch should be turned under and the soil cultivated as much as possible to mix fertilizing ingredients, air, and water evenly throughout the soil. Some seeds and bare-rooted perennials may be planted as soon as the soil can be worked. Seedlings of annuals and container-grown perennials are usually planted after the last frost. Spring is, of course, the busiest and most urgent time for a gardener.

Summer maintenance can be a pleasant pastime because now the plants are putting on their displays. It is important to keep the gardens weeded and cultivated for better water absorption and aeration. A layer of mulch will help retain soil moisture and prevent a crusty surface from forming. It will also produce an attractive, uniform appearance on the soil between plants. The whole garden can be fertilized lightly and frequently (follow package directions), and heavy feeders may need individual treatment. Stake plants that cannot support themselves, and remove dead or diseased plant tissue as they appear.

As fall progresses many of the plants will gradually fade into dormancy. Some new perennial plants can be placed now that are robust enough to withstand winter conditions. Spring-flowering bulbs should be planted. A thorough weeding now will mean an easier weeding job next spring. Remove all dead plant material that might rot or provide a shelter for insect eggs or pupae. A few of the sturdier stems of plants left in place will trap snow to form a protective blanket over the surface. This is an excellent time to analyze your garden and make notes for a more successful and attractive garden next year.

VEGETABLE GARDENS

Planning and Planting

A sunny spot is a requirement for a good vegetable garden. In addition, good drainage is especially important, since so many of the vegetables are underground crops such as carrots and radishes. Some of the suggestions given previously concerning garden sites are especially applicable to vegetable gardens.

Fig. 9-11. A raised vegetable garden reduces the bending and stooping necessary while tending the plants and provides convenient seating as well.

The size of your vegetable garden will depend on the quantity of produce suitable for your needs as well as the space available and your own ambitions. If it is too large it will be burdensome and you will soon lose interest. Therefore, at first, your garden should be somewhat small (probably no larger than 15 x 25 feet, approx. 8 x 8 m.) and then enlarged in later years as your skills and knowledge of the plants develop. It is best to start with the old standbys such as lettuce, tomatoes, carrots, radishes, and beans (see Fig. 9-12). In subsequent years expand the garden to include the more exotic or space-requiring varieties after you have mastered the easier ones.

Perennial crops such as asparagus and rhubarb should be planted where they won't interfere with turning the soil in the rest of the garden. Either place them along one side of the garden or plant them apart from the main garden. Tall plants should not be southward of short plants since they will shade them.

If planting times are staggered it is possible to extend a harvest of some vegetable plants through the growing season. Radishes and spinach, for instance, can be planted in three plantings to have crops from early summer into fall.

Always orient rows of plants perpendicular to the direction of drainage to minimize erosion, especially if the garden is on a slope. A winter planting of rye grass or other such cover will secure the bare soil between seasons and can be turned under as green manure in the spring.

Because some crops are heavier feeders than others it is practical to rotate or alternate the location within the garden of the various kinds of vegetables from one year to the next. Rotation maintains more uniform soil fertility and in addition helps to control insect pests and diseases.

N

Potential Yield
A 15′ x 25′
Vegetable Garden

70 cucumbers

tomatoes
80 lb.

zucchini squash
40 lb.

peppers
40 lb.

cabbage
20 heads
(2 plantings)

lettuce
44 heads
(2 plantings)

string beans
25 lb.
(2 plantings)

chard
48 lb.

beets
36 lb.
(2 plantings)

carrots
36 lb.
(2 plantings)

spinach
12 lb.
(2 plantings)

onions
24 bunches

radishes
24 bunches
(2 plantings)

parsley
48 bunches

celery
18 bunches
(2 plantings)

cauliflower
24 heads
(2 plantings)

peas
15 lb.

Fig. 9-12. Crow's-eye view of vegetable garden. Good vegetables for the family garden and an order to plant them in.

Maintaining the Garden

There are only a few differences in the basic care of vegetable and flower gardens. Mainly the vegetables need more feeding and less weeding than do the flowers. Germinating seeds and transplanted seedling plants should be kept weed-free, but generally once the vegetable plants are easily distinguished by their growth from the weeds, intensive weeding is no longer necessary. Occasional shallow cultivation will control weeds sufficiently and not harm vegetable roots.

Vegetables grown from seeds or transplants will respond to most of the same conditions and care as flowering plants, but spacing is especially important. Plants that produce fruits (including those considered to be vegetables from a culinary viewpoint) that require ripening before harvest must get maximum sunlight. When the plants are mature the removal of some of the leaves and stems may even be necessary, because tomatoes, green peppers, and eggplants will be more uniformly colored and shaped if given plenty of room to grow and lots of sunlight. It has even been shown that the level of vitamin C is considerably higher in tomatoes in sunshine than in those in shade.

A vegetable garden should have spaces for walking between the rows of plants, and these are good places to mulch. The mulch will help retain overall soil moisture, reduce weeds, and provide a less muddy walkway. Grass clippings and last year's leaves, which may be much too compacted for flower gardens, are quite acceptable for mulch in a vegetable garden, as are most of the other organic materials. If plastics or other non-decomposing mulches are used it will be all the more important to dig compost into the garden before planting for the next season.

As with a flower garden it is important to remove all plant debris at the end of the growing season for insect and disease control.

The vegetable planting chart on page 250 will provide necessary information for planning your vegetable garden.

COLD FRAMES

If you would like to extend your outdoor growing activities into the winter months, you can do so with a cold frame. Locate a sunny spot in your yard that faces south and has a weather-protective wall or hedge behind it. The frame is simply a bottomless wooden box, usually about 2 x 4 (75 cm. x 150 cm.) or 3 x 6 feet (1 m. x 2 m.), with a glass or plexiglass cover over it.

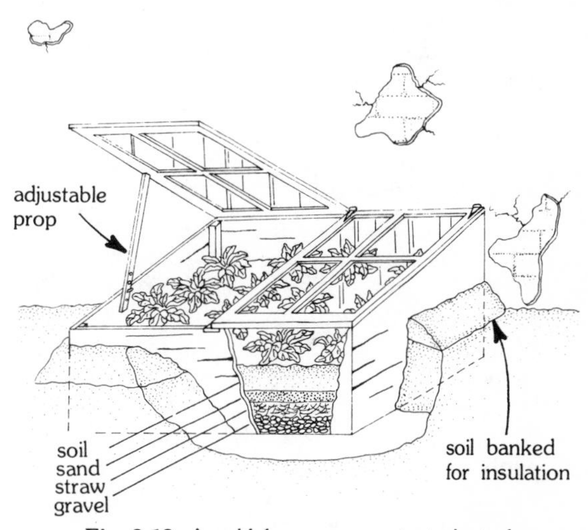

Fig. 9-13. A cold frame can protect plants from harsh weather or provide a cool place for forcing bulbs.

(An old glass storm door or window is an inexpensive source for a cover, and a sheet of heavy plastic is an especially safe cover if children are around.) The back of the box should be higher than the front so that it slants toward the mid-day sun and allows rain and snow to run or slide off. The box can be insulated by heaping soil around the outside of the box. The depth of the box and the soil within it will depend on what you plan to do with your cold frame. Most cold frames are between 12 and 30 inches (31 cm. and 77 cm.) deep at the back. (Flats of seedlings will do fine in 12"; dormant plants and pots of bulbs may need more.)

You might try the following suggestions for using your cold frame:

Pot up some bulbs for forcing in the fall. Place them inside the box either buried to the pot rim in the soil within the box or with wet leaves or peat moss stuffed between the pots. The purpose is to avoid widely fluctuating temperatures and prevent the pots from drying out. A leaf mulch can be laid lightly over the surface of the pots for additional protection. Cover the cold frame with its pane of glass. A little before Christmas you can begin to bring the pots into the house for welcome mid-winter blooming.

A cold frame is an excellent means of protecting the plants that may not be quite hardy enough for your winter climate but still desirable in your garden. When they are dormant, simply dig them up and replant them into the soil in the cold frame, or bundle them together and cover the roots with moist soil or leaves. The plants will be cold enough to remain dormant but not so cold as to be killed.

Stem and root cuttings of woody plants can be carried over the winter and gradually exposed to outdoor conditions by propping the lid of the cold frame open on mild days until they are large enough for transplanting. transplanting.

Best of all, though, a cold frame can be used to get a jump on spring. Plant flats, Jiffy-7's, or peat pots with seeds of hardy perennials or vegetables that can be placed in the garden later, or sow seeds of vegetables such as lettuce or radishes directly into the soil of the cold frame, thin and water properly, and you can have an early crop. These plants can also be sown again in the fall for an after-the-frost crop.

If you find the cold frame useful, you may wish to expand your skills by using a hotbed. This is simply a cold-frame structure with a heating coil within a layer of sand in the bottom. It will be warm enough to be used like a miniature greenhouse so that many of the tender seeds and plants can be started long before garden soil is warm enough. As the weather gets warmer the plants can be gradually exposed to it until they are transplanted into the garden. Garden annuals, some biennials, and vegetable seeds can all be started readily in a hotbed.

Houseplants from cuttings or seeds may also be grown in your hotbed until they are attractive enough for display or gifts.

A heating cable adequate for an average-sized cold frame costs around $10.00, and for the first-class gardener self-ventilating frames are available from garden-supply houses. These structures open and close automatically with temperature changes and are especially useful to the gardener who must be away from home during the day or for extended periods.

A FINAL WORD—ABOUT ORGANIC GARDENING

Organic gardening is the practice of growing plants using only natural nutrients and controls. No synthetically manufactured chemicals or other substances are used. If this is to your liking—fine. You should be aware, however, that a molecule is a molecule and, quite frankly, a petunia cannot in any way distinguish between a phosphate ion from a compost pile and a phosphate ion from a 5-10-10 synthetic fertilizer. Furthermore, the results of a number of tests have shown no differences that could be attributed to organic methods when organically grown flowers and vegetables were compared with plants fertilized with manufactured substances.

This does not mean of course that you should not garden organically. It takes a bit more effort, but you may well benefit from the additional exercise, and you will have the satisfaction of using a method harmless to the environment. The use of botanical pest controls is often all that is needed for sufficient control in an average home garden. Hopefully not all of your neighbors will be using only botanicals also. (See Chapter 5, page 142 for recipes for botanical insecticides.)

In the meantime, be tolerant of the fact that without the assistance of synthetic chemical fertilizers and pesticides our great "breadbasket" in this country would soon shrivel to supporting only a fraction of the population now fed by it, and food prices would soar. Efforts are being made to reduce the need for these synthetic chemicals; their adverse effects on the environment are well recognized and they are expensive. But until research brings us better methods, we have to feed our millions, and this can only be done with the help of synthetic compounds.

HELPFUL HINTS FOR BETTER GARDENS

General

1. Planting-zone maps showing frost dates and low winter temperatures (page 232) for the various areas of the United States are useful as indicators of what you can grow in your gardens. However, for the most reliable local information consult your county extension office.
2. Be prepared for unexpected freezes early or late in the growing season. A few stakes taller than the plants can be quickly pounded into the soil to support an old sheet over the plants. This covering is usually sufficient.
3. You can measure the amount of water (in inches) applied to your garden by placing a can or jar between plants.
4. Vegetables and sun-loving flowering plants need at least 6 hours of sun each day for best growth. For a shady corner or any spot that doesn't get its 6 hours, use sheets of aluminum foil as a mulch beneath and between the plants to act as reflectors to increase the amount of available light.
5. Fireplace ashes spread thinly in your garden will act as a barrier to snails and slugs, reduce soil acidity (don't overdo it!) and serve as a source of potash.
6. Organic gardeners can use rotenone- and pyrethrum-formulated pesticides and still conform with the organic methods, since these products are derived from natural plants without synthetic additives.
7. Where root-rotting microorganisms are a problem, use composted hardwood bark instead of peat moss in the soil.
8. Always wash your hands after handling diseased plants before touching others.
9. Scientific names can be important in buying seeds or plants to get the plants you intend to have. Every known plant has only one scientific name, but some flowering plants may be known by as many as 200 common names throughout the world.

Seeds and Transplants

10. The best plants sprout from seed no more than one year old. Older seeds germinate poorly, and sometimes poor-quality plants develop from the seeds that do germinate.
11. Some seeds sprout only at specific temperatures, so be sure to check the seed package. Those needing cool temperatures can be stimulated by placing a planted flat in a refrigerator for 5 days or in a cold frame before bringing it into a warmer greenhouse or hotbed.
12. Most flower seeds germinate best at a temperature of 75°F. (24°C.). Half-strength complete fertilizer (6-12-12) applied to new seedlings will give sturdy plants.
13. Don't confuse true leaves with seedling or cotyledon leaves which appear first on a seedling. Only the true leaves will look

like the leaves characteristic of the plant.

14. When planting transplants in peat pots into the garden, tear out the bottom of the peat pot without damaging roots, for strong root growth and development.
15. To protect transplants from spring cold and possible frosts, buy hot caps or hot tents from the garden store, or use cut-off milk cartons, bleach bottles, flower pots, cardboard tubes, or loose straw mulch around plants.

Perennials

16. Plant tender "bulbs" such as *Dahlia* and *Gladiolus* where they can be removed in the fall without disturbing other perennials.
17. Be prepared to reduce the size of most perennial flowering-plant clumps from time to time, generally every 3-5 years.
18. Late-fall applications of fertilizers to perennials will not force new growth but will increase the nutrient content of plants which in turn can enhance growth the following spring.
19. White flowers and those of the warm colors of red, yellow, and orange are the most easily visible and stand out well in a garden. Flowers of the cool colors of blue, lavender, and purple are harder to distinguish from the green foliage and appear to recede from our view.
20. To make cut flowers last longer, fill the vase with a mixture of Sprite or 7-UP with an equal amount of water and a spoonful or so of bleach. The carbonation and citric acid will control development of microorganisms that might block xylem vessels, the sugar provides nourishment, and the bleach kills bacteria that rot tissues.

Vegetables

21. In some mid-western communities it is traditional to plant pea seeds on St. Patrick's Day, March 17, a good way to jog the memory that the gardening season is about to begin.
22. Establish straight rows with string between pegs. Plant seeds along string and then remove it so it won't interfere with the growth of the seedlings.
23. Rows of slowly germinating vegetables such as melons or onions can be marked by planting a few radishes in the row. Radishes germinate fast and can be harvested before the other plants require much growing space.
24. Notch a spade or hoe handle for distances useful in spacing your garden. Then you won't need a yardstick.
25. A good starter fertilizer solution for vegetables is 10-55-10. The high phosphate insures strong root growth.
26. Sheets of black plastic 3 to 4 feet wide can be spread over the vegetable garden after the soil is fertilized and spaded. Anchor them in place with soil put over the edges. Cut holes for transplants, and water through the holes. The plastic kills weeds, warms the soil, and hastens growth of your plants.
27. Supports of stakes, nets, or cages will save space and expose tomatoes, cucumbers, melons, and pole beans to good air circulation and sunshine for better produce.
28. Cabbage, radish, lettuce, beet, onion, pea, and broccoli are called cool-season crops because they are not injured by frosty temperatures. We eat the roots, stems, leaves, and young flowers of these plants.
29. Tomatoes, peppers, squash, cucumbers, corn, and pumpkins are warm-season crops. They are planted after danger of frost is past. We eat the fruits of most of these plants.
30. Seeds from crosses between pumpkins and squash will not produce desirable plants the following year, but this hybridization will not affect the quality of the pumpkin or squash containing the hybrid seeds.
31. A 15 x 25 ft. vegetable garden with 18 varieties of vegetables will provide a fam-

ily of four with fresh vegetables with plenty left over for preserving and storing, for a value of more than $325 (1979 prices).

32. Gardening is a popular pastime. In Michigan along 78% of all families buy plants, and this seems to be a nationwide trend that is increasing. Petunias are the most popular bedding plants in the United States.

DIAGNOSTIC SERVICES

Every state has a state college or university with an agriculture department that offers diagnostic services for any farmer or gardener with insect, plant, or soil problems. These services are offered directly or through your county cooperative extension service.

Submitting Plant Material

Describe the problem as fully as possible. Enclose photos if appropriate. Send **whole plants**, if possible, or several samples of plant parts. Label and wrap each specimen individually.

To send small **potted plants**, secure with packing material in a sturdy box and place a plastic bag over the top to reduce desiccation. Do **not** water plant, and avoid crushing the plant.

Pressed leaves may be enclosed between cardboard sheets, wrapped in plastic or waxed paper, and sent in an envelope or box. Do **not** pack in moist paper towels.

Diseased twigs just beginning to show symptoms and a healthy sample from the same plant should be wrapped in newspaper or waxed paper. If foliage is attached, wrap the twig end in a damp paper towel and seal it in a plastic bag.

Fruits, vegetables, tubers, etc. should show early, not advanced, signs of disease. Include leaf and/or twig specimens if possible. Wrap specimen in dry toweling or newspaper and send in a crush-proof container.

Lawn turf samples should measure about 6" x 6" x 3" deep (13 cm. x 13 cm. x 7 cm.) showing symptoms but not dead. Place specimen in a plastic bag and sturdy box.

Submitting Insects

Try to send at least 10 insects if possible. Include plant material associated with the insect. Do **not** mail live insects—it is unlawful!

Soft-bodied insects should be preserved in alcohol in a plastic vial. Grubs and caterpillars should be dropped into boiling water and then placed in rubbing alcohol or 100-proof liquor. These treatments will cause color loss or change, so describe the original color.

Hard-bodied insects may be packaged between layers of cotton or tissue and sent in a crush-proof box or tube.

Enclose a letter stating location of pests, estimated size of population, extent of damage, attempts at control, and any other information you think may be useful.

Submitting Soil Samples

Remove surface litter and about 1/2" (1 cm.) of top toil. Insert a soil auger to its full length. If a space or trowel is used, remove a slice of soil by making two vertical cuts 6" deep, about 1/2" apart. Take several samples from the area of concern, mix together, and remove rocks and debris. Do **not** dry soil. Select about 1 cupful to be sent.

For soils beneath trees and shrubs, deeper soil samples are needed. For garden soils take samples before spring planting. A second sampling can be taken part way through the growing season after the application of recommended treatments to show if the treatments are effective.

A soil diagnosis will show soil pH and major nutrient contents. Recommendations will provide information for developing a soil appropriate for its intended use. The nutrients usually tested for are P, K, Ca, and

Mg. Include a letter with information about drainage, soil additives or fertilizers used, plants or crops in the soil, etc., for a reliable diagnosis. If you have trouble interpreting the recommendations sheet sent by the soil laboratory, consult your county agricultural agent or a fertilizer dealer.

COMPOSTING

Garden compost is prepared by fermenting or decomposing such materials as grass clippings, leaves, sod, table scraps, straw, vegetable refuse, manure, mushroom soil, corn stalks, asparagus stems, sunflower heads, weeds, and other easily decomposable plant material. The odor of a compost pile is sometimes objectionable, so select a site near the back of a lot and away from neighbors.

Recommended Procedure—Compost can be prepared in any quantity using the following method (from *Home Vegetable Garden*, Extension Bulletin E-529, Cooperative Extension Service, Michigan State University):

1. Spread a portion of materials to be composted in a layer 6 to 8 inches thick.
2. Sprinkle this layer with a small amount of complete fertilizer, such as 5-20-20 or 6-12-12 (3 cups per bushel of compost) or mix in some animal manure. Adding small amounts of colomitic limestone (2/3 cup per bushel of compost) will counteract excessive acidity and hasten decomposition.
3. Wet the layer thoroughly, but not enough to wash the fertilizer away.
4. Add a small amount of fertile soil to each layer to hasten bacterial action and decomposition and reduce odor.
5. Form additional layers 6 to 8 inches thick on top of the first one until all material is in the pile. Repeat Step 2 for each layer.
6. Add additional layers on top as new material becomes available.
7. Keep the pile moist.
8. The length of time for decomposing materials varies from 6 weeks to 6 months or more.
9. The rate of decomposition can be hastened by turning the pile over 2 or 3 times at 2-week intervals. Turning also reduces odors. When the compost material is light and crumbly and can be put through a coarse sieve it is ready to use.

Short-Order Compost—To make a small amount of compost in just 3 to 4 weeks, the American Association of Nurserymen recommends the following method:

1. Crumble one bushel of leaves into a clear plastic bag.
2. Add 2 gallons of water, and 5 pounds of high-nitrogen commercial fertilizer.
3. Place the bag in a sunny spot and turn every other day. It can be stored until needed.

". . . but what about the city? Who's got the room to build a compost heap, let alone two? Obviously, anything on a large scale is out of the question—but on the other hand, not as much leaf mold is needed, so it's all in proportion. What can be done is to use two large plastic garbage cans (two small ones will do if the space is really limited) as containers for compost piles. Since good drainage is important, drill holes at the bottom of the cans, Swiss-cheese fashion. Don't keep lids on, as the compost must be kept moist . . ..

If yours is a backyard garden, you'll have leaves to put into the containers as well as weeds, clippings, dead annuals, tops of bulbs and vegetable scraps . . .. Stick to this kind of "clean" organic garbage, do not use the compost container as a regular garbage can . . ..

Since the compost piles are so small, you can do the layering bit. Start the piles with a couple of inches of soil at the bottom, alternating soil and compost material until you reach the top. Make it a ratio of about three to one: three inches of compost to one inch of soil. It's also a good idea to sprinkle on some all-around fertilizer and a bit of limestone each time you add the soil . . .. In six months you should be able to use these two compost piles to enrich the soil in your garden. Afterward, get on the schedule of letting one pile "cook" while the other is being used." (From *Gardening in the City* by Carla Wallach, pub. Harcourt Brace Jovanovich, N. Y., 1976).

Table 9-1 - Garden Flowers

ANNUALS FOR SPECIFIC CONDITIONS

Annuals for Beginners

Scientific Name	Common Name
Ageratum houstonianum	Ageratum
Phlox drummondii	*Annual Phlox
Coreopsis tinctoria	Calliopsis
Celosia spp.	Cockscomb
Cosmos hybrids	Cosmos
Tagetes spp.	Marigold
Tropaeolum majus	Nasturtium
Petunia hybrida	Petunia
Portulaca grandiflora	Portulaca
Cleome spinosa	*Spider Flower
Lobularia maritima	*Sweet Alyssum
Zinnia spp.	Zinnia

Annuals for Poor Soil

Scientific Name	Common Name
Impatiens balsamina	Balsam
Coreopsis spp.	Calliopsis
Celosia argentea and *cristata*	Cockscomb
Centaurea spp.	Cornflower
Mirabilis jalapa	Four-o'clock
Godetia grandiflora	Godetia
Mesembryanthemum spp.	Ice Plant
Amaranthus candatus	Love-lies-bleeding
Mentzelia lindleyi	Blazing Star
Ipomoea spp.	Morning Glory
Verbena hybrida	Moss Verbena
Tropaeolum majus	Nasturtium
Perilla frutescens	Perilla
Papaver rhoeas	*Poppies, Shirley
Portulaca grandiflora	Portulaca
Cleome spinosa	*Spider Flower
Lobularia maritima	*Sweet Alyssum

Annuals for Dry and Hot Conditions

Scientific Name	Common Name
Phlox drummondii	*Annual Phlox
Gypsophila elegans	*Baby's Breath
Eschscholtzia californica	*California Poppy
Coreopsis spp.	Calliopsis
Celosia spp.	Cockscomb
Sanvitalia procumbens	Creeping Zinnia
Centaurea spp.	Cornflower
Mirabilis jalapa	Four-o'clock
Mesembryanthemum criniflorum	Ice Plant
Papaver rhoeas	*Poppy, Shirley
Portulaca grandiflora	Portulaca
Salvia splendens	Scarlet Sage
Euphorbia marginata	*Snow-on-the-mountain
Cleome spinosa	*Spider Flower
Limonium spp.	Statice
Kochia scoparia	Summer-cypress
Helianthus annuus	Sunflower
Zinnia spp.	Zinnia

Annuals for Moist and Cool Conditions

Scientific Name	Common Name
Campanula medium	Annual Canterbury Bells
Dianthus chinensis	Annual Pink
Nemophila menziesii	*Baby Blue-Eyes
Trachymene caerulea	Blue Laceflower
Anchusa capensis	Bugloss
Iberis umbellata	Candytuft
Nicotiana alata	Flowering tobacco
Myosotis scorpioides	Forget-me-not
Alonsoa spp.	Mask Flower
Mimulus variegatus	Monkey Flower
Nemesia strumosa	Nemesia
Polygonum orientale	Polygonum
Calendula officinalis	*Pot Marigold
Lathyrus odoratus	*Sweet Pea
Verbena hybrida	Verbena
Torenia fournieri	Wishbone Flower

Annuals for Shade

For Shade *or* Full Sun

Scientific Name	Common Name
Impatiens balsamina	Balsam
Myosotis palustris semperflorens	Forget-me-not
Begonia semperflorens—Pink Profusion, Lucifer, Stuttgart, Organdy Mixture, Paris Market Begonias	Fibrous-rooted Begonias
Vinca rosea—Little Bright Eyes, Coquette, Rose Carpet	Madagascar Periwinkle

Scientific Name	Common Name
Viola tricolor hortensis	Pansy
Lobularia maritima—Little Gem, Carpet of Snow, Royal Carpet, Rosie O'Day	*Sweet Alyssum
Viola cornuta—Chantreyland, Chinese Blue, Jersey Gem, Lutea Splendens, White Perfection	Tufted Pansies

For Shade *Only*

Scientific Name	Common Name
Browallia demissa—Major Blue, Sapphire	Browallia
Coleus blumei	Coleus
Begonia semperflorens—Snowbank, Carmen, Indian Maid, Sparkler, Blushing Baby	Fiberous-rooted Begonias
Fuchsia hybrida	Fuchsias
Impatiens sultanii—Dwarf Bright Rose and Orange, Salmon Jewel, Pink Sprite, Pixie White	Impatiens
Lobelia erinus compacta—Crystal Palace, White Lady, Sapphire, Rosamond	Lobelia
Torenia fournieri grandiflora and *Torenia fournieri compacta*	Wishbone Flower

Annuals for Edging

Scientific Name	Common Name
Ageratum houstonianum	Ageratum
Phlox drummondii	*Annual phlox
Iberis species	Candytuft
Dianthus species	Dianthus
Cineraria maritima	Dusty Miller
Myosotis sylvatica	Forget-me-not
Lobelia erinus compacta	Lobelia
Tagetes species	Marigold
Mesembryanthemum sciniflorum	Ice Plant
Viola tricolor hortensis	Pansy
Anagallis indica	Pimpernel
Portulaca grandiflora	Portulaca
Lobularia maritima	*Sweet Alyssum
Verbena hybrida	Verbena

PERENNIALS ARRANGED BY FLOWERING TIME

Name	Color	Height (inches)	Characteristics
		April	
Alyssum *(Alyssum saxatile)*	yellow, gold	15	'Basket of gold', good with tulips.
Anemone, Pasque Flower *(Anemone pulsatilla)*	purple, red, white	12	Cool location, moist soil.
Dwarf Bleeding Heart *(Dicentra eximia)*	pink	18	Lacy, attractive foliage and flower.
Dwarf Iris *(Iris reticulata)*	purple	8	Vigorous, blooms into May & June. Bulbs.
English Daisy *(Bellis perennis)*	red, pink, white	6	Cool, moist soil, does best in partial shade.
Primrose *(Primula spp.)*	violet, yellow, red	8-15	Best in cool, moist, partial shade, continues to bloom into May.
		May	
Bleeding Heart *(Dicentra spectabilis)*	rose, and white	30	Lacy foliage, foliage gone by mid-summer.
Coral Bells *(Heuchera sanguinea)*	red, pink, white	18	Small flowers, excellent cut flowers from May to October.
Dwarf Phlox *(Phlox subulata)*	pink, blue, lavendar, white	6	Vigorous, covers rapidly, full sun.
Geum or Avens *(Geum coccineum)*	red, orange, yellow	4-36	Attractive double flowers, sensitive to excessive soil moisture.
Bearded Iris *(Iris germanica)*	yellow, pink, blue, white, purple, bronze, etc.	36	May-June flowering, some again in fall.
Iceland Poppy *(Papaver nudicaule)*	red, orange, pink, white	18	Delicate, will self-seed, flowers from May to October.
Oriental Poppy *(Papaver orientale)*	orange, red, pink, white	36	Vigorous with single or double flowers, foliage gone by mid-summer.
Violets *(Viola cornuta)*	red, violet, yellow, purple, white	8	Excellent in shade, cool moist soil.
		June	
Columbine *(Auilegia hybrida)*	blue, white, pink, *purple, yellow*	36	Long spurred hybrids are best and most *vigorous.*
Daylily *(Hemerocallis* spp.)	yellow, orange, bronze, pink	12-50	Some are scented, flower May to October, vigorous plant.
Foxglove *(Digitalis purpurea)*	lavender, purple, white, pink, rose	24-60	Self-seeding biennial, best in partial shade.
Gas Plant *(Dictamnus albus)*	pink, white	36	Unusual flower, gives off an inflammable gas.
Peony *(Paeonia lactiflora)*	pink, white, red	30	One of the best perennials, single and double flowered, double require staking.
Painted Daisy or Pyrethrum *(Chrysanthemum coccineum)*	red, pink, white	30	Single and double daisy-like flowers until August with fern-like foliage.

Name	Color	Height (inches)	Characteristics
Shasta Daisy *(Chrysanthemum maximum)*	white with yellow center	36	Large daisy flower, single and double flowers into July.
Sweet William *(Dianthus barbatus)*	red, pink, white in combinations	24	May to early July flowering, biennial, reseeds itself yearly.
Siberian Iris *(Iris sibirica)*	rose, blue, violet, white	36	Vigorous, should be divided every 3 to 4 years.
Veronica or Speedwell *(Veronica longifolia)*	purple, blue, pink, white	36	Long spikes of flowers from June to September.
Yarrow *(Achillea filipendulina)*	yellow, pink	36	Large, flat flowers with gray-green fern-like foliage from June to September.
July			
Babysbreath *(Gypsophila paniculata)*	white, pink	36	Fine lacy foliage and flowers through August.
Beebalm *(Monarda didyma)*	red, pink, white		Scented foliage, spreads rapidly.
Coreopsis *(Coreopsis grandiflora)*	yellow, bronze	30	Very prolific flowering plant.
Delphinium *(Delphinium hybrids)*	blue, white, purple, pink	40	Tall spikes, new hybrids best.
Gaillardia *(Gaillardia x grandiflora)*	yellow, red, orange	36	Daisy-like flowers.
Hollyhock *(Althea rosea)*	white, pink, rose, maroon, red	60-100	Good background plant, double and single flowered, self-seeding biennial.
Astilbe *(Astilbe x arendsii)*	white, pink, red	28	Lacy foliage, flowers last into August.
Phlox *(Phlox paniculata)*	white, pink, salmon, purple	36	Large, showy, flowering into August.
August			
Black Eyed Susan *(Rudbeckia speciosa)*	yellow with black center	24	Rough leaves, good cut flower.
Cardinal Flower *(Lobelia cardinalis)*	red	48	Spectacular red flowers, does well in wet soils.
September			
New England Aster *(Aster novi-angliae)*	pink, white, violet, blue	48	Large plants with considerable variation in flower size.
Plantain Lily *(Hosta plantaginea)*	lavender, white	24	Large attractive leaves, grows in shade.
Stonecrop *(Sedum spectabile)*	red, white	18	Fleshy, gray-green foliage, flowers August to frost.

*Should be planted in cool soils, after danger of heavy frost in North, in late summer or winter in South and Pacific Coast areas. All others should be planted in warm soils or started indoors for transplanting later.

Table 9-2. Vegetable Planting

VEGETABLE	Planting Times*	Weeks from Seeding to Transplanting	Depth to Plant (inches)
Asparagus	April		6 to 8
Beans, Lima	May 20-Jun. 1		1 to 2
Beans, Snap	Apr. 20-Jun. 30		1 to 2
Beets	Mar. 20-Apr. 20, Jun. 20-30		½ to 1
Broccoli	Mar. 20-Apr. 20, Jun. 20-30	4 to 6	(plants)
Brussels Sprouts	Apr. 1-20, Jun. 20-30		(plants)
Cabbage	Apr. 1-20, Jun. 20-30	4 to 6	(plants)
Carrots	Apr. 1-10, Jun. 20-30		½ to 1
Cauliflower	Apr. 1-20, Jun. 20-30	4 to 6	(plants)
Celeriac	Apr. 1-20		½
Celery	Apr. 1-20, May 20-30	10 to 12	(plants)
Chinese Cabbage	Jun. 20-Jul. 30		½
Collards	Apr. 1-20	4 to 6	½
Cucumbers	May 20-Jun. 20	4	1 to 2
Eggplant	May 20-Jun. 1	8 to 10	(plants)
Endive	Mar. 20-Apr. 20, Jun. 20-30		½
Garlic	Mar. 20-Apr. 20		1½
Kale	July		½ to 1
Kohlrabi	Mar. 20-Apr. 20, Jun. 20-30	4 to 6	1 to 1½
Leeks	September		½
Lettuce (head)	Mar. 20-Apr. 20, July	4 to 6	¼ to ½
Lettuce (leaf)	Mar. 20-Apr. 30, July		¼ to ½
Muskmelon	May 20-Jun. 1	4	1 to 2
Mustard	Apr. 20-30		½
Okra	May 20-Jun. 1		½
Onion (sets)	Mar. 20-Apr. 20		1 to 2
Onion (transplants)	Mar. 20-Apr. 20		(plants)
Onion (seeds)	Mar. 20-Apr. 10	4 to 6	½
Parsley	Mar. 20-30		¼ to ½
Parsnips	Apr. 1-20		½
Peas	Mar. 20-Apr. 20		1 to 2
Peppers	May 20-Jun. 1	8 to 10	(plants)
Pop Corn	May 20-Jun. 1		2 to 2½
Potatoes	Apr. 20-Jun. 1		4
Potatoes, Sweet	May 20-Jun. 1		(plants)
Pumpkins	May 20-Jun. 1		½
Radishes	Mar. 20-Apr. 20, July		½
Rhubarb	Mar. 20-Apr. 30		(plants)
Rutabaga	Jun. 20-30		½
Salsify	Apr. 1-20		½
Spinach	Mar. 20-Apr. 20, July		¼ to ½
Squash (Summer)	May 20-Jun. 1		1 to 1½
Squash (Winter)	May 20-Jun. 1		1 to 1½
Sweet Corn	Apr. 20-Jun. 20		2 to 2½
Swiss Chard	Apr. 1-20		½
Tomatoes	May 20-Jun. 1	4 to 6	(plants)
Turnips	July		1 to 1½
Watermelons	May 20-Jun. 1	4	1 to 2

* Planting times are based on conditions at East

Chart. (Michigan Extension).

Amount of Seed	Days to Maturity	Planting Distance (inches) In Rows After Thinning	Planting Distance (inches) Between Rows	Row Length (feet)	Estimated Production
12 plants	2 to 3 yrs.	12 to 18	48 to 60	20	6 pounds
½ pound	65 to 88	6 to 8	24 to 30	50	4 pounds shelled
⅛ pound	50 to 68	3 to 4	18 to 24	15	7 pounds
¼ ounce	60 to 65	2 to 3	18 to 24	25	25 pounds
12 plants	55 to 74	16 to 20	30	25	10 pounds
15 plants	90 to 95	18 to 24	30	25	8 pounds
6 plants	63 to 100	15 to 24	24 to 30	12	6 heads
½ pkt.	68 to 85	1	18 to 24	15	15 pounds
5 plants	60 to 85	18 to 24	30 to 36	10	5 heads
⅛ pkt.	120	6	30	10	6 pounds
30 plants	100 to 125	4 to 8	30 to 36	15	30 stalks
¼ pkt.	70	12	24 to 36	10	12 heads
½ pkt.	75	6 to 8	24 to 30	25	20 pounds
½ pkt.	50 to 72	12	48 to 72	10	6 pounds
3 plants	72 to 80	30	36	6	12 fruits
10 plants	85 to 98	8 to 12	12 to 18	6	10 heads
4 cloves	115	3	12 to 18	1	4 bulbs
6 plants	55	8 to 15	24	6	6 heads
24 plants	55	4 to 8	18 to 24	12	24 stems
1 pkt.	130	1 to 2	15 to 18	10	30
18 plants	73	8 to 15	18 to 24	15	15 heads
1 pkt.	45 to 50	6	12	5	2½ pounds
½ pkt.	82 to 90	36 to 48	48 to 60	16	18 fruits
¼ pkt.	35 to 45	1	18 to 24	10	5 pounds
¼ pkt.	55 to 58	12 to 15	36	8	5 pounds
½ pound	90	2	12 to 18	10	5 pounds
120	90 to 115	2 to 3	12 to 18	30	25 pounds
1 pkt.	105 to 130	2 to 3	12 to 18	30	25 pounds
⅛ pkt.	75 to 85	6	12 to 18	3 plants	3 bunches
⅓ pkt.	130	3 to 4	24	15	15 pounds
1 pound	62 to 69	2 to 3	18 to 24	100	28 pounds
6 plants	62 to 80	14 to 18	24 to 30	10	6 pounds
⅓ pkt.	90 to 120	10 to 12	30 to 36	25 — 2 rows	1 peck
5 pounds	100 to 120	10 to 12	24 to 36	50	50 pounds
25 plants	150	12 to 18	36	25	10 pounds
⅓ pkt.	100 to 120	60 to 72	72 to 96	3 hills	30 pounds
1 pkt.	24 to 28	1 to 2	6 to 12	12	8 pounds
3 plants	1 to 2 yrs.	36 to 48	48	9	8 pounds
½ pkt.	90 to 92	6 to 10	18 to 24	15	15 pounds
⅓ pkt.	120	3 to 4	15 to 18	5	15 pounds
⅛ ounce	46 to 70	3 to 6	12 to 18	10	5 pounds
½ pkt.	49 to 55	36 to 48	36 to 48	2 hills	24 fruits
1 pkt.	85 to 110	48 to 60	60 to 72	4 hills	10 fruits
¼ pound	63 to 94	10 to 12	30 to 36	25 — 2 rows	40 ears
¼ pkt.	60	6 to 8	18 to 24	8	7 pounds
10 plants	62 to 83	24 to 36	48 to 60	40	3 bushel
⅛ pkt.	58 to 60	4 to 6	18 to 24	20	20 pounds
½ pkt.	88 to 90	72 to 96	72 to 96	2 hills	4 melons

Lansing. Change these times to suit your location.

10. PRUNING AND GRAFTING

Pruning of trees and shrubs is the art of judiciously removing parts to make the plants healthy, beautiful, and productive. One of the worst threats to the attractiveness of a garden is a reluctance to prune, since pruning is the key to the perpetuity of well-formed and appropriately-sized trees, shrubs, and vines. A good set of gardening tools should include good-quality clippers, loppers, shears, and saws to prune with clean, sharp cuts, thus avoiding problems caused by ragged or torn wood (Fig. 10-1).

Why Bother?

There are a number of specific reasons to prune, such as the following:

1. For normal maintenance. This is the reason most gardeners buy pruning tools in the first place. If a plant is to retain the beauty for which it was planted, it is necessary to remove spent flowers, excess shoot growth, basal suckers, broken or damaged twigs, and any unsightly growth.
2. To keep to desired size. The wise gardener will plant a tree or shrub whose natural growth habit is appropriate to its location. NEVER try to change the natural shape or form or size of a plant radically, but routine pruning will keep plants from outgrowing their sites. The prompt removal of any weak, gangly,

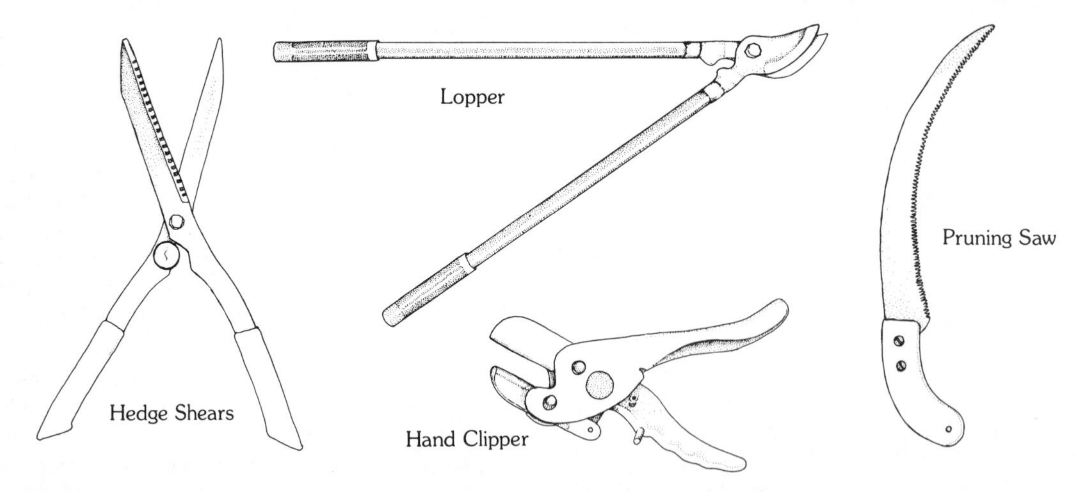

Fig. 10-1. These tools are needed for good pruning. **Hand clippers** serve for removal of spent flower heads and stems of small diameter. **Lopping shears** will cut off stems of larger diameters; the extra length of the handles provides increased force. The **pruning saw** is used for very large branches that are too big to fit within lopping shears. **Hedge shears** will remove a wide swath of stem tips to keep hedges smoothly trimmed.

or malproportioned growth will produce sturdier, healthier plants as well.

3. To increase or decrease flowers and fruits. Young branches are more productive of flowers and the resulting fruits than older ones. However, if the growth of branches is too dense or twiggy, flowers and fruits do not form properly. Remove the oldest nonproductive branches or pinch out tips to promote new growth with flower buds. Opening up the tree by removing excess growth will increase the sun and air necessary for good flower and fruit production.

   An overabundance of flowers often results in flowers and fruits that are smaller than desired. Fruit trees may produce so many fruits that their weight threatens to break branches. Early spring removal of some of the smaller branches results in larger but fewer flowers and fruits. If branches are so heavy with fruit that they may break, prompt removal of the excess may be necessary.
4. To remove injured or diseased parts. When winter storms or carelessly used lawnmowers break branches or wound plants, usually the best way to repair the damage is to prune off the injured part. Cankers and other disease-caused growths and fungus-damaged tissue should be promptly pruned off, if possible, to keep the organisms from spreading. Secondary infection in broken or wounded wood can be prevented by removing damaged parts before the diseases can become established.
5. To rejuvenate old plants. Overgrown, neglected, disease-ridden, tangled shrubs are often on the grounds of older homes. Can they be saved? Often—yes. First remove old, broken, or diseased wood, then cut back the tips of healthy stems to stimulate branching, and finally remove suckers and weak stems. If the job is rather extensive, it may be wise to accomplish it over a two- or three-year period to avoid bare-looking shrubbery.
6. To produce special forms or effect. When plants are allowed to grow naturally, they take the line of least resistance and are not always as handsome and well-formed as we might like them to be. Boston ivies must be pruned away from the windows, hedges must be kept confined and thick growth encourged by regular shearing, and trees along a tree-lined lane must have their lower branches removed. Espaliers, topiaries, and bonsai are produced by special kinds of pruning and training that result in artificial forms and fanciful figures. The gardener and his pruning tools are essential for producing these novelties. (See page 252.)

UNDERSTANDING PLANTS FOR PRUNING

To understand the methods employed in effective pruning it is important to recall (see Chapter 1, p. 10) and be aware of certain characteristics and functions of plants. The stems of woody plants consist of a thick core of water-conducting xylem tissue surrounded by a thinner layer of food-conducting phloem tissue, and between them lies a sheet of dividing cambium from which new xylem and phloem cells are derived. All of these tissues can be affected by disease and injury, and pruning (or grafting, as discussed later) may alleviate a plant's distress.

Xylem tissue is not easily damaged, since it is relatively abundant and deep within the stem. If some xylem cells are destroyed, sufficient water and minerals can usually be carried by alternate xylem routes so that the plant is not seriously affected. **Phloem** cells, however, are much more susceptible to damage since they are close

Fig. 10-2. Many years ago this wire was wrapped around a then-slender tree trunk. The bulging circumference above the wires indicates that the flow of materials through the phloem is being interfered with. The tree will probably not survive to old age.

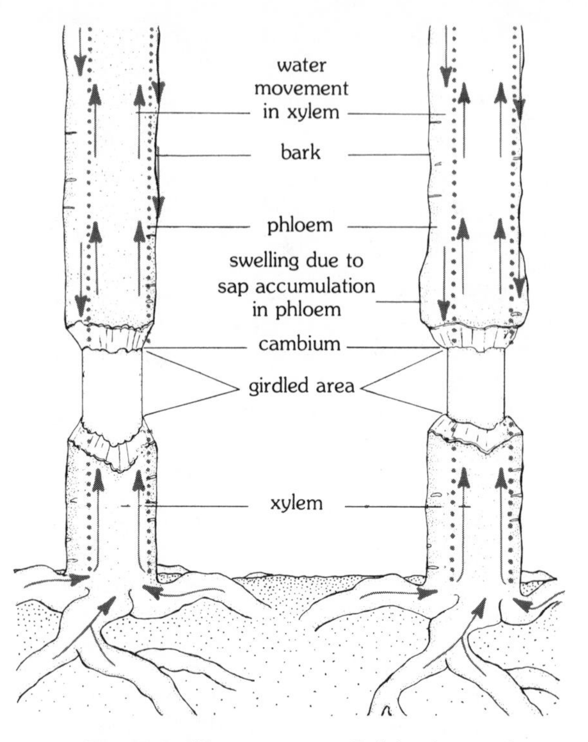

Fig. 10-3. When a tree is girdled deeply enough to remove phloem tissue, nutrients will be carried downward only as far as the girdled area. The roots below will be starved, and the nutrients will accumulate in the tissues above the wound. The flow of water upward in the xylem will not be affected.

to the outside of the stem. If the phloem is constricted with wires or girdled by gnawing rodents the downward flow of sap will be interrupted. This will cause an accumulation of sugars in the top of the plant and deprivation or starvation of the roots. If the main trunk is completely encircled by such injury, the plant will eventually die unless remedial measures (such as grafts) are taken promptly. Injured branches may either be repaired by grafts or pruned off, depending on the extent of the damage and the effect on the health and appearance of the plant. **Cambium** cells are small, thin-walled, undifferentiated, and extremely fragile. Since the cambial tissue lies just inside the phloem, it too is easily injured. In fresh wounds the exposed tissue is damaged by drying out, and the cells are readily invaded by disease organisms. When plants are pruned, clean, sharp cuts will minimize the exposure and possible damage of cambial cells.

Understanding the nature of woody tissue is important in effectively pruning or grafting to improve plant appearance and, especially, to repair wounds. Often, however, we prune to induce new growth as a means of producing plants of a desired size or shape, or to stimulate a plant to recover more readily from a damaged condition. Plant hormones are involved in the growth or suppression of bud development, and pruning, which removes parts of plants, may bring about physiological responses influenced by the presence or absence of certain hormones.

You will recall from Chapter 1, (page 24) that auxins are produced in the shoot apex and stimulate cell elongation and differentiation. Cytokinins, derived from the roots, induce cell division. When both hormones are present in the bud, shoot development occurs. Apical buds will have high concentrations of both hormones but lateral buds will show only low levels of concentration. When the apical bud is present, apical dominance occurs in which the terminal bud develops but the lateral and axillary buds are suppressed. The buds closest to the apex are most strongly inhibited, while those farther down the stem are less likely to be affected and may develop into shoots. If the apical bud is removed, the nearby lateral and axillary buds develop and a bushy plant is produced. It may be of interest to note that while most plants produce obvious buds, a few plants have latent buds hidden as tiny lumps beneath the bark that nevertheless break (i.e., emerge) easily if the apical bud is removed. Holly, *Azalea*, and *Taxus* produce new growth from seemingly bare stems.

This phenomenon of apical dominance is well known but not completely understood. Part of the explanation for dormancy or suppression of buds is that the vascular tissues of the lateral buds do not connect directly with the xylem and phloem of the stem, and therefore these buds do not receive the nutrient molecules necessary for development. In addition, the growth-stimulating hormones are also transported as vascular solutes, and, again, these buds are isolated from the main vascular stream. The vascular tissues, however, terminate at the apical bud so that this bud is in a favorable location to receive vascular solutions containing abundant nutrients and the cytokinins (and perhaps other hormones) which will interact with auxins to promote growth.

In some species of plants, the auxins, as they migrate downward in the stem, produce an inhibitory effect on lateral buds. In these cases high concentrations of cytokinins are required to counteract the suppressing effects of the auxins.

When the apical bud is removed, it will no longer exert an inhibiting influence on

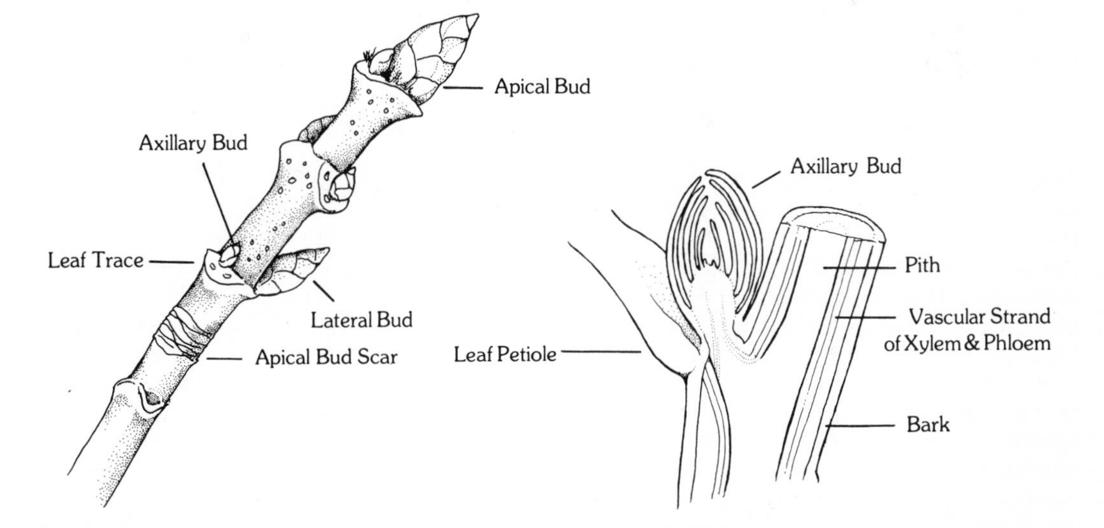

Fig. 10-4. An early spring stem, showing apical or terminal, lateral, and axillary buds. Diagram at right is a longitudinal section through an axillary bud; vascular tissue extends to the apical bud but not directly to the axillary bud.

the buds below. Suppressing concentrations of auxins are reduced, and the lateral buds of susceptible plants will no longer be restrained. The vascular flow into the apical tissues is now interrupted, and the solutes tend to accumulate somewhat in the lower regions of the stem with a resulting increase in cytokinin levels there. This increase stimulates completion of the vascular connection between the bud and the stem. Now the bud receives nutrients, more cytokinins, and other necessary substances. Growth-promoting auxins are produced, and as all of these substances interact, new shoot tissues result. Current and future research will someday establish more completely the roles of growth regulators in the complex process of bud development.

LET'S PRUNE

Proper pruning requires that an assessment be made of what the finished plant should look like before any cutting is done. Balance and uniformity are important goals in most shrubs, flowers and fruits in orchard trees, and strength and vigor in shade trees. Certain generalizations about pruning woody ornamentals (i.e., trees and shrubs) can be made, but because there are so many species that vary in so many ways, each must be treated in a way suitable to it. We'll discuss some of the major considerations in pruning in the following sections. For advice on a specific plant variety it would be wise to consult a pruning manual.

Shrubs

All garden shrubs benefit from regular pruning to maintain their health, size, shape, and beauty. And the gardener will spend most of his pruning time on shrubs and hedges rather than on trees.

Two terms are used to describe the kind of pruning routinely performed on shrubs and other plants as well:

1. Heading back. This is the removal of growing tips back to buds that will then give rise to new shoots. Shearing hedges is an extreme form of heading back, since by shearing young growth is removed and new shoots farther down the stems are encouraged to grow. Pinching back, often done with the thumb and forefinger, of stray or unwanted growth or as a means of spot-pruning a shrub to fill in a damaged area, is also a form of heading back. To establish the direction of the growth of a new shoot, prune down to

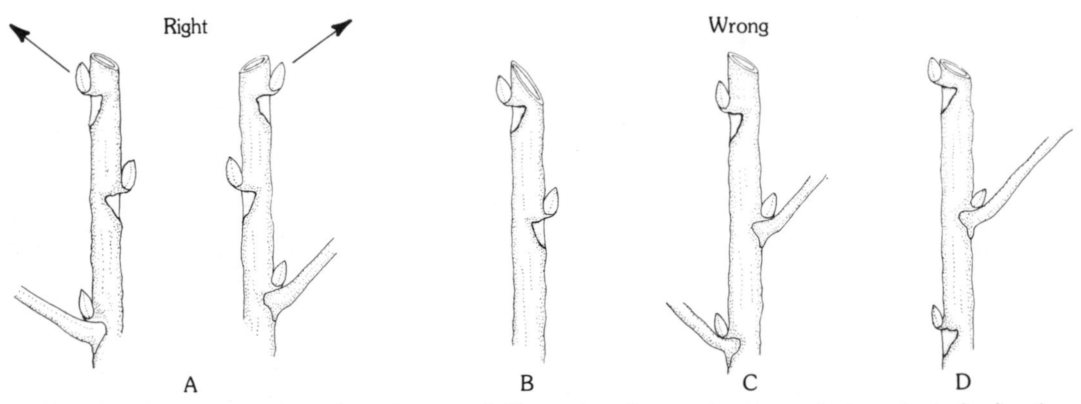

Fig. 10-5. The direction of growth can be controlled by cutting off a stem just above a bud growing in the direction desired for the new branch. The two stems at left illustrate the best cut in relation to a bud. The third stem is cut too steeply, exposing more surface than necessary; the fourth has too long a stub; and the stem at right is cut too close to the bud.

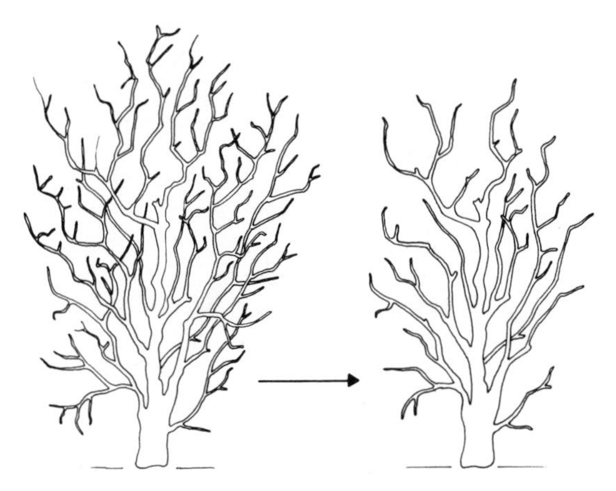

Fig. 10-6. Heading back. To maintain shrub size and prevent overgrowth, remove excess lengths of stem down to a vigorous lateral bud. New growth will produce a compact, well-shaped plant.

a bud growing in the direction desired. Always make a clean cut just 1/4" (6 mm.) above the desired bud slanting slightly away from the bud. This type of cut will minimize the area open to possible decay and provide enough stem tissue to promote healing and shoot development (Figure 10-5). Heading back should be done a branch at a time, stepping back to assess the results as the work progresses to be certain that the desired form will be achieved (Fig. 10-6).

2. Thinning out. This involves the removal of whole branches at their source, either at a main trunk or at ground level. Shrubbery that is so dense that air circulation and sunlight are restricted will soon weaken. Removing whole branches does not stimulate the production of new shoots as heading back will do, and the shrub will benefit from its new openness for several years at least (Fig. 10-7).

Deciduous shrubs should never be allowed to develop stems with mature bark, since these tend to be less graceful and produce fewer flowers and leaves than younger branches. When such branches appear, or if you are faced with the job of rejuvenating neglected shrubs, these gnarled old woody branches should be taken out.

Shrubs with multiple stems sprouting from the ground, such as *Forsythia*, Japanese flowering quince, and the honeysuckles, and whose beauty lies in their graceful arching branches, may be pruned every year or so by thinning out the oldest canes

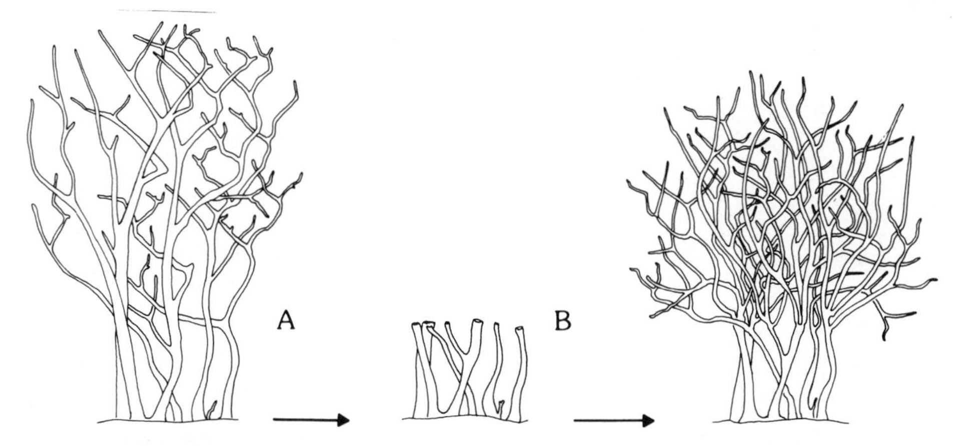

Fig. 10-7. Thinning out. Multiple-stemmed shrubs should be thinned out by cutting old mature stems off as close to the ground as possible. They will be replaced by new and vigorous shoots. If the shrub A is pruned as shown at B, lateral buds will be stimulated to grow into the tangle of branches above bare stems.

right at ground level. Heading back should be used only if more bushiness is needed.

A damaged or diseased shrub should be pruned at any time such problems occur in order to maintain the health of the plant.

Flowering shrubs must be pruned at specific times if they are to bloom:

1. Those that bloom in the spring or early summer form their buds on wood of the previous season. They should be pruned *after* they bloom, before new buds begin forming for next year. Examples—*Forsythia, Spiraea, Viburnum.*
2. Those that bloom between mid-summer and late fall form buds on the wood of the present season just before they bloom. These shrubs should be pruned in the winter or early spring. Examples— *Hypericum, Buddleia, Hydrangea.*
3. If you are in doubt, most shrubs can be pruned in any year *after* they bloom and will produce flowers in the next year. The accompanying list on page 270 recommends the best pruning time for a number of ornamental trees and shrubs.

Hedges

Hedges are simply shrubs placed closely together for their particular effect and should be given the same care all shrubs benefit from. Hedge plants can be planted in a single row, evenly spaced, or for more density, in a double row with the plants of one row alternating with those of the other, zigzag fashion.

X X X X X X X X X X X

or

X X X X X X X X X X X

X X X X X X X X X X

Informal hedges develop when shrubs are allowed to grow in their natural form and are pruned only when necessary for normal maintenance. Formal hedges are produced by shearing certain shrubs such as privet, box, and barberry to create living walls of plants.

The pruning of formal hedges should be begun when the plants are young. Prune the newly planted shrubs back severely so that a dense, bushy plant is produced right from the beginning. Cut back about half of the new growth each year so that the desired hedge height is reached slowly and the plants become full and beautiful as they mature. Depending on the species, most formal hedges need at least two trimmings a year to keep them from becoming shaggy-looking. To produce neat, straight-sided hedges, drop a plumb line from a ladder placed next to the hedge and/or sight along a straight line tied between two stakes. In any event, the shrubs should be pruned so that the direct sunlight can reach leaves all the way through the shrubs, and to the lowest levels. A hedge should never be wider at the top than it is at the bottom. (See Figure 10-8).

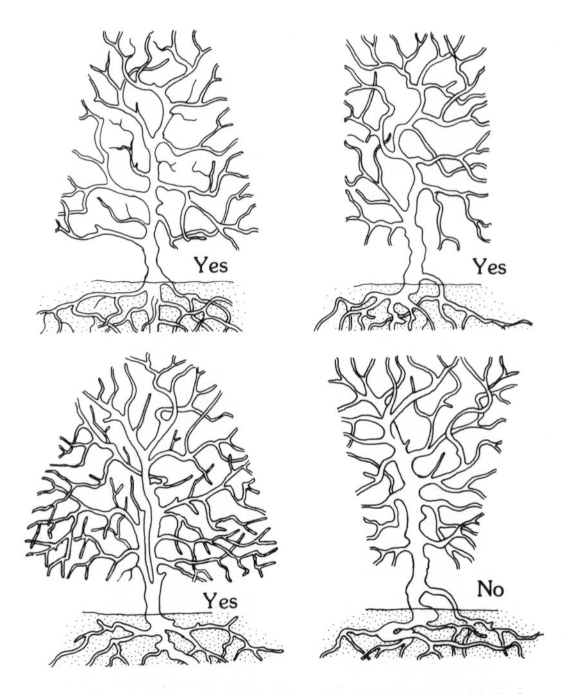

Fig. 10-8. A dense hedge will develop only if the leaves are sufficiently exposed to light. The hedges in the lower figures are shaped so that the lower branches are shaded and growth will be poor.

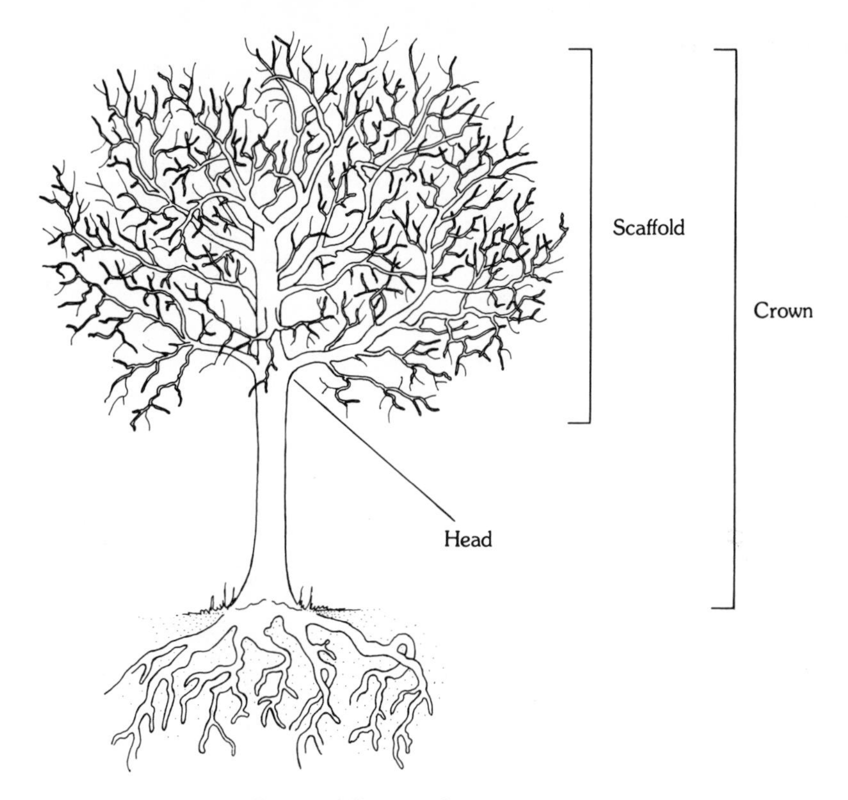

Fig. 10-9. Regions of tree.

Trees

Large healthy trees are likely to require very little pruning. Old damaged trees suffering from internally diseased tissues, broken limbs, and injured roots may require the attention of a professional tree worker and special equipment to reach repairable areas with safety.

To remove a large injured, diseased, or obstructing limb, first remove most of the branches and stems from the length of it, then suport it by tying the end of the limb to another strong branch overhead or to a nearby tree. Cut the limb off by using the "jump cut" which is illustrated in Fig. 10-10. This method allows the branch to be removed safely without tearing into the bark below.

It is always important to repair an injured plant promptly if the damage is likely to be compounded by further injury or disease. But when repairs or pruning are optional, they are best done just before the normal growth period begins in the spring. At this time a good assessment can be made of the branches before the leaves conceal them, and the tissues will heal quickly and new growth will cover the wounds.

However, trees that "bleed" should not have any cuts made on them in the spring. Sugar maple, birch, and walnut ooze sap profusely from cut surfaces which attracts insects and disease organisms and delays healing. Such trees are best pruned in midsummer when the sap flow has slowed.

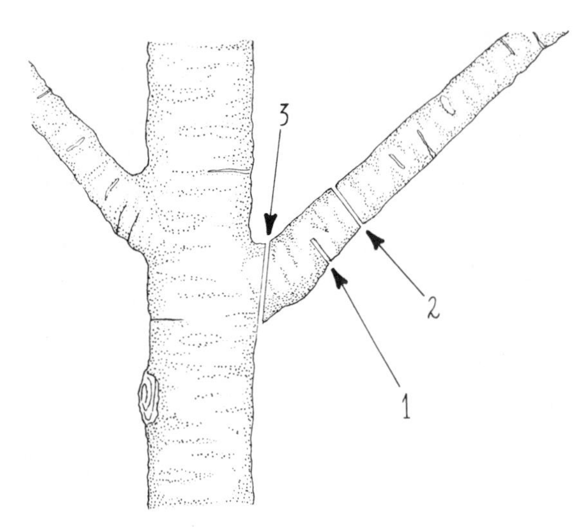

Fig. 10-10. The "jump cut" method for removing large branches from trees. Make the initial cut upward partway through limb. Then cut downward to remove most of limb. If the weight causes the limb to break off, the break will "jump" to the first cut, thus not tearing into the main trunk. Finish with a smooth cut close to the trunk.

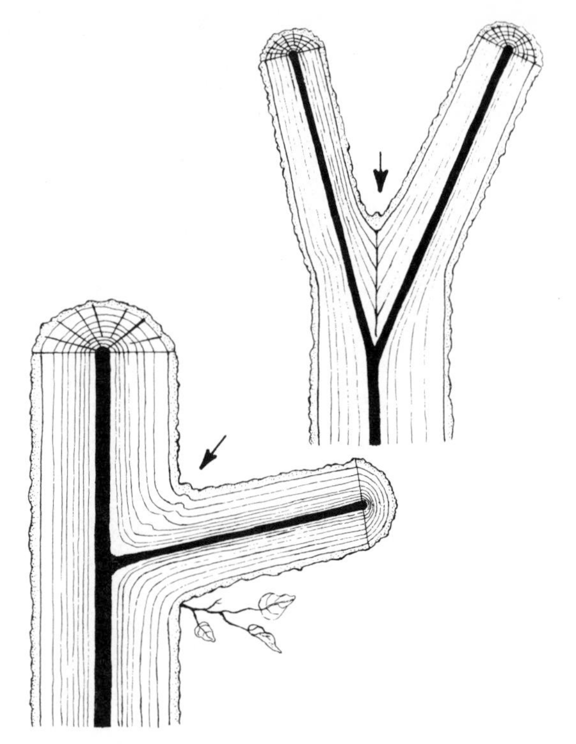

Fig. 10-11. Branches forming a narrow angle split apart readily along the grain of the main trunk. Branches forming a right angle are held strongly together by their grain.

Stems, of course, increase in diameter over the years, and because of this lateral expansion, branches may interfere with each other's growth. It may be difficult to imagine two finger-sized branches a foot apart ever crowding each other, but in time they may, if remedial measures are not taken. Branches forming a right angle to the stem from which they are derived will produce strong growth, while narrow angles cause weak joints where bark tissues meet, and stem growth becomes uneven. With pressure the layers are easily separated and the wood splits. For this reason, when thinning out trees (and some shrubs as well) remove those branches forming the narrowest angles with other stems. If both branches are nearly equal in size, remove the branch with fewer leaves on it since it will have less effect on the health of the tree in comparison with a branch with many productive leaves on it.

If, for any reason, one branch cannot be removed, several other solutions are possible. If one branch is cut partially back, its growth will be slowed and the other branch will become stronger. Growth may also be slowed down by weighting down one branch. The flow of food and water is generally in a vertical direction, and the weighted, horizontal branch will not develop as rapidly as its more vertical companion. If both branches must be allowed to grow equally, a board or stick with notched and padded ends may be placed between them to spread them far enough so they don't meet. Caution must be taken not to injure the bark in doing this.

If it should be necessary to support a weakened branch or tree trunk, never use naked guy wires or any other material wrapped around the branch in such a way

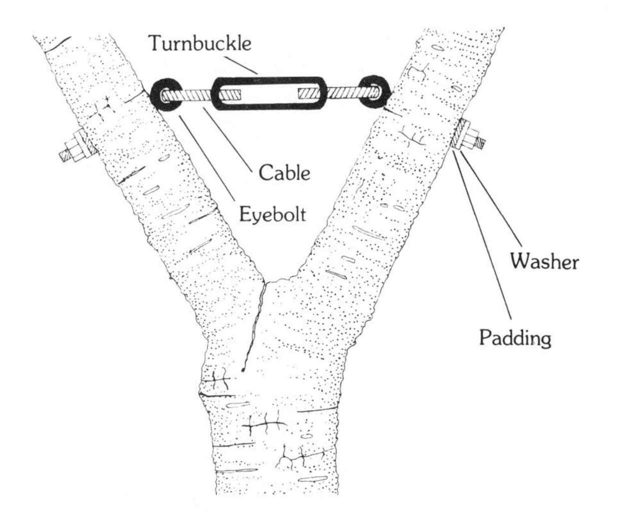

Fig. 10-12. When narrow crotches threaten to split, they can be supported by rods or eyebolts in holes drilled through both members.

that the phloem may become constricted. Use padded supports or old pieces of garden hose to prevent supporting lines from cutting into the wood. If a permanent support is required, such as might be needed to prevent a narrow crotch from splitting, drill eyebolts directly into the wood or run a bolt all the way through both branches. (See Fig. 10-12.)

If a wound is caused in the tree trunk by mechanical injury or disease, healing can be encouraged even though pruning off the damaged part is not possible. Make a longitudinal oval cut around the wound to smooth out torn edges. Tissue repair tends to be most easily accomplished in the longitudinal direction corresponding to the direction of flow of vascular fluids. Bark tissue often dies along horizontal cuts in the area of a wound, and square wound areas allow

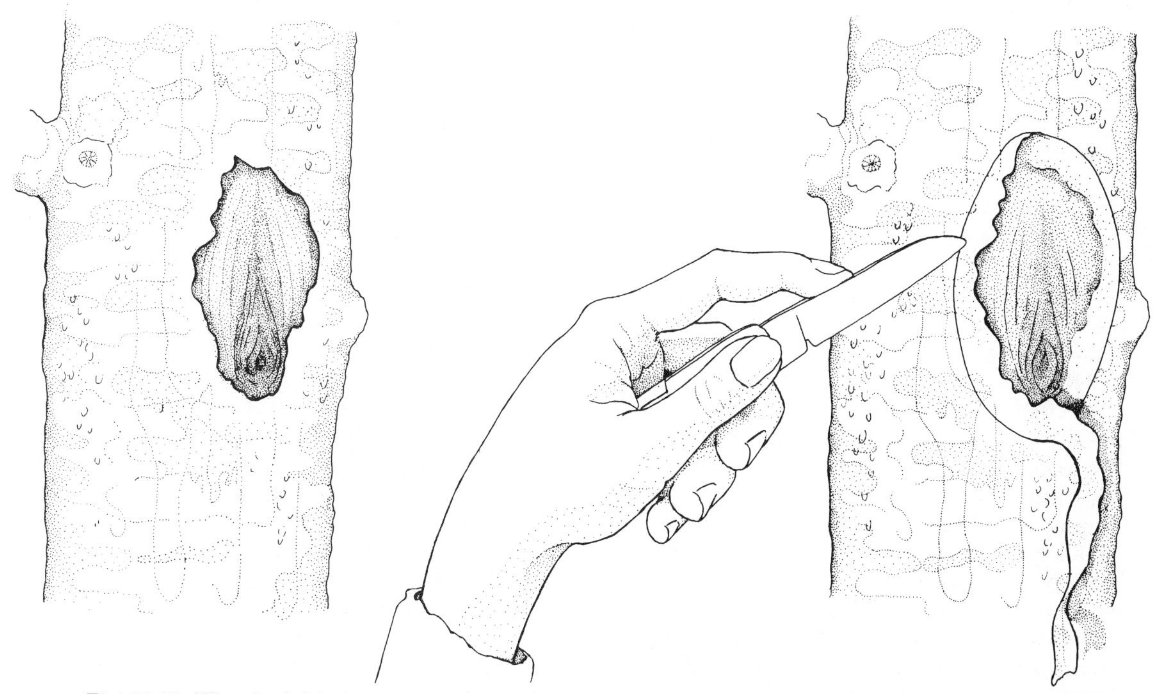

Fig. 10-13. When bark injuries occur on the trunks of large trees, it is important to speed healing and prevent further damage by insects or a disease. The edges of the wound should be trimmed as shown (it is no longer considered necessary to cover with wound paint).

insects and diseases easy entrance, inviting secondary damage. It will not be necessary to apply wound sealers because the wound will heal more rapidly if they are not used.

Recent studies have shown that after a tree is wounded, the living cells beneath the wound react by forming compounds that serve as inhibitors to a wide variety of potentially damaging organisms. In a sense, a tree produces its own wound dressing, and the natural defense reactions within the tree will keep out decay and facilitate rapid healing. The application of wound sealers may interfere with these natural reactions, and the use of them is now being discouraged by the U. S. Forest Sciences Laboratory.

If you feel that you must apply a wound dressing, use a commercial wound paint which is a harmless asphalt material with a fungicide added. Rubber-based or linseed-oil paints, shellac, and spar varnish also will not harm plants, but turpentine and creosote will kill cambium and should never be used.

Flowering trees should have only the pruning necessary for maintaining a strong tree. Many branches mean many flowers and maximum beauty. Any necessary pruning should usually be done after blooming.

Shade trees allowed to keep their lower branches will provide more ample shade. Cut back the tips of lower branches to increase lateral branch growth. Street trees often need to be pruned to permit walking and parking beneath them. They should also be thinned to allow light to pass through them, especially that from street lights at night. Power lines passing through trees can be cleared without chopping ugly holes in the tree. Remove just enough small branches so that wires can pass through the tree inconspicuously or so that the weight of the main branch is reduced enough that it rises above the wires. Future growth can then be guided away from the wires so that the tree remains attractive.

Dwarf trees are usually produced by grafting onto slow-growing rootstock, by root pruning, or by pot confinement. Dwarf fruit trees, which produce normal-sized fruit, can be a home gardener's delight because of the manageable size of the trees. They do however require careful pruning to open them up to provide air and sun for good ripening of the fruit. A book on dwarf fruit trees should be consulted for advice on specific trees.

Coniferous Evegreens

Evergreen trees such as pines, firs, and spruces are not easily rejuvenated or repaired. Old wood does not give rise to new shoots, so lost lower limbs cannot be replaced. Top repairs only are possible.

Most conifers exhibit strong apical dominance so that lateral branch development is uniformly inhibited near the top and the familiar cone-shaped tree results. If an evergreen tree loses its leader shoot, all of the surrounding laterals will begin to develop, and a multi-headed, often misshapen tree is produced. To avoid this, tie a splint to the trunk of the tree and then draw up one of the side branches into a vertical position. Tie it to the splint. It will replace the missing leader in about a year and the splint can be removed.

Although most single-trunked conifers will grow into attractive trees with little or no pruning, trees that will be used as Christmas trees or special ornamentals often have their new growth (called candles) pruned back part way to keep these trees especially well formed and to increase their bushiness. This kind of pruning is best left to the experienced experts in the nurseries, and pruning on the home property should usually be limited to curtailing a twig or two that may stray from the desired form.

The multi-stemmed conifers such as arborvitae, hemlock, cypress, yew, and juniper will respond to light pruning at almost any time of the year. Such plants will recover most quickly from their "haircut" if the trimming is done in the spring just

Fig. 10-14. Evergreens produce a leading shoot with several other shoots clustered below it. If the leader is damaged by insects or weather, one of the nearby shoots can be tied into an upright position along a wooden splint, as shown. This shoot will replace the damaged leader in a year or two.

before new growth starts. Carefully selecting individual branches and nipping them out at a point beneath an overlying branch will maintain the radiating, feathery appearance of conifers such as junipers and keep them to an appropriate size for many years. Badly overgrown conifers of these multi-trunked varieties can be cut back as much as 5 or 6 feet (±2 meters) and still produce an attractive plant. Work over the whole plant to restore its symmetry and form, but never cut beyond **green** foliage. Old wood without foliage will not produce new shoots.

For hedges of hemlock, yew, or arborvitae, shearing produces a more formal appearance and a dense, impenetrable wall. Such evergreens easily give rise to new shoots from near ground level to the top. Since these conifers are not harmed by between-season pruning, mid-December pruning can provide boughs for Christmas decorating.

FANCY FUN

Training plants into artistic forms called espaliers, topiaries, and bonsai is an ancient skill that is still popular in many parts of the world. Since producing these delightful creations requires quite a bit of time and effort, in the United States, in these days of high labor costs, we are most likely to see them as special attractions in botanical gardens, at flower shows, and in the private home or garden of the hobbyist who finds them a unique and satisfying challenge.

Espaliers

Espaliers are trees and shrubs trained to grow two-dimensionally (i.e. in one plane only) and are usually supported against walls, fences, or trellises. This form of trained plant originated in southern Europe and was commonly used for fruit trees against walls in the Italian and French Renaissance palace gardens. The trees were pruned into intricate and rather formal patterns so that they not only produced good fruits in a limited space but added to the artistic quality of the gardens as well. Espalier fruit trees were commonly grown by our early colonists, and today we still see some very old espaliers that have been maintained in some of the colonial gardens

Fig. 10-15. An espalier-trained pear tree produced by selectively pruning out young branches to maintain a pattern.

that are popular tourist attractions.

Dwarf peaches and pears are the easiest fruit trees for espaliers but other trees and shrubs may also be used. *Pyracantha*, *Lantana*, flowering quince, *Cotoneaster*, privet, and some of the viburnums are good choices. It is important to choose plants that put out buds readily and can tolerate severe pruning.

Begin with young plants, plan a pattern, and then prune individual branches to produce shoots to follow the pattern. Supports are often needed to hold the shoots in place against a framework.

Topiaries

Topiaries are horticultural sculptures. The exact origin of this art is a bit obscure but the word is derived from the Latin *topiarius* that the Romans defined as "ornamental gardener" (which, incidentally, was the highest class of slaves). Descriptions of the beautiful gardens of the private estates and the imperial palaces of ancient Rome make it clear that topiaries were commonplace among the many ornamental plants and that the art of topiary was already well advanced during the days of the Roman Empire.

The Italian, French, and English garden of the 1600's and 1700's found topiary very accommodating to the ornate and geometrical designs that were so popular. These were gardens designed exclusively for pleasure, and toward the end of this period topiary work reached ridiculous extremes. French and English gardens in particular exhibited whimsical animals, people, chessmen, birds, and creatures-of-the-imagination as well as elaborate geometric shapes. In the more serious times that followed, top-

Fig. 10-16. A turban-shaped boxwood topiary. The plant receives a weekly shearing to maintain its design.

iary lost favor and became a much more subdued garden form, if used at all. Many of the surviving gardens of Europe still exhibit some of the topiary work that characterized them so many years ago.

In colonial America, topiaries, along with espaliers, were occasionally used in gardens, but they were never as elaborate as the earlier European versions. The formal topiary garden at Longwood Gardens near Philadelphia and the amusing figures along the road to Disneyworld in Florida are two notable exhibits of topiary work in the United States today.

A topiary can be formed with one or more plants depending on the size and shape of the intended form. The plants should be trimmed as soon as they are planted to establishe directions of growth. As the scaffold of branches grows, it is gradually sheared into shape and then pruned regularly (perhaps weekly) to maintain the shape once it is attained. Plants that produce leafy twigs are most suitable. Privet, *Rhamnus*, *Buxus*, and the evergreens *Taxus*, juniper, and arborvitae will prune easily into shape in a few years.

Bonsai

Bonsai, which is the art of growing miniature trees in pots, originated in China and was perfected by the Japanese more than 1000 years ago. It is still a popular and much-used art form, not only in Japan but throughout the Orient, and has, in recent years, attained international favor.

Basically, a bonsai is a miniature, perfectly proportioned tree or shrub, not a malproportioned dwarf! One grower says that he tries to make his plants look as they would if you were viewing a normal-sized plant at a distance. In true Japanese fashion bonsai should suggest a landscape or an environment. Thus we see trees leaning as if windblown or gnarled and twisted as if clinging to a precipice. Sometimes a whole grove of trees in a pot is designed to simulate a forest.

Fig. 10-17. A pine tree miniaturized into a bonsai creation. Both roots and branches are pruned back severely to produce dwarfing.

Bonsai trees are cultivated in shallow pots to restrict root growth. Both roots and branches are pruned to create and maintain a miniature-sized plant in the intended form. Branches may be further shaped with soft copper wire.

Most woody bonsai are not suitable as houseplants, since the best subjects are normal "outdoor" plants and survive only if kept outdoors subject to the usual seasonal changes. Almost any tree or shrub (especially dwarf varieties) can be miniaturized for a bonsai and, given proper care, will survive for the normal life span of the species. A few tender, woody plants such as *Azalea*, cycads, and *Fuchsia* can be trained for indoor growing, and some herbaceous plants can be miniaturized effectively. Or you can get a bonsai effect with certain houseplants that resemble trees or have the appropriate appearance. Jadeplant *(Crassula argentea)*, ming aralia *(Polyscias fruticosa)*, or ming asparagus *(Asparagus densiflorus "Myriocladus")* are especially suitable.

Bonsai has become a popular form of plant growing in our country in recent years and many gardeners are becoming quite skilled at it. It is unlikely, however, that anyone will ever surpass the Japanese in producing truly exquisite bonsai. They are justly proud of their art, and, in commemortion of our nation's bicentennial anniversary in 1976, Japan presented a collection of ancient bonsai to the National Arboretum in Washington. The trees are attractively displayed in a lath-house-like building which one enters through a Japanese garden. The oldest tree in the collection is a magnificent 350-year-old pine that stands about three feet (one meter) tall.

Any of the trained plants are fun to produce but do require special skills and a great deal of patience. If you wish to develop these skills, specialized gardening books should be consulted for specific instructions for growing espaliers, topiaries or bonsai.

GRAFTING

Grafting is an ancient art that has served gardeners well in many ways. Grafts may be used as unique solutions to many gardening problems. In a graft, a piece of one plant, called a **scion**, is united with another, called a **stock** or **rootstock**, by the cambium layers of each. (It should be noted that since cambium is necessary for a successful graft union, monocotyledonous plants, which usually lack cambium, cannot be grafted. Only dicotyledons make good subjects for grafts.)

Some grafts are a form of asexual reproduction. If you don't have room for several apple trees, plant one tree with branches of several other varieties of apple grafted to it. Since a branch retains its own genetic make-up, each grafted scion will continue to produce its particular type of apple. Or graft a scion of a seedless variety of plant to a rootstock of a related plant and you will have a new plant of the seedless variety.

Grafts can be used to repair or to add structural strength to many plants. If your trees have been gnawed by rabbits, repair the wounded trunks with bridge grafts or replace whole branches with whip-and-tongue or side grafts (grafts are described below). Natural braces for trees can be formed using neighboring twigs as scions for branch-to-branch supports. You can even join weak, trailing forms of plants to sturdy stocks for support. You have probably seen some of the epiphytic cacti (e.g., *Rhipsalis* and *Zygocactus*) grafted onto stocks of *Pereskia*, a common item in the local plant stores.

Although no genetic changes occur in grafts, scions and stocks may impart certain beneficial effects to each other. For example, in roses, tough winter-hardy varieties of rootstocks are used to support the more delicate hybrid-tea relatives, or one variety of grape may be grafted to a mildew-or insect-resistant variety to produce a uniformly healthy plant.

For the home gardener, one of the most important uses of grafts is in the production of dwarf fruit trees. When a naturally dwarf variety of plant and a normal-sized variety are grafted together, the normal portion will be reduced in size. Dwarf fruit trees are produced by grafts made between normal, fruit-producing trees and dwarfing stocks (usually rootstocks). Actually the size of the dwarf is determined by the vigor of the scion variety as well as by the dwarfing effect of the stock. For example, most dwarf apple trees are grafted onto Malling crab apple (a natural dwarf) rootstocks, of which there are several varieties. The M-9 rootstock produces trees that are 6-9 feet (2-3 m.) tall, while those of the semi-dwarf varieties reach 12-15 feet (4-5m.).

Another way of producing a dwarf tree is by using three sections, instead of two, grafted together. In apples, a full-sized tree provides the roots, a portion of a Malling tree provides an interstem or mid-section of the trunk, and a standard tree provides the top fruiting portion. The dwarf trunk section reduces the height of the tree while allowing the desired variety of fruit to be grown, and there is an ample root system to support them both. Bartlett-pear dwarfs are produced in three sections using quince roots, which have a dwarfing effect but with which Bartlett is not compatible. An **interstem** of "Old Home" pear is used since it is compatible with both the quince and the Bartlett. Other kinds of pear may be grafted directly onto quince for dwarfing.

The best time for grafting is just before spring growth begins. To produce a graft, select clean, healthy scion and stock material that are closely related. The more closely related they are, the more likely the graft will be successful. Use a very sharp knife for cutting (professional horticulturists have an assortment of knives appropriate for the various kinds of grafts). Cut the piece or pieces so that there are no frayed or ragged edges. Always join the scion by its basal end to the stock. It is important that the apical and basal ends not be reversed. Since new cells in stems are produced by the cambium, match the pieces so that some of the cambium of the scion is in contact with some of the cambium of the stock. If this is done, new cells from each will be produced adjacent to each other and a firm union will be formed between them. Keep the cut surfaces moist while matching them. Tie or tape them together for support, and then cover the graft area with grafting wax or compound over the outside to prevent drying out and diseases. The tape should be removed after growth begins so that it does not cause a restriction.

The grafting of woody plants is commonly used as a means of obtaining plants more rapidly than possible from seed or of circumventing some of nature's hazards and quirks. Herbaceous grafts are likely to be performed only as demonstrations or novelties or used in scientific studies designed to seek an understanding of the transmission of hormones, nutrients, diseases, etc., between plants.

The following list describes most of the common types of grafts, and the accompanying drawings illustrate how the stock and scion pieces should be cut and matched for a successful union.

Approach graft—Remove a tiny slice from one side of each stem. Press the wounded areas together and bind firmly until healed. Use this graft to rescue trees with damaged roots. The damaged tree can be grafted to a sucker (if present) or to a seedling tree planted next to it. The root system of the support plant will replace or supplement the damaged roots. The approach graft may also be used on herbaceous plants to produce plants with more than one variety or color of flower, etc. For example, a geranium may be produced with some branches bearing white flowers and others bearing red. The root portion of one plant may be cut away after the graft union is healed.

Wedge graft—A tapered scion is fitted into a V-shaped cut in a stock of nearly equal diameter. Good for herbaceous plants (e.g., tomato and peony), soft-wooded shrubs (e.g., lilac and rhododendrons), and grape vines. This graft is very easy to do but not very strong, and until growth occurs the graft may need to be suported by a splint or stake.

Saddle or inverted wedge graft—Simply the reverse of the wedge graft in which the stock is tapered and the scion is notched with a V-cut. It may be used as an alternate to the wedge graft.

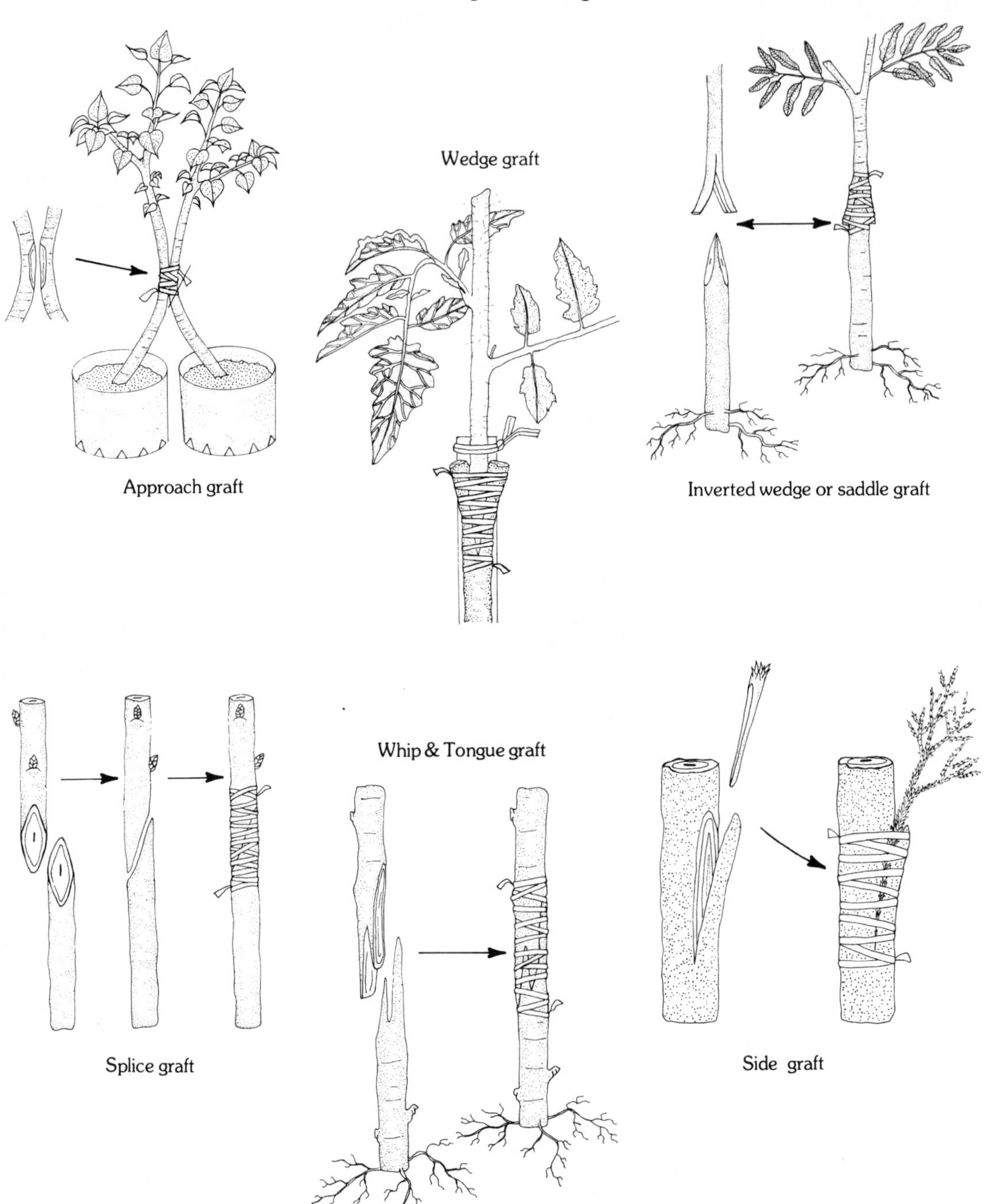

Fig. 10-18. Common types of grafts described in text.

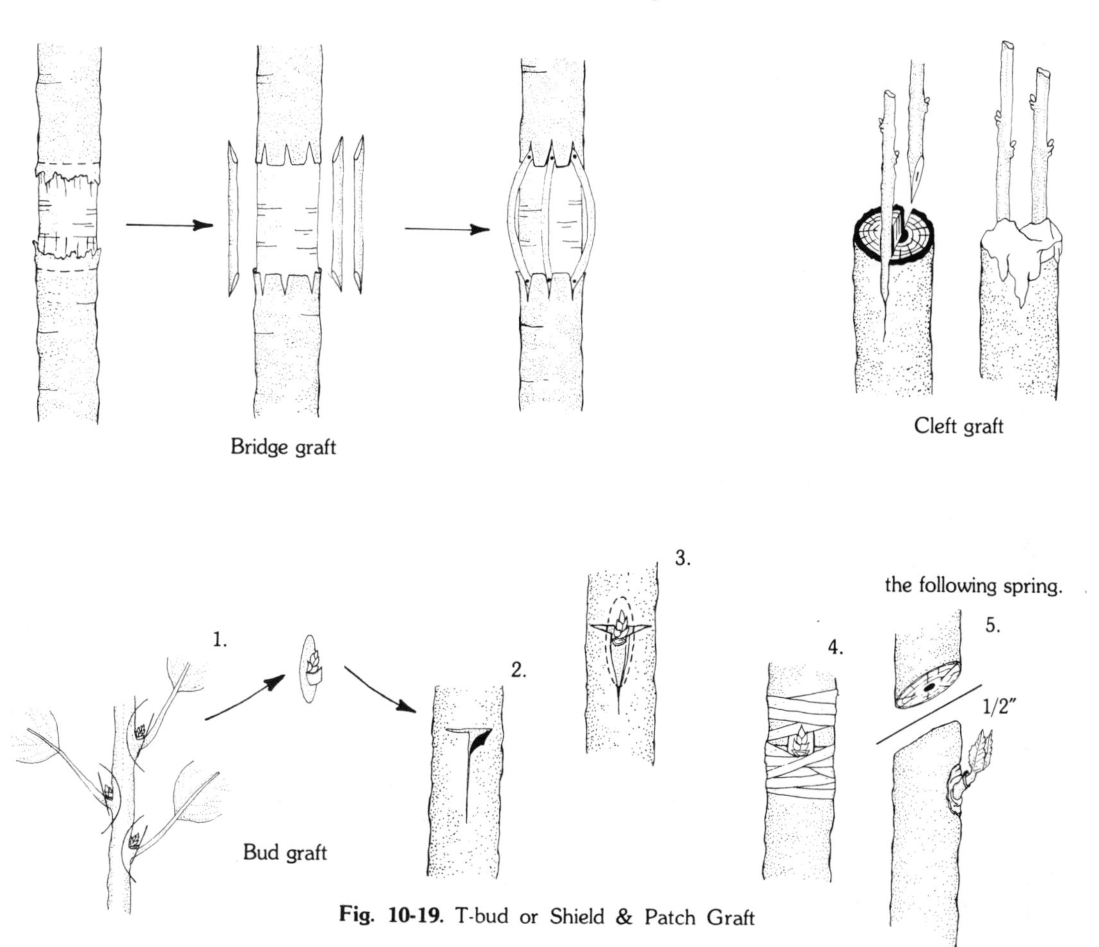

Fig. 10-19. T-bud or Shield & Patch Graft

Splice or English speed graft—Make a single oblique cut on a scion and on a stock of nearly equal diameter. Tie the cut surfaces tightly together. It is a rather fragile graft but suitable for plants in greenhouses where they are protected from hazardous weather conditions. It is easy to do and fast to heal.

Whip-and-tongue graft—Each year millions of fruit trees and ornamental flowering trees are propagated by grfting them on sturdy rootstocks with the whip-and-tongue method. A pair of oblique cuts produces a tongue-like cleft in small diameter scions and stocks that are then fitted together. The stocks are entire roots of young (1-2 year) seedlings that are collected in the fall and stored in moist sand. The scions are derived from young dormant twigs that are collected in winter. The grafts are performed indoors during the winter and then stored until they can be planted in the spring. Branch grafts can also be done by this method to produce trees such as apple and pear with branches of several related varieties.

Side graft—An excellent graft for scions and stocks of different diameters. A young vigorous branch is tapered to be used as a scion and is inserted into a slanting cut in a large branch or trunk. Broken junipers are often repaired with this graft, and the appearance of many of our favorite trees and shrubs can be improved with branches added by side grafting.

Bridge graft—Used to repair trunks or branches girdled by mice or rabbits, or restricting wires, or damaged by lawnmowers. The injured area is trimmed to remove ragged edges, and several young branches are removed for scions. The scions are tapered at both ends and the ends are then inserted under the bark of the trimmed edge of the wound area, often being nailed in place. Vascular flow is restored by means of the scions, and as they increase in diameter the wound area will be recovered.

Cleft graft—Generally used for "top-working" a tree. If the scaffold is broken off the trunk, as in a storm, the remaining stump can serve as a rootstock to restore the tree. Or a "scrub" apple or other fruit tree can have scaffold branches of a desirable variety. Vigorous young scions are usually cut while dormant and tapered at one end. In early spring the selected stock tree or damaged stump is cut squarely across

Table 10-1. When to

Spring flowering trees and shrubs which should be pruned after flowering.

Common Name	Scientific Name	Common Name	Scientific Name
Shadblow	*Amelanchier*	Honeysuckle	*Lonicera*
Azalea	*Azalea*	Magnolia	*Magnolia*
Barberry	*Berberis*	Crabapple	*Malus*
Sweetshrub	*Calycanthus*	Mock orange	*Philadelphus*
Peashrub	*Carangana*	Andromeda	*Pieris*
Bittersweet	*Celastrus orbiculatus*	Firethorn	*Pyracantha*
Redbud	*Cercis*	Flowering cherry and plum	*Prunus*
Flowering quince	*Chaenomeles*	Rhododendron	*Rhododendron*
Smoketree	*Cotinus coggygria*	Black jetbead	*Rhodotypus scanden*
Flowering dogwood	*Cornus florida*	Climbers and shrub roses	*Rosa*
Hawthorn	*Crategus*	Mountain ash	
Deutzia	*Deutzia*	Spirea	*Sorbus*
Forsythia	*Forsythia*	Japanese snowball	*Spiraea*
Mountain laurel	*Kalmia latifolia*	Lilacs	*Styrax japonica*
Beautybush	*Kolkwitzia amabilis*	Viburnum, Wayfaring tree,	*Syringa*
Privet	*Ligustrum*	European cranberrybush	*Viburnum*

and a wedge-shaped cut made in it with a chisel. Two scions are then placed in the cut on opposite sides of the stock so that the cambium of the scion matches that of the stock. If both unions are successful one scion is usually removed to allow better development of the other one.

Bud grafts—There are several kinds of bud grafts but each utilizes a dormant bud removed from one plant and grafted onto another in an incision in the bark of the stock plant. These are good grafts for producing several varieties on one plant, such as apples, or with plants that produce excessive wound fluids from cut xylem, or when scion material is in short supply. Budding is commonly used for the commercial production of apple and other fruit trees and for tea and hybrid tea roses which are budded onto stocks with strong root systems. The bud should be wrapped firmly in place to apply adequate pressure for healing. These grafts are usually done in July and August when the bark separates most easily from the wood to facilitate bud insertion.

une Trees and Shrubs.

Summer flowering trees and shrubs which should be pruned before spring growth begins

Common name	Scientific name
Aralia	*Acanthopanax*
Glossy Abelia	*Abelia grandiflora*
Silk tree	*Albizia*
Butterflybush	*Buddleia*
Beautyberry	*Callicarpa*
Clematis	*Clematis*
Shrub-althea	*Hibiscus syriacus*
P.G. Hydrangea	*Hydrangea paniculata 'Grandiflora'*
Oakleaf Hydrangea	*Hydrangea quercifolia*
Goldenrain tree	*Koelreuteria paniculata*
Hybrid Tea	*Rosa*

Trees and shrubs which should be pruned both before and after bloom

Common name	Scientific name
Red Osier dogwood	*Cornus stolonifera*
Cotoneaster	*Cotoneaster*
Oregon hollygrape	*Mahonia aquifolium*
Anthony Waterer and Frobel spirea	*Spiraea bumalda*
Snowberry and Chenault coralberry	*Symphoricarpos*
Rose Weigela	*Weigela florida*

11. LANDSCAPE PLANNING

he history of gardens mirrors human history. Our early ancestors were, of necessity, nomadic, following meat-providing animals as they both searched for edible plants. With ingenuity these early humans soon learned to plant and harvest, and relatively permanent settlements became possible.

As civilization advanced and technological developments were achieved, time for leisure became available and with it the increase in artistic expression. Gardens were a way of experiencing peace and beauty while at the same time exhibiting personal wealth and achievements. There have been styles and fashions, fads and trends in the use of plant material just as there have been in art, clothing, buildings, literature, entertainment, and all other human endeavors.

To recount the entire history of gardens and landscaping would be unreasonable here, but such a history would reveal a long and fascinating past. The Bible records the idealized Garden of Eden, and in the literature of most other religions also, their deities are usually placed in garden-like settings. The grounds of temples, mosques, churches, and monasteries all over the world are attempts to simulate these imaginary gardens.

In many ancient cultures private and royal gardens were so important that they were pictured on the walls of stone temples and tombs. Some of these walls still remain though the homes and palaces are gone. In ancient Egypt, for example, shade for comfort and water to sustain life were essential, so the Egyptians built high-walled estates with water reservoirs in the form of decorative pools from which slaves dipped water for the plants.

Fig. 11-1. Monasteries such as this served to preserve knowledge through the Dark Ages, and every monastery had a cloistered garden where the monks preserved herbs and other plants for both intellectual and esthetic purposes.

In the emptiness of lifeless deserts, green oases were created and gardens in the centers of cities became commonplace. As civilization spread to Europe and throughout Asia, so too did the gardens. Later, as the walls of medieval cities were discarded, a new period of beauty and esthetic enlightenment emerged. The still-beautiful formal gardens of Versailles in France and the naturalistic wooded estates of Britain were envied the world over.

The designs of the past were usually on a large scale and reflected the social and economic moods of the time. While it is of value to know the past we must not copy it. We may be inspired by it, we may adapt from it, we may admire it, but we should create designs to express our own particular attitudes and life styles.

LANDSCAPING TODAY

Nature has always been the main inspiration and it is still the most satisfactory one, but we view nature today differently from the way we did in the past, and we must try to express that view in our surroundings. Landscape design brings art and nature together. We attempt to tame nature just enough to achieve a sense of orderliness and refinement.

There are today occasional individuals who want to "let Nature have her way" with a piece of property. It must be admitted that this is certainly an energy-and time-conserving method of designing but one which would hardly evoke much admiration for a harmonious and satisfying effect. Ideally a good design "stirs the mental process, is stored in the mind and becomes a pleasant memory long after the design is no longer being observed" (Ortloff and Raymore, *The Book of Landscape Design*; Morrow and Co., N. Y., 1975).

Fig. 11-2. A well-designed landscape has served this house well for over 50 years and is still attractive and functional.

Unlike static forms of art, landscape compositions are strongly inspired by the "genius of the site", that is, the topography and other characteristics of the particular piece of land, and by the dynamic, ever-changing nature of the living materials being used.

A good designer must first of all develop a "seeing eye" and this is emphatically stressed in design courses. Whether a plan is formal, informal, or natural, the trained and experienced eye will observe the subtle, lasting features that lead to an appreciation of the art and which are unrecognized by the untrained or indifferent observer.

Landscape architecture is constantly changing. New plants are developed, new styles emerge, new social and economic ideas prevail. Changes are desirable and necessary but should be based on the basic principles of good design common to all forms of art.

Why Should You Landscape?

In addition to being an artistic creation a landscape should be utilitarian, practical, and economical, as well. There are, in fact, numerous functional and decorative purposes for landscaping your property:

Fig. 11-3. A house without landscaping looks forlorn and unloved.

1. To give a home a suitable setting
2. To extend living space
3. To preserve and enhance beauty
4. To take advantage of a natural setting
5. To increase property and home value
6. To make surroundings functional
7. To provide easy maintenance
8. To create a more pleasant place to live

Whether the various ideas of a plan are for utilitarian or soul-restoring purposes, if the landscaping is well designed the results will be most satisfactory.

Landscaping Professionals

It is generally estimated that 10-30% of the cost of owning a residence is for outdoor development. This includes driveways, sidewalks, patios, walls and fences, lawns, plants, and the related professional services. With this much investment required the services of trained individuals are well worthwhile.

Ideally a landscape architect will be employed to work with the building architect and the contractors to produce a harmonious result for the house and site beginning with the initial plans. Whether the property already has a home built upon it or not, a landscape professional can contribute significantly.

Landscape plans are available from several kinds of qualified professionals. A landscape architect has a degree from an accredited college or university and qualifies for membership in the American Society of Landscape Architects. A few states grant licenses to those passing special board examinations. A landscape architect, like a building architect, provides plans for a fee. He (she) does not sell plants or do constructions but may, upon request, contact and work with those who do.

A landscape designer is professionally trained in specialized design and horticulture courses, usually being granted a certificate by the institution attended. A designer will provide plans and do the follow-up work required to transform the property. He grows many of his own plants and has the heavy equipment (and strong backs) necessary to develop even the most difficult property. The recommended plans will, of course, reflect the plants he has available and the construction skills at which he is most adept.

Nurserymen may or may not be professionally trained, but they are primarily horticulturists. They usually have an enviable sense of the effective use of plants to create attractive gardens and pleasant yard plantings of trees and shrubs. They will provide plans of limited scope emphasizing the plants which they would hope to sell to the homeowners who carry out their plans. Very often nurseries employ landscape designers for customers interested in more extensive plans.

LANDSCAPE PLANNING

A good set of plans for landscaping is a very worthwhile investment in money, if you hire a designer, or in time, if you should do them yourself.

1. It utilizes the land to the best and fullest extent.
2. It avoids costly mistakes.
3. It prevents unsatisfactory appearances.
4. It provides an opportunity to devise and try solutions to problems on paper before executing changes on the land itself.
5. It can break a scheme down into smaller, manageable sections.
6. It allows for easier budgeting of time and money.
7. It discourages haphazard and impulse plantings.
8. It results in a unified, harmonious whole toward which one can proceed over the years with confidence.

It is at least as bad to overlandscape as to

do nothing at all, but most attempts are better than none. Landscape designs are not only for public parks and wealthy estates (of which we have fewer every year) but are even more important for the average to small property. The smaller the site, the less room there is for error and the more carefully it should be planned.

It is also true that a "difficult" piece of property can benefit from professional planning more than a less challenging site since it requires more extensive modifications to make it attractive.

A landscape plan must begin with a feeling of respect for the land itself. Just as people have requirements, so also does the land. Upon this land you will impose structures, plants, and modifications to fit these requirements that will be evaluated in terms of the natural features and assets of the land. As Aldo Leopold said, ". . . that land is a community is the basic concept of ecology, but that land is to be loved and respected is an extension of ethics . . ." (Leopold, Aldo, *A Sand County Almanac.* New York: Oxford University Press, 1977) . Land is a heritage that we pass on to those who come after us. We dare not abuse it.

A house should be attractive and suitable to the family. The garage and service and storage areas should be conveniently situated. Children's play areas and spaces for outdoor sports and social activities should be considered. The need for privacy and seclusion will depend on both personal desires and on the surrounding environment. Gardens for flowers, vegetables, herbs, or fruit trees must provide for the needs of the plants. Finally, the time, money, and effort required to establish and maintain the landscaping must be considered.

A landscape design should be thought of as a collection of three-dimensional spaces; the earth is the floor, the walls are shrubs, fences, or other indicators of vertical boundaries, and the ceiling is the canopy of trees and/or the sky.

The Elements and Principles Design

It is within our property spaces that we attempt to create pleasant and livable extensions of our indoor spaces. We do this by applying the basic elements and principles of design. Although all of our five senses may be stimulated by our environment, only our eyes are perceptive of design. As our eyes wander constantly, we seek patterns and organization that satisfy an inner feeling. A successful design will utilize the following universally recognized component parts in its creation:

Elements of Design

1. Point or accent. A spot or focal point to which our eyes are drawn. This can be achieved in a garden with a flowering tree, a gazebo, a piece of fine sculpture, or other such singular item.
2. Line. Two spots connected so that the eye moves from one to the other. The line may be straight so that the eye moves quickly along it, or it may be a simple or compound curve that leads the eye more gently. Effective lines are strong and definite with sharp, clean corners and graceful, firm curves.
3. Form. Defined by line, form is the shape of things. The form of a plant or structure will often determine the function or appropriate use of it.
4. Color. The hues and intensities perceived under various light conditions.
5. Pattern. Repeated shapes producing a design.

The above elements are utilized in establishing the **Principles of Design:**

6. Proportion and scale. The proper relation between all parts of the design with respect to sizes, extents, masses, etc.
7. Balance. A visual feeling of stability. If the objects in view project a feeling of equal weights such as one would find on a see-saw, the balance is symmetrical or formal. Asymmetrical balance

Fig. 11-4. Elements of Design: **Accent** (upper left): The dogwood at the entrance to this house attracts the eye and is a focal point or accent in the surroundings. **Line** (upper right): The curved line for these foundation plants guides the eye and adds interest and attractiveness. **Form** (lower left): This radiating brick walk leads to the several entrances of this house and creates an unusual shape. **Pattern** (lower right): The small rectangles of various sorts reduce the scale of this tiny garden and create an interesting pattern of living and non-living elements around the accent tree.

using dissimilar components is more subtle and difficult to achieve and is usually associated with informal plans.

8. Repetition. The repeated use or a sequence of objects and/or plants to provide emphasis or a feeling of movement (rhythm). It is related to balance to introduce order and interest into the design. If overdone, monotony may result.
9. Unity. The feeling that all the parts fit together. This can be created by enclosing or enframing an area with plants and structures or tying all parts

Fig. 11-5. Principles of Design. **Proportion** (upper left): The small dogwood trees reduce the apparent height of the house and are in proportion to the height of the line over the carriage drive. **Balance** (upper right): Even in winter the balanced placement of plants with the oriental arch make an attractive appearance. **Unity** (lower left): This fenced dooryard unifies the U-shaped house and the various features and plants within the entrance. **Combination** (lower right): Most landscape plans utilize most, if not all, of the elements and principles of design in the various features. In the last picture above, the brick walk produces a form and a pattern, the lamp is an accent, the well-proportioned dwarf flowering trees are repeated across the front of the house, and the clumps of bright red tulips beneath them add a touch of color. Do you see other elements?

together as with a continuous lawn.

Other terms may be substituted for or added to those above, but these will do for a discussion of landscaping ideas. It is more important to discover ways of applying the elements and principles of design than it is to know what to call them.

Formal and Informal Designs

Today there are two major themes in American design, the formal and the informal. The formal emphasizes the supremacy of Man over Nature. Such plans are geometrical and symmetrically divided by two axial lines.

The classical formal gardens of the past were rectangles based on the "Golden Mean", a reference to Horace's exhortation to avoid extremes and maintain balance. Mathematically the Golden Mean or Golden Section is a relationship of 1.6 to 1.0, a proportion that occurs commonly in nature. When applied to the rectangular formal garden it meant that if the width were 10 feet the length must be 16 feet, or if the width were 16 feet the length had to be 1.6 times that or 25.6 feet. Today our residential gardens are not so mathematically precise, although it will be found that most nicely proportioned landscaping forms will fit or simulate the Golden Mean proportion (see Figure 11-6).

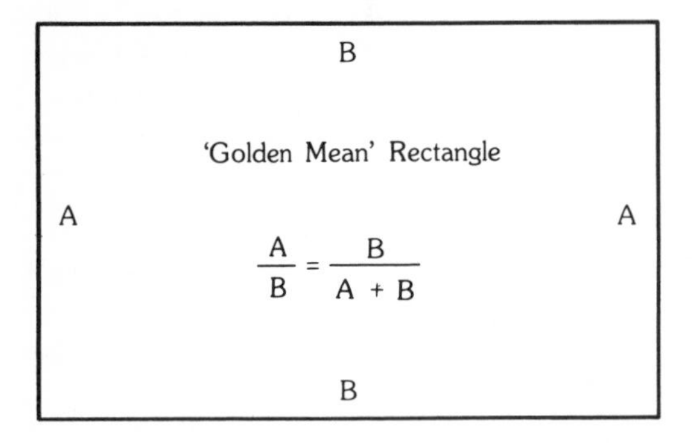

Fig. 11-6. Rectangle illustrating ratio of the Golden Mean.

In the traditional formal garden the main axis terminates in a focal point and all other components are subservient to it. Frequently there is a central accent such as a fountain or statue where the long and short axes cross.

Although straight lines, circle segments, and angles are clearly perceived in a formal plan, it need not be stiff or uncomfortable. Small rectangular lots dominated by a rectangular house and framed by straight streets or sidewalks lend themselves to a formal rather than an informal approach to landscaping, although the proportions are not usually too precise, and bits of informality should be juxtaposed for relief.

The informal or naturalistic mode recognizes our desire to live with nature but not dominate it. The lines are relaxed and free-flowing and arranged to accommodate plants with interesting combinations of colors, textures, and forms. The strong geometry and defining walls are gone and the design is closely related to the natural surroundings. It is difficult to produce an informal design without letting it dissolve into a meaningless jungle of unorganized growth. The informal can be most readily applied to a site that is a bit larger than the normal, with interesting topographical features, and with a house that is rustic or of an irregular shape.

Actually most American homes today exhibit a combination of the formal and the informal, often with emphasis on the formal because it is easier to achieve. If a plan is well executed there will be little awareness of any transitions between the two. The choice of formal of informal or the degree of each should depend upon personal preferences, the surrounding neighborhood, and

Fig. 11-7. The rigidly geometric lines of the formal English garden contrast sharply with the freely flowing lines of an American informal garden.

Fig. 11-8. An informal, rustic house.

the "genius of the site." What is to be avoided at all costs is an unyielding application of either of these modes to the point of stiltedness or artificiality. We Americans are a relaxed, comfort-loving people, and our surroundings should reflect this.

Selecting and Analyzing the Property

When you decide to acquire a home you should begin by selecting a piece of property that fits your personal needs and is economically appropriate. It should be located conveniently close to the places you usually go. Public transportation may be required and, if so, should be within easy walking distance. The surrounding neighborhood should be attractive, with a minimum of heavy traffic and an absence of unsightly junk yards, heavy industry, and the like. Public utilities and services and the presence or absence of paved streets and sidewalks must be considered.

Once a site is selected it should be analyzed for its specific characteristics. Observe the topography and the orientation of the lot. What views does it provide? Is the soil suitable for constructions and growing plants? Of what value are any existing trees? Is the architecture of the planned or existing house suitable for the property? If greenhouses or outbuildings are planned it is important to check on zoning regulations and deed restrictions.

Preliminary to the actual planning must be the recognition of the major features, assets, and problems of the property. The major considerations include the following:

1. View of the property from the street and neighboring properties. The house and its surroundings must blend into, not clash with, the rest of the neighborhood. There can be a world of difference between being distinctive and being a "sore thumb." Landscaping improvements can make any house more attractive and, unless you own the most expensive house on the block, will add quite a bit to its value. Consider the kinds of landscaping that will incorporate some of the better features of the surroundings and make your home an asset to the neighborhood.
2. Grade of the land. Only rarely is a piece of land truly flat. Most lots will have some changes in elevation that will accommodate interesting lines, allow for steps and terraces, and in general improve the effectiveness of a plan. Do not, however, force extensive grade changes onto a piece of property. Natural grades flow easily into the surrounding properties, while created grades, aside from being costly, usually look contrived and tense.

Fig. 11-9. The diagonally positioned steps effectively and attractively cope with the problem of a lot level that is higher than the street level.

Water runs downhill. It would seem that this fact should be universally known, but more than one owner of a suburban home will tell you that his builder did not. The result is wet basements, flooded patios, and soggy gardens. So check the drainage patterns around a house before you buy or build.

Land can be classified by grade into three categories:

Flat land. Less than 5 feet of vertical fall for 100 feet of land.

Rolling or sloping land. About 10 feet of vertical fall for 100 feet of land.

Steep slope. More than 10 feet of vertical fall for 100 feet of land.

Flat land requires little or no grading unless special features are to be added. Rolling hills are generally a welcome feature on acreage, but a sharp drop in land (slope) next to the house can cause problems, and such an area may need to be regraded.

Steep slopes frequently call for construction of steps and terraces, requiring the use of heavy equipment. When regrading for terraces, care should be taken to avoid dividing the land so evenly that it takes on the appearance of gigantic steps.

3. Wind direction. Of all the climate factors, wind is the least constant or predictable. In much of our hemisphere the prevailing winter winds are out of the north and northwest, and these can be very damaging to rose bushes and evergreens in particular. The more gentle winds of summer can sometimes cause damage by snapping off the tender stems of annuals or excessively drying the soil. Plantings of wind-resistant trees and shrubs or the construction of deflecting walls and fences may be called for if these or similar problems occur on your property.

Check snow patterns in the winter for the garden areas that may be deprived of protective snow cover by strong winds. Plant only the hardiest plants in these areas or provide extra mulching in winter. Be particularly aware of tunnel effects between closely placed structures as well as of flat walls that cause winds to ricochet. Both of these situations may cause unexpected damage to plants from strong winds and overdrying. Contrarily, areas that are totally enclosed often suffer from high humidity and other ill effects from the lack of circulating air.

Fig. 11-10. This enclosed sculpture garden allows enough circulation for the widely spaced plants but would create serious humidity problems for more conventional plantings.

4. Orientation of sun. Horizontal sun angles vary considerably through the year. The noon sun in December will flood the interior of a house with bright sunlight through its south exposed windows and shadows will be long, while in June at the same hour the sun will be high overhead and the shadows will be short.

We welcome the warmth of the sun in winter but try to protect ourselves from it in the summer. Sitting porches and patios can be protected from the sun by the well-planned use of roof overhangs, screens of shrubs or lattice work, and large trees. Especially try to avoid using heat-absorbing materials such as brick and concrete in landscaping structures where they will radiate back a day's accumulation of heat from the sun on persons trying to relax in the evening shade.

5. Natural features such as rock outcrops, hills and slopes, streams, ponds,

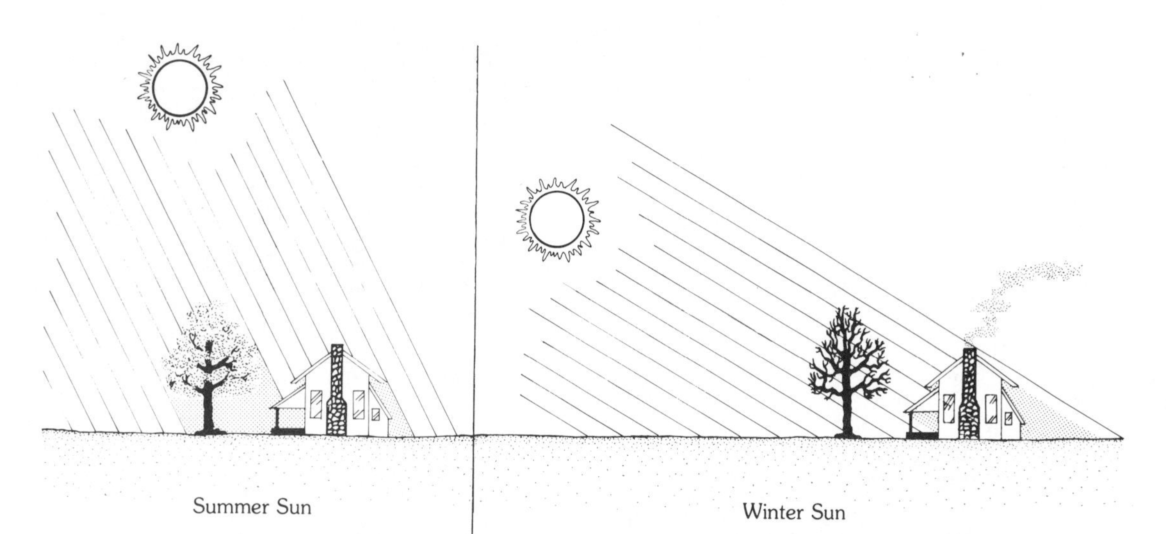

Fig. 11-11. The noon sun in mid-summer (left) is high in the sky, but in mid-winter (right) it is at a lower angle and shines into south-facing windows.

Fig. 11-12. This lovely swan pond was produced from a weedy marsh on the lake side of the property.

Fig. 11-13. This lovely old oak tree was saved by building a retaining wall around a portion of its growing space before other parts of the property were leveled for building.

marshy areas, and existing trees. While large individual rocks and boulders can be moved, albeit with difficulty, to more attractive locations, a rocky outcrop must usually be worked around. Such an outcrop appears hard and cold, and plants around it have a softening effect. Perhaps the area can be turned into a rock garden or, with properly concealed pipes, into a waterfall. In any case it cannot be ignored and will have to be worked into a plan somehow.

Hills and slopes, aside from the drainage problems they may present, can be fine assets for avoiding monotonous landscaping, as can water in ponds and streams. How difficult it would be to create an oriental garden, for example, without a swan pond or an arched bridge over a flowing stream.

Trees are definite assets on a site, and in general they should be retained when possible if they are healthy, not too old, and reasonably spaced. They, along with the other "permanent" features, should be incorporated into the final landscaping design.

6. Good and poor views from the house and yard. A good design does not end at the property line; only the tax assessment does. So consider how you can frame an attractive view as it will be seen from the patio, or design your yard so that it seems to flow pleasantly into the neighboring property. Where unsightly views intrude, screen them out or direct attention in another direction.

Fig. 11-14. If the confinement of animals or children is not required, let backyards flow into each other without obstructing fencing for a pleasant expanse.

Basic Space Planning

Most residential properties have three kinds of outdoor space utilized by its human occupants. (See Measurements for the Home Landscape Plan at end of chapter.) These are the public or foreground space, the service and storage space, and the private or outdoor-living space.

The **public space** should blend house architecture, ornamental plants, and lawn to present a warm, friendly appearance compatible with the general character of the neighborhood. The front door should be the focal point so that it is readily located by visitors. The overall design should create an impression of tasteful simplicity.

Fig. 11-15. A short walkway from the house to the driveway is preferred over a long sidewalk that divides the front lawn out to the street. Notice the convenient broad apron adjoining driveway.

Houses with long straight front walks between the street and the entrance often give an appearance of being perched like a bird house on a pole. To avoid this look, use a gently curved or extra-wide walk or, preferably, place the walk so that it runs parallel to the front of the house. A parallel walk also permits an attractive expanse of uninterrupted front lawn. For suburban homes, the walk generally connects with the driveway which then serves as a walkway to the street. When designing any walk, keep in mind that humans do not turn easily at right angles and tend instead to cut across them. Try to expand or curve a walk where it joins the driveway or connecting sidewalks.

The driveway is a utilitarian feature. If short, it should be as straight as possible, and if long, should not have tight or hazardous curves. It should be simple, direct, and wide.

The entrance is the area first seen by visitors and establishes the mood for the rest of the property. Unfortunately, it is often overdesigned, overplanted and overdecorated. Sometimes it is cluttered up with tasteless statuary and accessories and planted with flower-show-like beds of blooming annuals requiring a great deal of tedious maintenance. Driveways and walks that are outlined with uniform rows of flowers or shrubs create a feeling of rigid formality and detract from other more interesting features. The plants may also make it difficult to alight from cars without stepping into them and can be a problem when clearing snow in the winter.

Fig. 11-16. Carefully selecting plants and positioning them in interesting lines and arrangements adds beauty to the appearance of this home and avoids the rigidity of more formal arrangements.

Plants should be used to perform functions, enhance structure, and create interest in the design. Foundation plantings tie the house visually to the ground and soften its lines, but they are the one feature that, more than anything else, identifies the budget development home. The builders all too often "landscape" all the houses identically with the atrocious "ball and spear" arrangement of low rounded evergreen shrubs (usually beneath the picture window) between tall, pyramidal arborvitae at the corners. While the house should not appear to be floating on a sea of green, generous groupings of plants at corners, entrances, and other strategic places rather than dot-dash lines of them will present a natural transtition between the house and the land.

Fig. 11-17. Typical example of ball-and-spear landscaping.

Tall houses need tall trees and rounded, softening foundation plants. Long, low houses need some vertical plants to interrupt the strong horizontal lines. In both cases the trees and shrubs should frame the front view of the shouse to complement the house and site, not compete with it. Design elements and principles will give guidance and reassurance in planning this front view.

Fig. 11-18. Lovely trees, but in this case they have been allowed to grow out of proportion to the house so that they compete with it for the viewers' attention.

Service and **storage areas** are necessary evils in a landscape design. Garbage and trash cans, clothes poles, gardening tools, lawn mowers, wheel barrows, recycling containers for glass and metals, badminton sets, bicycles, compost piles, and firewood are just some of the items an ordinary family may need to cope with.

Accessibility and concealment are the prime factors to consider when working service and storage areas into a plan. Larger-than-normal garages or additions of the lean-to type may provide an ample amount of storage space. Trash containers and potting sheds, etc. can be screened with hedges, vines, or decorative fencing.

Play areas for small children should be placed where they can be easily observed by adults inside the house but away from windows and flower beds that may be damaged by children's activities. A nice approach is to place the play area where it can be appropriately converted into a garden when the children are grown.

Fig. 11-19. This patio on the east side of the house is warmed by the early morning sun but protected from hot afternoon rays as the sun moves around to the other side of the house.

Nowhere is it more important that the indoors and outdoors complement each other than in the **private** or garden portion of the property. Basically the outside areas should be extensions of the inside rooms of the house. Patios for dining should be placed near the kitchen for easy service. Sunning terraces and swimming pools should have dressing rooms nearby, and flower and vegetable gardens should not be too far from necessary tools and water.

A good backyard will look organized and logical if it utilizes design practices to direct the eye and develop the space effectively. Too often poor planning results in a mere collection of plants distributed here and there, whereas with a bit of consideration a secluded, restful haven of satisfying composition could be achieved.

Drawing the Plan

To produce a landscape design, start with the outline of the property drawn to scale on a piece of graph paper. To this add the house plan, driveway, utility lines, trees to be retained, and any other "unchangeable" features. Structures on neighboring property should be noted and may be added.

Over this graph paper place a sheet of tracing paper and indicate with bubble- like circles the public, service, and private areas of the property. Gardens, play areas, swimming pools, and other large features will also be added to this initial "bubble plan". Use several sheets of tracing paper to try various arrangements until a satisfactory one is achieved.

Now add suitable constructions and groups of plants. It is easy to arrange and rearrange on paper, so doodle away trying various possibilities. You need not think of species of plants or of exact kinds of construction materials at this time. Keep in mind that the house is the most important feature of the design and simply work with masses, spaces, and patterns that will enhance its appearance. Then, when you are satisfied with your basic plan, choose the specific plants and materials that will accomplish your design.

STRUCTURES AND PLANTS FOR YOUR LANDSCAPE PLAN

Man-made constructions and natural materials such as water, rocks, and plants are used in landscaping. The hard, geometric nature of structures contrasts sharply with the free-form naturalness of the others, and together they can be used to carry out a landscape plan whether it is very simple or elaborate.

Structures in the Landscape

Structural features include such things as some surfacing materials, walls and steps, planting beds, edgings, and fences.

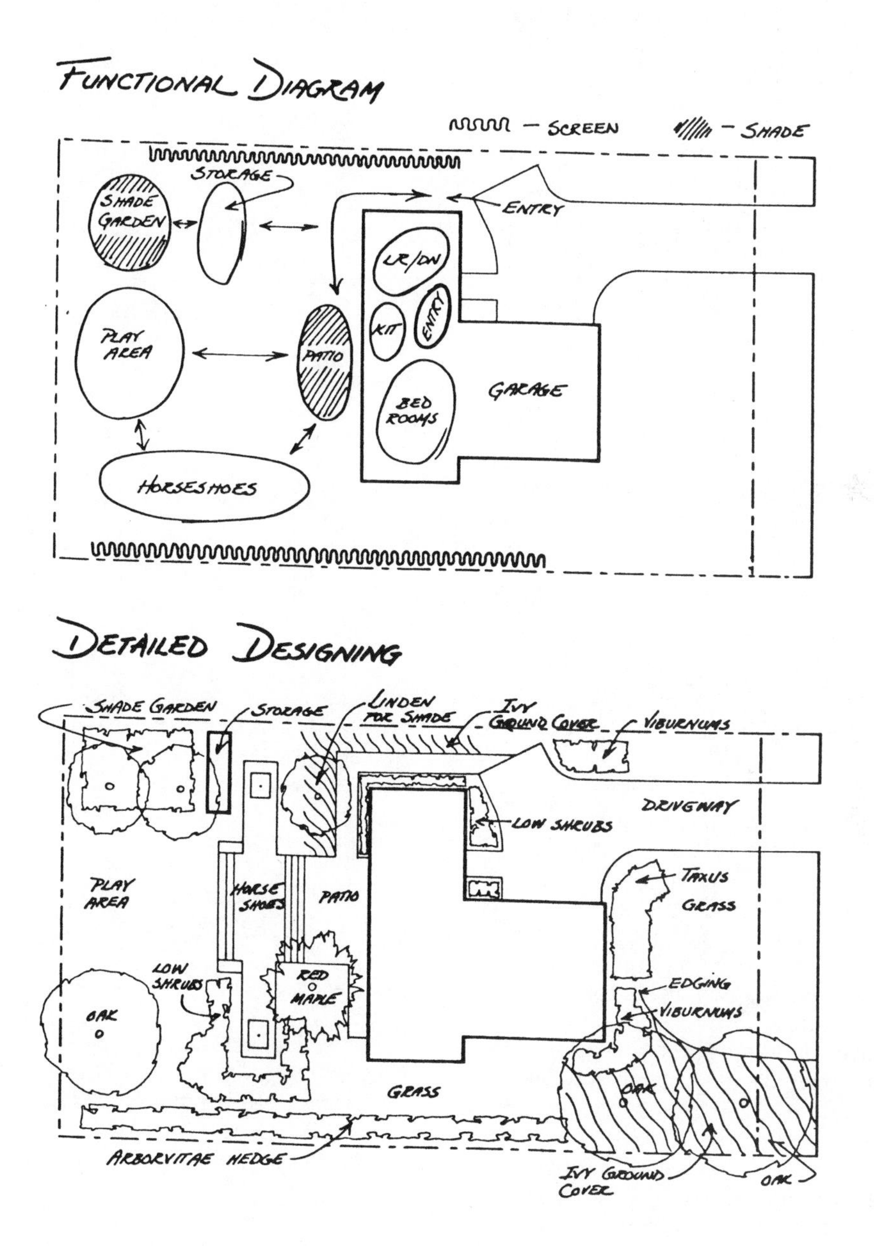

Fig. 11-20. A "bubble plan" is the preliminary plan that will be developed in detail for the final design. (USDA).

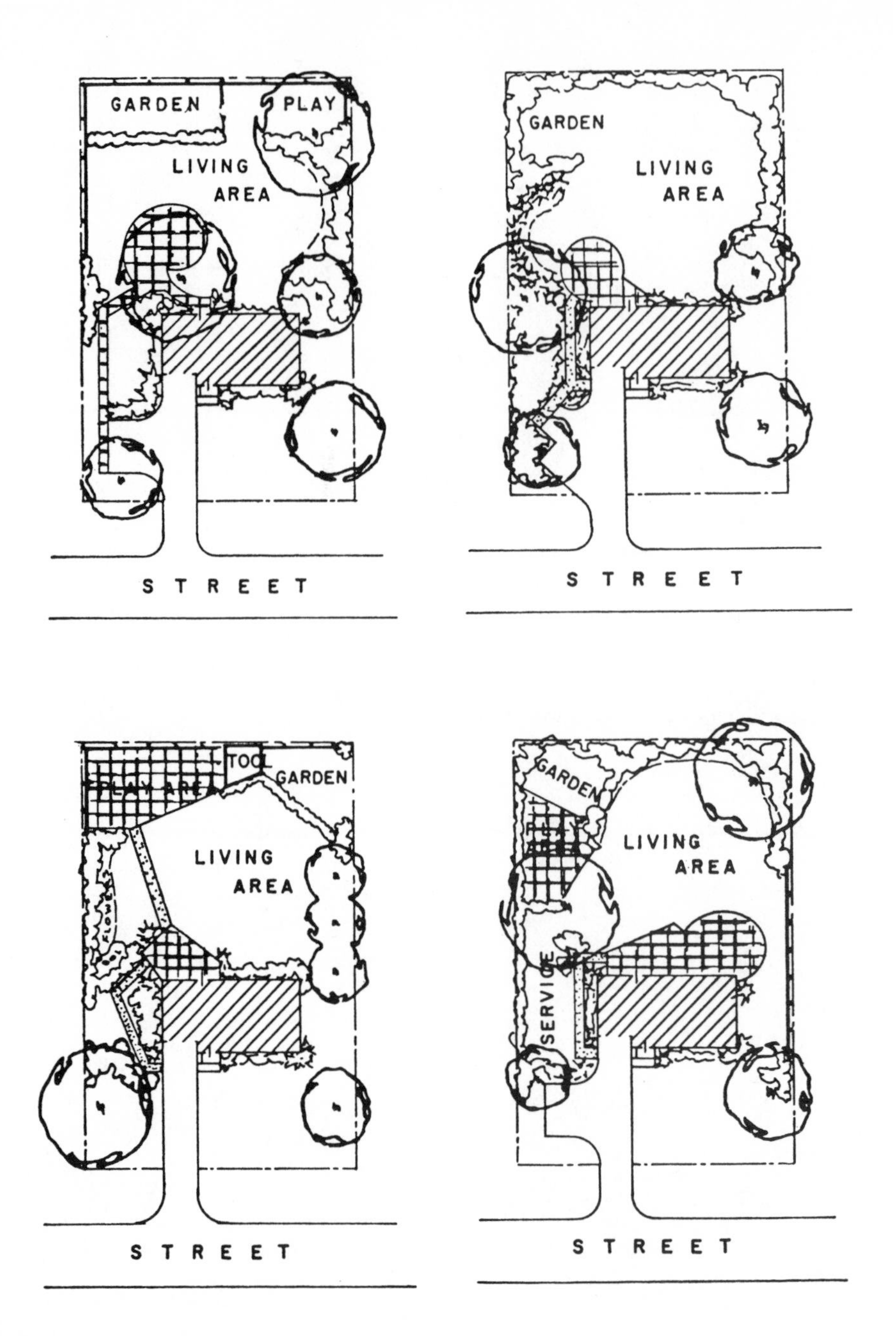

Fig. 11-21. There are always a number of landscape designs possible for any property depending on family interests and desires. Samples of four alternate schemes for a particular piece of property are shown above. (Cornell Extension).

They serve not only to achieve interesting and creative designs but also to cut maintenance efforts and costs.

Surfacing materials in general may be considered either hard or soft. Blacktop and concrete are the most utilitarian of the hard materials. Commonly used for driveways, they may also be used for walls and patios. Blacktop is the less attractive, and it is heat-absorbing and subject to cracking and pitting. Concrete is more expensive but longer lasting. It is easy to clean but can be slippery and reflects glare and heat in the sun. Expansion lines will help to prevent cracks.

Fig. 11-22. Hard surfacing materials may be attractive pre-formed squares (top) or rounds, brick or stone walkways (middle), or loose aggregates such as the limestone gravel (bottom).

Bricks, flagstones, and synthetic slabs can be used for interesting patterns in walks and patios. Bricks are usually set in sand, and since water drains down between them, they tend to heave in winter and may occasionally need to be reset. Their colors and textures, however, make them well worth a few minor problems.

Crushed stone and gravel are fine materials to fill irregularly shaped areas, although confining them can be troublesome. Epoxy coatings are available to bind them together if necessary. Such materials produce interesting colors and textures but tend to sink into the ground after a time, and new material may need to be added. Weed growth can be prevented by laying heavy black plastic sheets under the gravel. Shrubs planted in crushed limestone, if not lime-tolerant (e.g. acid-requiring broad-leaved evergreens such as *Euonymus radicans*, etc.), will probably need to be acidified occasionally as decomposition of the rock releases alkaline carbonates. Chlorosis (a yellowing of the leaves) is a symptom of too much soil alkalinity, and healthy color reappears within a few days after the addition of iron or acidifying agents. (See Chapter 2 for discussion of soil acidity and alkalinity.)

Soft surfacing materials are tanbark, wood chips, tree-trunk rounds (slabs), ground covers, and lawn. They are particularly appropriate for enhancing a quiet, relaxing atmosphere. Wood-chip pathways add rustic colors and textures while providing a soft, non-muddy medium to stroll along. Large tree trunks sliced into flat rounds can often achieve a softer effect

Fig. 11-23. Lawn grass provides a soft pathway to stroll along while admiring the flowers in a garden and the myrtle (*Vinca minor*) provides interesting texture along a garden stairway.

than that of flagstone or concrete stepping stones while still providing firm material underfoot. Wood materials are, of course, subject to decay, and will have to be replaced as they break down and become slippery.

Living plants may also be used as surfacing materials. Lawns, which can be walked upon, and ground covers, which usually cannot, serve to cover large areas of surface inexpensively and unify all the various elements of the design composition. Used together they can produce pleasing contrasts of textures and colors and direct the eye along lines and forms or to accents. The result can be immensely effective.

Walls may be used to form an enclosing shelter from wind, block an unsightly view, retain a steep slope, or provide garden seating. They can provide actual or psychological boundaries, or they may be merely decorative. They can be open or solid, rough-hewn or ultra-modern. With them you can form graceful lines, create a unified enclosure, or focalize a view. The height, width, and materials of the wall should be in proportion to the space involved. If a psychological, rather than actual, barrier is needed, a low wall will restrict movement but not limit the view. A waist-high wall will not only block movement but will alter and limit the view. A high wall will block all movement and vision. Since such walls may appear formidable, they are often softened with plantings of small trees or climbing vines.

Often a retaining wall can be extended into or be combined with a set of steps to make a grade change more negotiable. Good outdoor steps will add drama and flair to the most commonplace lot and, coordinated properly, will unite a house and its site.

Steps should be simple, attractive, and in good proportion. Since human feet and legs are to be accommodated, the riser (vertical portion) of each step should be of a consistent height for the set, but not higher than the tread (horizontal portion) is wide, and the tread must be large enough for the feet to step securely. A relaxed, easy-going appearance to garden steps will emphasize the outdoor mood. A long flight of steps should be broken by landings, and if the direction can be changed at these points the flight will appear less arduous.

Raised **planting beds** can be both labor-saving and decorative. A wide retaining wall along a side or two of a flower bed

Fig. 11-24. Walls serve many purposes, both decorative and practical. The limestone wall above (left) controls drainage on a slope, while the brick wall on the right becomes a front door planter.

Fig. 11-25. Garden steps should be comfortably low and broad. Steep slopes appear less arduous if the stairway turns or curves.

can provide seating while gardening chores are done. Artistically, raised beds are especially effective in creating strong patterns or in breaking the monotony of feature-poor sites.

When clean, sharp lines are required, **edgings** will establish and retain the lines and prevent grass and cover plants from obscuring them. Six-inch-high strips of metal edging will keep grass out of gardens, but railroad ties, flagstones, bricks, or cedar posts can add much to the charm of a garden and its design. Since bricks absorb water readily, they should be glazed or painted when bordering areas of growing plants to keep the soil from losing water through them. If the edging material can be placed level with the soil surface of a lawn it will act as a mowing strip and eliminate the need for grass trimming.

Finally we come to **fences**, more often than not the most poorly used structure on a piece of property. Fences provide privacy and protection but, for some reason, are too often considered apart from, rather than a part of, a landscape design. If they are included in the composition they can be both utilitarian and beautifying. A few appropriate do's and don'ts may emphasize how to use fences to their best avantage.

DO select materials that are durable and sturdy, and use weather-resistant paint and wood preservatives.

Fig. 11-26. Raised planting beds define lines, prevent the invasion of grass, and provide seating while performing garden tasks.

DON'T use chain-link or other strictly utilitarian fencing except where required by a need (e.g., animal enclosures, swimming pools, etc.) and for which other kinds of fencing are not adequate.

DO use plantings to hide or screen functional but unattractive fencing.

DON'T expect vines to grow successfully over metal fencing if it gets too hot in direct sun.

DO avoid sharp junctures. Soften necessary ones with plants.

DON'T mix types of fencing without considering the effect very carefully, especially around a small, wholly visible yard.

DO keep fence heights and designs uniform and coordinated with color and style of house.

DON'T fence across the front property line unless protection and privacy are needed. A formidable and unfriendly appearance may result.

DO try fencing off a dooryard or other small cozy area rather than the whole yard for a pleasant effect.

DON'T mix a fancy or elaborate gate with plain fencing.

DO consider whether the view beyond the property is to be retained or excluded before selecting an open or solid type of fencing.

DON'T use an unnecessarily high fence. You may offend your neighbors.

DO allow for good air circulation through and around fencing to avoid plant diseases.

DON'T impede movement around a house with fences. Use gates on both sides if necessary.

DO use the horizontal and vertical lines of fencing to draw the eye to desired

Fig. 11-27. Bricks make excellent edging materials, but to avoid absorbing water from the soil they should be glazed or painted.

Fig. 11-28. Fences should add to the attractiveness of a home, not merely confine it. The dooryard fence contributes charm to an otherwise ordinary home. Wherever fences and plants combine be sure to provide for the circulation of air as in the alternating plank fencing in the backyard above.

objects or areas and to create interesting patterns.

DON'T fence directly on or past the property line. Fencing placed a foot or so within the line allows for plants on the outside and secures neighborliness.

DO seek to achieve a secure, sheltered feeling with the fencing and not one of confinement.

Plants in the Landscape

Before selecting the plants for a landscape, consider the setting and environment in which they will be placed. Plants around a one-story house must be slow-growing, smaller varieties, while a two-story house will permit and/or require larger plants. Avoid evergreen monotony with pleasing combinations of coniferous and deciduous plants. Using evergreens exclusively produces an overly rich and heavily geometric appearance often funereal in impression.

In harsh or environmentally challenging locations it is wise to use the well-adapted native varieties of plants, since they will have natural characteristics that make them undemanding of special treatment. Exotic selections, however, can add interest and artistry to any plan. A good way to determine which plants grow well in a town to which you are new is to visit the local parks and cemeteries. Such places are usually planted by professionals well acquainted with plants suitable to the area.

Avoid rigidly aligned "foundation" plants, striving instead to arrange artistic and balanced groupings. And, while small plants are usually less expensive than large ones, one large plant, equal in cost to several small ones, is usually attractive for a longer period of time. Small ones soon crowd each other as they grow. It is wise to plan and allow for growth when planting.

Efforts made to plant and maintain plants properly pay off in healthy, long-lived specimens that will effectively carry out a plan. However, plants over the years often lose their attractiveness for any number of reasons, and the homeowner should expect to replace some of them from time to time.

Although structures are of importance in a design, it is basically the plants that will be used to frame a house or a view, provide attractive backgrounds and foregrounds, and serve as well-placed accents. Therefore, before planting, consider the immediate environment for the intended plants and the individual characteristics of each.

Environmental Effects on Plants

The environmental factors that are most likely to affect the health and suitability of the trees, shrubs, and flowers used in a landscape include soil, air circulation, air and soil temperature, sun, and wind. Each factor will, of course, interact with the others in an actual situation, but we'll discuss them individually in the sections that follow.

Soil. Soil type varies in different parts of the country, of course, but, aside from that, the homeowner should be aware that the soil around a house is usually considerably modified in the process of building. Topsoil may or may not have been preserved. Added soils may be of good or poor quality (see Chapter 2, Soils and Plants). Soils high in lime content will be alkaline while peaty soils will be acidic. Broad-leaved evergreens such as *Azalea*, *Rhododendron*, and holly as well as most conifers will appreciate acidifying agents or acid composts as supplements to their soil. Average neutral soils will support the other usual trees and shrubs used on home grounds, with the exception of the desert or near-desert plants of the West.

Air Circulation. Evaporation affects plants adversely if it causes wilting. The greater the amount of air circulation, the faster the rate of evaporation. However, if circulation is restricted, the humidity of the air may become high enough to encourage molds and mildews.

Air Temperature. Plants generally prefer rather moderate air temperatures, and tender plants breaking dormancy in the spring are frequently damaged when cold air flows downhill and collects in pockets or produces frost where they are growing. Conversely, sun-heated radiating surfaces of stones, brick, and concrete can heat the air around plants to a damaging level.

Soil Temperature. Soil temperatures of 50-65°F. (10-18°C.) are ideal for plants. At lower temperatures roots are chilled or frozen and water uptake is retarded. At higher temperatures root metabolism is rapid but inefficient and damage may occur.

Sun. Sun can either damage plants or promote their growth. Evergreens in the winter suffer when the direct rays of the sun speed loss of moisture from the needles. This loss is especially detrimental when the ground is frozen and the plants can't get moisture. Evergreens with the resulting brown tips are said to have "winter burn".

There are some plants that have specific sun or shade requirements. The annual flowering *Impatiens* does best in shady gardens, but lilac *(Syringa vulgaris)* bushes produce little or no bloom unless they are in full sun. Variegated shrubs may lose their variegation, and red barberry *(Berberis thunbergi*, Crimson Pygmy) and Vicary golden privet *(Ligustrum vicaryi)* may lose

Fig. 11-29. A bed of shade-loving *Impatiens* grows nicely in the shade of this old pine tree, while the fibrous and tuberous begonias tolerate well the shady side of a house.

their distinctive colors if any of them are planted in shade, since sun is needed to induce their coloring. Although most plants are less fussy, it is always wise to look up the sun or shade needs of any plant to determine the best location for it. (Some trees and shrubs are listed on page 302 that are especially adapted to shade.)

Winds. Winds can be delightfully gentle on a summer afternoon or destructively fierce in a storm. How well your plants fare in the wind will depend partly on where they are placed. For example, the flowering dogwood *(Cornus florida)* forms buds in the fall that will open in the spring only if they are protected from the sharp northwest winter winds. Russian olive *(Eleagnus angustifolia)*, on the other hand, thrives readily in a harsh, windy environment and is frequently used as a windbreak or screen. (See list of wind-tolerant trees on page 300.)

Site variations in wind, light, and temperature should always be considered. If you select plants according to their needs and how well you can match them, your time and money wil usually be well spent.

Plant Characteristics

Plants, like all living things, have distinguishing features, a given growth potential, and reproductive capabilities. Select plants that are in good health so that they are reasonably disease- and insect-resistant. Never buy a plant suffering from disease or insect attack and hope to cure it. The results are generally unsatisfactory.

After determining that the selected plants are healthy, consider the individual characteristics of each that will make it suitable or not for a given site. These characteristics are:

1. mature size, growth rate, and life span,
2. natural form, texture, and color,
3. special foliage, flowers, fruit, or bark.

All of these characteristics must, of course, be considered with respect to the environmental conditions (discussed above) to which they will be subjected. The special and desirable traits of plants can only be realized if the environments in which they are growing are favorable and the needs of the plants are met. (See lists of plants beginning on page 300.)

Mature Size, Growth Rate, and Life Span: A large shade tree, such as sugar maple, will easily exceed 80 feet (25 m.) in height at maturity, while a naturally small variety, such as a dogwood, may

Fig. 11-30. All plants have expected mature sizes. (above) The oaks and hickories in this small woodlot compose an upper canopy of tall protective trees for the naturally smaller dogwoods (*Cornus florida*) that form a lower canopy of trees that need shelter from the cold and winds of winter. (below) Sugar maples (*Acer saccharum*) are climax forest trees that grow into very large, full trees needing a great deal of both horizontal and vertical space. The trees are overly large for the house on this property and appear to be engulfing it.

grow to only 20 or 30 feet (6 or 9 m.) Watch for power lines, overhanging roofs, and nearby trees and shrubs. Be sure that they do not interfere with the vertical space required by the mature height of the tree.

It is also important to allow for potential spread of both shrubs and trees. A row of shrubs, each of which has a mature spread of 5 feet (1.5 m.), must be planted 5 feet (1.5 m.) apart even though at the time of planting they are only tiny plants. Deciduous shrubs usually have a mature spread equal to their height or nearly so. Trees close together in a forest will be tall and narrow, their lower limbs reduced or eliminated by the lack of light. The same species of trees will have dense, full scaffolds of branches when grown in an open lawn where their light needs are more adequately met.

Because of the slower growth rate of trees it is usually wise to plant them first on a new property. Shrubs and flowers can be added later and will rapidly catch up with the trees for a uniformly aged appearance. But, be careful! It is so tempting to plant a fast-growing tree for quick results that we often overlook the fact that such trees are also usually short-lived. While it's easy enough to plant a young tree, removing it 20 years later when both you and the tree are no longer young is quite a different story.

The silver maple *(Acer saccharinum)*, the poplars (*Populus* spp.), and the willows (*Salix* spp.) are all very fast growers but are not long-lived and have other objectionable traits as well. Silver maples are disease- and insect-prone, shed copious amounts of winged seeds, and are so shallow-rooted that it is difficult to grow lawns beneath them. Poplar wood is weak and breaks easily in storms, and both the poplars and the willows have water-seeking roots that work into water drains and septic tanks.

If fast growth is needed, sycamore *(Platanus acerifolia)*, honey locust *(Gleditsia triacanthos)*, tulip tree *(Liriodendron tulipifera)*, and pin oak *(Quercus palustris)* are much better choices than the above-named trees. They grow only slightly more slowly, and their growth can be speeded up with heavy fertilizing and watering.

If the rate of growth is of little consequence, or if larger trees can be afforded to begin with, then the oaks *(Quercus* spp.), sweet gum *(Liquidambar Styraciflua)*, hard maple *(Acer saccharum)*, and beech *(Fagus grandiflora)* are long-lived choice selections suitable to a variety of locations. (See page for list of trees especially recommended for urban grounds.)

Natural Form, Texture, and Color: The characteristics of form, texture and color are of importance in selecting the right plant for a particular site. The natural form of silhouette of a shrub or tree should be retained. Pruning a plant into any other shape produces little more than an artificial appearance and a great deal of frustration (see Chapter 10 on pruning). Good selections of shrubs and trees are available in a wide variety of shapes: columnar, broad, spreading, round, oval, horizontal, cone, weeping, etc. Select the shape you want and buy a plant that fits it.

Texture is the general appearance of

Fig. 11-31. These three conifers, a pine, a spruce, and a cedar, exhibit notably different textures, sizes, shapes, and lines.

the component parts of an object. It will vary with the distance from which objects are viewed and by the interplay of light and shadow upon them. Leaves especially determine the texture of a plant, although twigs, flowers, and other parts may also contribute. Depending on the effect to be achieved there are many textures available. For example: coarse (the large, thick leaves of the sycamore) or fine (the divided compound leaves of the honey locust), rough (the exfoliating bark of the shag-bark hickory) or smooth (the close gray bark of the beech), dense (the long, thick clusters of needles on the ponderosa pine) or thin (the spaced fascicles of sprays of dainty needles on the tamarack), and heavy (the ponderous clusters of horse chestnut flowers) or light (the delicate sprays of *Amelanchier* flowers).

Similarly color will also vary with distance. Background colors appear less vivid and more blue than they do when viewed close up. If we want cool, receding effects we choose the blue-violet-green range of colors. For warm, cheerful, and exciting colors the yellow-orange-red range is effective, and plants of these colors will seem to be closer to us as well. Choose plants with colors of leaves and flowers that blend or contrast attractively with each other and avoid those combinations that compete with each other (see tree and shrub lists on page 300).

More colors can be tolerated in a flower bed, since flowers are small and live briefly, than in the colors of tree and shrub leaves which last throughout the season. The many hues and intensities of normal *green* leaves cover a wide range from light to dark as well as occuring as blue-greens, yellow-greens, silver-greens, etc. As a rule of thumb, most designers limit the use of strongly colored foliage to an apparent or visual ratio of 1:9 with more normally green foliage. With foliage colors less is better. If a blue spruce, for example, is used as an accent plant for its attractive color, the eye must not be diverted to several other contrastingly colored trees such as Japanese red maple *(Acer palmatum)* or the silver-green Russian olive *(Eleagnus angustifolia)*. Too many colored trees or shrubs, especially if randomly planted, tend to produce a patchwork-quilt effect.

Special Features: The special features of flowers, fruits, and bark can contribute delightfully changing effects at different times of the year. The spring blooming period can be considerably extended with the right selection of plants, and the whole garden can take on a new look in the fall with trees especially selected for their brilliant fall colors. (See list on page 301.) Textured or colored bark can add contrast and variety at any time but is especially visible after the leaves have fallen. (See list on page 302.)

The long life spans, large sizes, and extensive variations in form, color, and texture of trees and shrubs should make it quite apparent that planning ahead is essential for a beautiful garden. Keep the design simple, balanced, and in good proportion, and select trees, shrubs, and vines carefully according to the special charms that will contribute to your overall, season-long plan.

Kinds of Plants

When producing a landscape plan you should draw the mature sizes of the trees, shrubs, and vines to scale and indicate their exact locations. Flower beds, borders, and areas of ground cover can be outlined, but it is not necessary to locate individual plants since only collectively do they form a part of the design.

Trees are the major plants to be used. The larger the tree the more it will influence accompanying plants. Not only must the spread of the branches be allowed for, but the tree must be placed where its roots too can grow where there is room for them and so that they don't deprive other plants of water and nutrients. The root spread usu-

Fig. 11-32. (left) The bright bittersweet-like fruits of *Euonymus radicans vegetus* and the green and white color combination in the variegated form add color and interest through the growing season into early winter. (center) The unusual bark of the paper-bark maple, *Acer griseum*, is best appreciated if the tree is planted where it can be observed closely. (right) The Japanese yew (*Taxus cuspidata*) bears fleshy, scarlet, berry-like fruits that surround an inner seed.

ally equals (or exceeds) the branch spread. Avoid planting sun-requiring plants within the "drip line" or outer circumference of a tree, although you may plant shade-tolerant plants with shallow roots.

New trees or 8-10 foot (2.5-3 m.) height are small enough to handle easily and large enough to establish quickly and well. Do consider how the tree will look in winter as well as in its leafy covering. Watch for hazardous or undesirble characteristics. Spines and thorns that can snag passersby, nuts, pods, and fruit juices that drip onto patios or attract insects are problems that are best avoided by not planting such trees. (See list on page 302.)

Shrubs can be used as individual accents, in groups, or as hedges, trimmed or untrimmed. They can hide service areas, be combined with fencing, soften constructions, or break monotonous ground covers. they can be grown for their foliage alone or for an annual floral or fruit display. (See lists beginning on page 300.) Many undesirable traits can be found in shrubs. Because of their smaller size the problems too are usually smaller.

Above all, don't plant too many shrubs or too many varieties. To avoid an exhibit-hall effect, select shrubs whose prime display periods don't interfere with or detract from each other. When placing either trees

Fig. 11-33. A combination of coniferous and flowering shrubs creates an interesting corner planting.

or shrubs avoid "scatter-planting" them. Carefully arranged groups, keeping design principles in mind, will please the eye and avoid a haphazard or polka-dot appearance.

Vines are not generally appreciated for their many uses and therefore are not planted as much as they could be. Vines attach themselves with holdfasts or tendrils or by twining so they need supports upon which to climb, but they require little ground space and fit nicely into spots too small or narrow for shrubs or trees. They can provide screening, overhead protection, or attractive foliage and flowers. A patio can be roofed with grapevines climbing over minimum and inexpensive supports. The dappled bits of penetrating sunlight and the clusters of hanging grapes create a charming atmosphere. A large expanse of bare wall can appear etched with the creeping branches of *Euonymus radicans*, or a bower of purple panicles of *Wisteria* flowers can grace a garden alcove. Vines should be selected, like trees and shrubs, for the special characteristics that they can contribute to a garden plan. Equal consideration should also be given to the structures that will be required to support them.

Ground Covers: Lawn grasses are the most universally used coverings for expanses of land. Early humans evolved in the grassy African savannah, and reputable scientists in anthropology, biology, and psychology have written papers exploring our present apparent need to recreate, in our own backyards, the habitat of our evolving ancestors. The lawn-scape with its assortment of shrubs and trees is prevalent the world over.

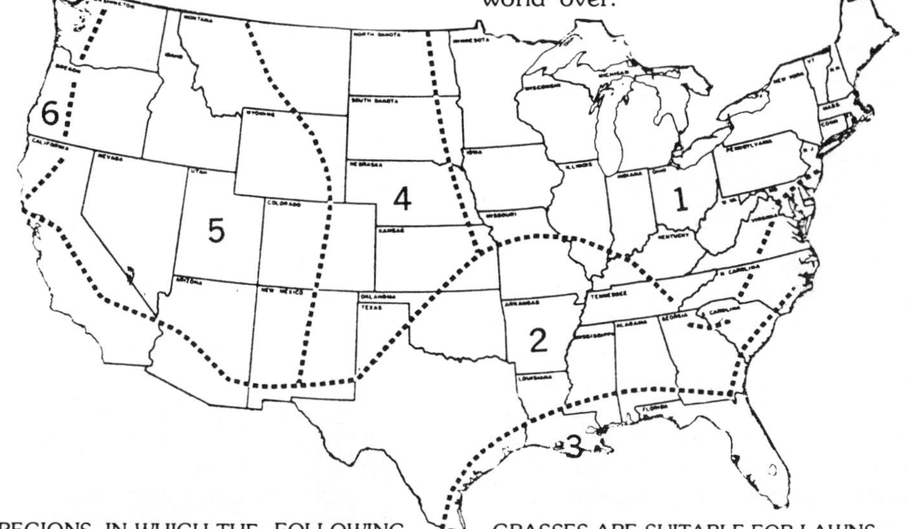

CLIMATIC REGIONS, IN WHICH THE FOLLOWING GRASSES ARE SUITABLE FOR LAWNS:

1. Kentucky bluegrass, red fescue, and Colonial bentgrass. Tall fescue, bermuda, and zoysiagrasses in the southern part.

2. Bermuda and zoysiagrasses. Centipede, carpet, and St. Augustine grasses in the southern part; tall fescue and Kentucky bluegrass in some northern areas.

3. St. Augustine, bermuda, zoysia, carpet, and bahiagrasses.

4. Nonirrigated areas: Crested wheat, buffalo, and blue gramagrasses. Irrigated areas: Kentucky bluegrass and red fescue.

5. Nonirrigated areas: Crested wheat-grass. Irrigated areas: Kentucky bluegrass and red fescue.

6. Colonial bent, Kentucky bluegrass, and red fescue.

Fig. 11-34. Grasses for different regions of the U.S. (USDA).

However we often find it desirable and even practical to use ground covers other than lawn in some situations. Myrtle *(Vinca minor)*, pachysandra *(P. terminalis)*, ivy *(Hedera helix)*, and even colorful native violets are only a few of the many plants suitable to create interesting areas or cover difficult slopes.

Fig. 11-35. This ground cover of *Vinca minor* on the slope resists erosion and eliminates a mowing problem.

Flowers add color and accent to any landscape but are a special feature in themselves requiring their own coordinated plan. Flowers are usually in beds or borders, the main difference being that a border is backed by a shrub row, fence, or wall and thus is viewed from the front side only. The sizes of both beds and borders should be in proportion to their surroundings, and the colors should blend or contrast according to the normal principles of using color, although because they are natural materials and are on display for a relatively short time, a bit more leeway may be permitted.

The frontyard flower gardens of Europe and early America were made necessary by the lack of backyards or because backyards were occupied by livestock and vegetable gardens. Today, however, we are likely to place our loveliest flower beds where we can view them at leisure in the relaxing atmosphere of a secluded patio or back porch. Since flower beds require a lot of work to produce and maintain, we are not likely to have elaborate ones in the front yard where they are usually only fleetingly glimpsed by passersby and may leave unattractive bare spots when the growing season ends. Flower beds in the public space are usually seen as delightful color accents, often in monochromatic or related-color masses, or as part of the design on very large properties.

Fig. 11-36. (above) Even a second-floor porch may be turned into an area for beauty and relaxation. (below) This tiny condominium garden extends the living space for several months of the year and adds a pleasant charm to the unit.

For seasonal accents or in areas difficult to plant try using container plants. There are attractive planters in any size and of many materials available to enhance a patio corner, doorway, terrace or any other situation imaginable.

Regardless of what plant material is being used, if the elements and principles of design are appropriately applied the garden will be economically and esthetically satisfying to the homeowner for many years to come.

MEASUREMENTS FOR THE HOME LANDSCAPE PLAN

Driveways

Single-lane drive, width = 9 feet
Double-lane drive, width = 18 feet
Minimum length to accommodate average car (wheel base) =10 feet
Minimum turning radius for small car = 15 feet
For head-in parking allow space 9-10 feet wide x 18-20 feet long + 25 feet for backing and turning.

Walkways

Entrance walk to front door, width = 4 feet
Walk for two people, width = 4 1/2 to 6 ft.
Maximum width for appearance = 8 feet
Garden work path = 18 inches
Stepping stones = approx. 18 inches square, 24 inches center to center.

Steps

Indoors = 6-7 inch riser x 10-12 inch deep tread
Outdoors = 4 inch riser x 14 inch tread
= 5 inch riser x 13 inch tread
= 6 inch riser x 12 inch tread
= 7 inch riser x 11 inch tread
Width for two people = 5 feet

Terrace or Patio

Minimum width = 16 feet
Golden Mean for rectangular, elliptical or oval = 16 x 25 feet

Miscellaneous

Gates, for wheelbarrow width = 4 feet
Gates, for deliveries = 8 feet

Walls, sitting height = 15 inches
Walls, retaining, should lean into bank by 5° angle for stability.
Hedge, width, minimum = 12 inches pruned
Fence, width, minimum = 6-8 inches

Slopes, for lawn, ideal = 7:1
Slopes, for lawn, maximum = 4:1
Slopes requiring ground cover = 3:1 or less

Games

Badminton = 54 x 24 feet (Orient courts N & S to avoid sunset)
Deck tennis = 39 x 18 feet + side space
Volleyball = 30 x 60 feet + 10 foot bounce space all around
Croquet = variable 30 x 60 feet
Swimming pools, flat edge all around = minimum 8 feet
Swimming pools, area at diving board end = 15 feet
Swimming pools, security fence, height = 4 feet
Seating at shallow end to avoid splash.

Plants

Soil depths, lawns and annuals = 6 inches
Soil depths, ground covers = 12 inches
Soil depths, flowers, perennials = 18 inches
Soil depths, shrubs = 24 inches x 24 inches diameter planting hole
Soil depths, trees = 36 inches x 36 inches diameter planting hole

Planting distances, small trees = 10 feet apart (or away from buildings)
Planting distances, medium trees = 12-15 feet apart
Planting distances, tall trees = 20-25 feet apart
Planting distances, shrubs, all alike = mature or pruned diameter apart
Planting distances, shrubs, varied = 1/2 mature diameter of each of two adjoining

LISTS OF ORNAMENTALS FOR HOME GROUNDS

The following lists of plants suggest only a few of the many trees, shrubs, and vines for landscaping that may be available for your area. For more complete lists and descriptions, consult the references in the bibliography beginning on page 318. The starred (*) references are especially helpful for selecting plants for particular purposes and regions.

Trees for Urban Grounds

(See Chap. 9, p. 232 for U.S. Hardiness Zone Map.)

Pin Oak *(Quercus palustris)*	4-8
Ginkgo *(Ginkgo biloba)*	4-8
Green Ash *(Fraxinus pennsylvanica)*	3-8
Norway Maple *(Acer platanoides)*	3-10
Japanese Pagodatree *(Sophora japonica)*	4-7
Thornless Honeylocust (*Gleditsia triacanthos* var. Shademaster)	4-8
Serviceberry *(Amelanchier canadensis)*	4-8
Horsechestnut *(Aesculus hippo-* castanum)	3-10
Grecian Laurel *(Laurel nobilis)*	7-8
Modesto Ash *(Fraxinus velutina)*	5-8
Eucalyptus or Lemon-scented Gum (*Eucalyptus citriodora* and others)	6-10
White Fir *(Abies concolor)*	5-10
Norway Spruce *(Picea abies)*	2-7
White Pine *(Pinus strobus)*	3-6
Scotch Pine *(Pinus sylvestris)*	2-8

Wind-Tolerant Trees

Russian Olive *(Elaeagnus angustifolia)*	2-9
Bur Oak *(Quercus macrocarpa)*	4-10
Hackberry *(Celtis occidentalis)*	5-10
Austrian Pine *(Pinus nigra)*	4-7
Green Ash *(Fraxinus pennsylvanica)*	3-8
Ponderosa Pine *(Pinus ponderosa)*	5-10
Eucalyptus, various spp.	6-10
Juniper, various spp.	5-10

Trees with Colored Foliage

- Cherry Plum *(Prunus cerasifera)* — purple
- Japanese Maple (*Acer palmatum* 'Atropurpureum')—red
- European Beech (*Fagus sylvatica* 'Copper' or 'Purple')—copper and purple
- Russian Olive *(Elaeagnus angustifolia)* — silver
- Blue Spruce (*Picea pungens* 'Glauca') — gray-blue

Shrubs with Colored Foliage

- Red Japanese Barberry (*Berberis thunberi* 'Crimson Pygmy') —red
- Golden Privet *(Ligustrum vulgare 'vicaryi')* — yellow
- Wormwood *(Artemisia arborescens)* — silver
- Smoketree (*Cotinus coggygria* 'Foliis purpureis') — purple
- *Actinidia kolomikta* — mixed green, pink, white

Many other species of both trees and shrubs have horticultural varieties exhibiting red, golden or variegated foliage. Check garden catalogues for those that interest you.

Trees with Conspicuous Flowers

Mountain Ash *(Sorbus aucuparia)*	3-8	White
Horsechestnut *(Aesculus hippocastanum)*	3-10	White
Japanese Pagodatree *(Sophora japonica)*	5-10	Yellowish
Goldenrain tree *(Koelreuteria paniculata)*	6-10	Yellow
Tuliptree *(Liriodendron tulipifera)*	5-10	Creamy
Mimosa or silk tree *(Albizia julibrissin)*	7-10	Pink
Crepemyrtle *(Lagerstroemia indica)*	7-10	Pink, red

Redbud *(Cercis canadensis)*	5-10	Purplish-pink
Dogwood *(Cornus florida)*	5-9	White, pink
Washington Hawthorn *(Crataegus phaienopyrum)*	5-10	White
Crabapple *(Malus floribunda)*	2-10	Pink
Pacific Madrone *(Arbutus menziesii)*	7-10	White
Saucer Magnolia *(Magnolia soulangeana)*	6-10	White

Shrubs with Conspicuous Flowers

Japanese Andromeda *(Pieris japonica)*	6-9	Creamy
Azalea and Rhododendron (Various species)	5-10	Various
Bridal-wreath *(Spiraea vanhouttei)*	5-9	White
Scotchbroom *(Cytisus scoparius)*	6-10	Yellow
Deutzia *(Deutzia gracilis)*	5-9	White
Forsythia or Goldenbell *(Forsythia intermedia)*	5-8	Yellow
Chinese Hibiscus *(Hibiscus rosa-sinensis)*	9-10	Pinks, yellow
Hydrangea *(Hydrangea macrophylla)* (Color depends on soil pH)	5-10	Blue, pink
Mountain-laurel *(Kalmia latifolia)*	5-8	White-pink
Lilac *(Syringia vulgaris)*	3-8	Purples, white
Oregon grapeholly *(Mahonia aquifolium)*	5-10	Yellow
Flowering quince *(Chaenomeles japonica)*	5-10	Apricot
Smokebush *(Cotinus coggygria)*	5-10	Purplish
Viburnums (*V. sieboldii, V. carlesii, V. plicatum*, etc.)	5-10	White

Trees for Fall Color

Sweetgum *(Liquidambar styraciflua)* — pink, red, yellow
Tuliptree *(Liriodendron tulipifera)* — yellow
Shagbark Hickory *(Carya ovata)* — yellow
Oaks (*Quercus* spp.) — red to dark red
Sugar Maple *(Acer saccharum)* — yellow, orange, scarlet
Serviceberry *(Amelanchier canadensis)*—pinkish red
Hawthorn *(Crataegus phaenopyrum)* — dark red
Tamarack *(Larix laricina)* — yellow
Sourgum *(Nyssa sylvatica)* — scarlet
Ironwood *(Carpinus caroliniana)* — orange, scarlet
American Yellowwood *(Cladrastis lutea)* — yellow
Dogwood *(Cornus florida)* — red
Winged Euonymus or Burningbush *(Euonymus alatus*, also *E. europaeus* and *E. atropurpureus)* — rose, purple, and scarlet, respectively

Plants with Conspicuous Fruits

Strawberry Tree *(Arbutus unedo)*	Red, large	7-10
Hawthorn, many varieties *(Crataegus* spp.)	Scarlet	5-10
English Holly *(Ilex aquifolium)* and American Holly *(Ilexopaca)*	Red	5-10
Mountain Ash *(Sorbus aucuparia)*	Orange-red	3-8
Japanese Aucuba *(Aucuba japonica)*	Scarlet	7-10
Glorybower *(Clerodendron trichotomum)*	Blue/Scarlet	8-10

Cotoneaster, most species	Scarlet	5-10
Firethorn *(Pyracantha coccinea)*	Orange-red	5-10
Snowberry *(Symphoricarpus albus)*	White	3-10
Viburnum, several species (esp. *betulifolium* and *opulus*)	Red or Black	5-9
Oregon Grapeholly *(Mahonia aquifolia)*	Blue-black	5-10
Bittersweet *(Celastrus orbiculatus)*	Red/Yellow	4-8
Evergreen Bittersweet (*Euonymus redicans* var. *vegeta*)	Red/yellow	3-8

Trees and Shrubs with Interesting Bark

Crepe myrtle *(Lagerstroemia indica)*	7-10
Amur corktree *(Phellodendron amurense)*	3-10
Ironwood *(Carpinus caroliniana)*	2-9
Sycamore *(Platanus acerifolia)*	5-10
Red-barked dogwood (*Cornus alba* var. *'siberica'*)	5-9
Paperbirch *(Betula papyrifera)*	2-7
Riverbirch *(Betula nigra)*	2-8
Paperbark Maple *(Acer griseum)*	3-10
Pacific Madrone (*Arbutus menziesii*, and others)	7-10
Burningbush *(Euonymus alatus)*	3-8
Gum tree (*Eucalyptus* spp.)	8-10

Generally ornamentals with unusual bark can be best appreciated if viewed at close range rather than at a distance so that their unique properties can be fully appreciated. Many of the above have both distinctive colors and other physical traits such as peeling, curling, splitting, etc.

Trees for Shady Places

Dogwood *(Cornus florida)*	5-10
Sorrel-tree *(Oxydendrum arboreum)*	6-9
Hackberry *(Celtis occidentalis)*	5-10
Arborvitae *(Thuja orientalis)*	5-10
Katsura tree *(Cercidiphyllum japonicum)*	5-8
Redbud *(Cercis canadensis)*	5-10
Japanese Yew *(Taxus cuspidata)*	5-10
Redwood *(Sequoia sempervirens)*	7-10
Hemlock *(Tsuga canadensis)*	3-7
Serviceberry *(Amelanchier canadensis)*	5-7

Shrubs for Shady Places

Witch-hazel *(Hamamelis virginiana)*	5-8
Inkberry *(Ilex glabra)*	3-8
Japanese Holly *(Ilex crenata)*	5-10
Mountain-laurel *(Kalmia latifolia)*	5-8
Oregon grapeholly *(Mahonia aquifolium)*	5-10
Japanese Andromeda *(Pieris japonica)*	6-9
Rhododendrons, in general	5-8
Arrowwood (*Viburnum dentatum* and others)	5-10

Plants with Possibly Objectionable Traits

Junipers—sharp needles
Japanese Barberry—sharp thorns
Firethorn—spines
Crabapple—fruits drop
Hawthorn—spines
Oaks—acorns drop
Maple—abundant samara production
Holly—dioecious plants (must have both a male and female tree for berries to form)

Horsechestnut—large nut-containing fruits that drop
Flowering Quince—fruits drop (they make good jelly, though!)
Japanese Pagoda Tree—seedpods that stain
Thornless Honeylocust—some produce large, twisted pods, 8"-18" (20-45 cm.) long
White Pine—large, sticky pine cones, 4"-6" (10 cm.) long
Ginkgo—female trees produce foul-smelling fruits

Trees and Shrubs Attractive to Birds

Autumn Olive *(Elaeagnus umbellata)*
Dogwood (*Cornus* spp.)
Mountain Ash (*Sorbus* spp.)
Russian Olive *(Elaeagnus angustifolia)*
Firethorn (*Pyracantha* spp.)
Sunflower (*Helianthus* spp.)
Crabapple (*Malus* spp.)
Elderberry (*Sambucus* spp.)
American Cranberry bush *(Viburnum trilobum)*
Cherry (*Prunus* spp.)
Cotoneaster (*Cotoneaster* spp.)
Tatarian Honeysuckle *(Lonicera tatarica)*
Redcedar *(Juniperus virginiana)*
Bittersweet *(Celastrus scandens)*
Holly (*Ilex* spp.)
Hawthorn (*Crataegus* spp.)

GLOSSARY

The numbers appearing in parentheses after each entry refer to a chapter or chapters in which the word or phrase is used, discussed, or more fully explained.

Abscission: Dropping of leaves or other plant parts usually after the formation of a special layer of cells (abscission layer). (1)

Abscisic Acid (ABA): A plant hormone that promotes leaf fall, dormancy, and stomatal closing, and interacts with other plant hormones in other plant responses. (1)

Acaricide: A substance for killing members of the order Acarina, which includes ticks, mites, and red spiders. (5)

Achene: A small, dry, hard, one-seeded fruit which does not split open naturally and in which the seed coat is not fused with the pericarp, or wall, of the fruit. (1)

Acid: A compound that reacts with a base to form a salt; a compound that releases (H+) hydrogen ions and has a pH lower than 7 when dissolved in water. (2)

Adaptation: Modification of an organism in structure or function in adjusting to a new condition or environment. (1, 3)

Adventitious: Plant structures, especially stems, roots, or buds, that arise from other than usual plant tissues. (4)

Aggregate Fruit: Fruit from a single flower with a number of pistils on a fleshy common receptacle, e.g., strawberry. Compare with multiple fruit. (1)

Air Layering: A procedure for producing a rooted cutting from an upright stem section while it remains attached to the parent plant. The cutting is separated from the parent plant after adventitious roots form. (4)

Alternation of Generations: A reproductive cycle consisting of a haploid gametophyte (gamete-producing) phase and a diploid sporophyte (spore-producing) phase. The gametes produce a zygote that develops into a sporophyte plant which, in turn, produces spores for a new gametophyte generation. (1, 7)

Angiosperms: Flowering plants that produce seeds within an ovary. (1, 3)

Anion: A negatively charged ion. It is attracted to positively charged ions (cations). See ion. (2)

Annual Plant: One which completes its entire life cycle in a single growing season, surviving until the next season by seed alone. (1, 3)

Annulus: In ferns, a row of special thick-walled cells around the sporangium that aids in the dispersal of spores, (7); in mushrooms, a ring of tissue that remains on the stalk as the cap tears loose as it expands.

Anther: Pollen-producing structure. (1)

Antheridium: Sperm-producing structure on the gametophyte of non-seed plants. (1, 7)

Apical Dominance: Suppression of growth of lateral buds and elongation of terminal bud as a response to auxins. (1, 10)

Apical Meristem: Dividing tissue at the tip of stems and roots that gives rise to new growth. Compare with lateral meristem. (1, 10)

Archegonium: Egg-producing structure on the gametophyte of non-seed plants. (1, 7)

Arthropod: An invertebrate of the phylum Arthropoda which includes the insects, spiders, arachnids, centipedes, millipedes, and crustaceans. (5)

Asexual Reproduction: The vegetative production of a new plant not involving the union of gametes.. (1, 4)

Autotroph: An organism capable of making its own food from simple inorganic molecules and some form of energy, most commonly sunlight. Green plants are autotrophic. (1)

Auxin (IAA): A plant hormone that regulates growth. (1, 10)

Axil: The acute angle formed by a leaf petiole and stem from which it grows. (1, 10)

Axillary Bud: The bud in the axil of a leaf. (1, 4, 10)

Bacteria: Minute, unicellular organisms without cell organelles. Most exist as parasites or saprotrophs and are the chief agents of fermentation and decay. Many are pathogenic. (1, 2, 3, 5)

Bark: The outer tissues of a stem, usually including those from the cambium outward. (1, 10)

Basal Plate: The reduced flattened stem at the base of a bulb. (4)

Base: A compound that reacts with an acid to form a salt; a compound that releases hydroxyl ions (OH-) when dissolved in water and has a pH higher than 7. (2)

Biennial Plant: One which completes its life cycle in two years. (1, 3)

Blade: The broad, expanded part of a leaf. The lamina. (1)

Bone Meal: Bones ground into a powder and used as fertilizer. A natural source of nitrogen and phosphorus. (2)

Bract: A modified leaf-like segment, usually one of several, that subtend a flower or inflorescence; in some species petal-like as in dogwood and *Poinsettia*. (7, 11)

Broadleaved: Possessing leaves that are thin and flattened as compared to the needle leaves of most coniferous plants. (1, 10, 11)

Bulb: An underground storage structure consisting of a short stem and fleshy scale leaves surrounding a bud. (1, 4)

Bulbil: A vegetative growth in the leaf axils of a lily stem that may be planted to develop into a new bulb. (4)

Bulblet: A vegetative growth from the basal plate of a bulb that will develop into a new bulb.

Callus: An undifferentiated mass of cells often formed in the area of a wound. (4)

Calyx: Composed of sepals; the outer whorl of flower parts. (1, 3)

Cambium: A thin layer of meristematic tissue that gives rise to new tissues. See cork cambium and vascular cambium. (1, 10)

Carbamate Pesticides: Toxic synthetic chemicals, less persistent in the environment than the organochlorines, which are used against insect and mite pests on plants. Examples: Carbaryl, aldicarb. (5)

Carbohydrate: An organic molecule of hydrogen, oxygen, and carbon; examples are sugars, starch, and cellulose. (1, 2)

Carnivore: A flesh-eating animal or one of a few especially adapted plants. (1, 8)

Carnivorous Plant: One which can utilize proteins from insects (usually) trapped by the plant. Sometimes referred to as insectivorous. (8)

Carpel: The ovule-bearing organ of a flower; a simple pistil or member of a compound pistil. (1)

Cation: A positively charged ion. It is attracted to negatively charged ions (anions). See ion. (2)

Cellulase: A plant enzyme that hydrolyzes cellulose. (4)

Cellulose: A complex carbohydrate, the main component of cell walls. (1, 4)

Chitin: A tough polysaccharide forming the outer covering (exoskeleton) of insects and other arthropods; also found in the cell walls of certain fungi. (5)

Chlorophyll: The green plant pigment that absorbs light energy for the process of photosynthesis. (1, 2)

Chloroplast: A chlorophyll-containing plant organelle located in the cell cytoplasm. (1)

Chlorosis: A condition in which chlorophyll does not develop properly and leaves are abnormally pale or yellow. (2)

Clay: An earthy material of colloidal-sized particles, usually associated with sand and silt as a component of soil. (2)

Climax Community: The final and most stable of a series of communities in a succession, remaining relatively unchanged as long as climatic and physiographic factors remain constant. (1, 2)

Clone: A population or group of individuals all derived asexually from one individual, e.g., all the bulblets from one bulb. (4)

Colloid: A suspension of small-sized particles of one immiscible substance in another, e.g., smoke in air, clay in water. (2)

Community: An association of plants and animals in a given area in which the various species are more or less interdependent upon each other, e.g., a desert community, pond community. (1)

Compost: Organic debris, often mixed with soil, lime, and fertilizers, that undergoes decay and is later used for garden mulch and fertilizer. (2, 9)

Compound Leaf: A leaf so deeply lobed that individual leaflets are formed, all attached by a single petiole to the stem. (1)

Cone: A unisexual structure found in some gymnospermous plants. In coniferous gymnosperms the seed-producing cones are characteristically large and woody, the pollen-producing ones are usually smaller and of brief appearance. (1, 3)

Conifer: A cone-bearing tree with needle-like leaves; one type of gymnosperms, e.g., pine, spruce, fir. (1, 10, 11)

Cork: A plant tissue in the outer layer of stems and roots that is impervious to air and water. It is produced outwardly by the cork cambium. (1)

Cork Cambium: Phellogen. A meristematic tissue that produces cork outwardly and phelloderm inwardly in stems and roots of vascular plants. (1)

Corm: An enlarged underground storage stem covered by leaf bases, like that of *Gladiolus*. (4)

Cormel: A vegetative growth offset derived from a corm. (4)

Cornell Mix: A potting medium of peat moss, perlite, and fertilizers developed at Cornell University. (2, 7)

Corolla: The petals of a flower collectively. (1)

Cortex: Unspecialized outer tissue in roots and stems consisting mainly of parenchyma cells. (1)

Cotyledon: Seed leaf; the single leaf of monocots remains below the soil upon germination, the pair of leaves of dicots usually emerging; temporary and distinctive from the plant's characteristic leaves. (1, 9)

Cross Pollination: The deposition of pollen from one plant onto the stigma of a flower on another plant; out-crossing. (1)

Crown: The above-ground portions of a plant, especially trees and shrubs. (10)

Crozier: The tightly coiled fiddlehead of a fern frond in its early development. (7)

Cuticle: The waxy layer over the outer surface of epidermal cells. (1)

Cutting: A part of a plant, most commonly a stem but also a leaf, bud, or root, used to grow a whole new plant. (1, 4)

Cytokinins: Plant hormones regulating cell division and other growth activities. (1, 3)

Cytoplasm: The protoplasm of a cell outside the nucleus in which other cell organelles are situated. (1)

Damping Off: A fungus disease of seedlings in which plants die just before or after emergence from the soil. (5, 9)

Deciduous: Shedding leaves seasonally; in contrast to evergreen. (1, 10)

Decomposers: Organisms (e.g., bacteria and fungi) capable of reducing complex organic molecules to simpler forms, often inorganic, that may be used as plant nutrients. (1, 2)

Dicotyledon (Dicot): A flowering plant possessing two seed leaves, net-veined leaves, vascular bundles in a ring arrangement, and flower parts in 4's or 5's; compare with monocotyledon. (1, 3)

Differentiation: The specialization of cells or tissues during development. (1, 4)

Diffusion: The dispersal of molecules from an area of greater concentration to one of less concentration until they are uniformly distributed. (1)

Dioecious: (Two houses) Unisexual, all the flowers of a plant being either staminate or pistillate only, e.g., willow. Compare with monoecious. (1)

Diploid: Possessing paired chromosomes, one set of which is normally derived from the male parent, the other from the female parent; the somatic chromosome number in sexually reproducing organisms; the sporophyte generation in plants. Compare with haploid. (1, 7)

Division: The act of dividing, or a section of a plant derived by dividing, a (usually) perennial plant into two or more portions. (4)

Doctrine of Signatures: An ancient mystical belief that plants reveal in their characteristics the uses to which they might be put; e.g., *Hepatica*, with its liver-shaped leaves, was used to treat ailments of the liver.

Dormancy: A usually prolonged state of physiological rest or inactivity; a resting or nonvegetative state, as in buds, seeds, spores. Compare with quiescence. (3)

Downy Mildew: One of a number of lower fungi that form a downy or moldy growth on the surface of a host plant or other substance. The causative agent of a number of plant diseases such as downy mildew of grapes and many other food crops. (5)

Ecosystem: An ecological system, a natural unit of living and nonliving components which interact to form a stable system in which a cyclic interchange of materials takes place between living and nonliving units, as in a balanced aquarium or in a large lake or forest. (1)

Edema: Excessive accumulation of fluid in tissue spaces. (7)

Electromagnetic Spectrum: The visible band of colors together with the nonvisible extensions at either end which are produced by radiant energy, as from the sun. (6)

Element: A substance in its simplest form, which cannot be broken down further, e.g., gold, carbon, oxygen. (2)

Embryo: A young sporophyte plant within the seed or gametophyte. (1, 3, 7)

Endosperm: Nutritive tissue formed within the ovule of a seed plant, constituting a reserve food in seeds of many plants, especially cereals, that is used by the embryo at germination. (1)

Energy: The capacity to do work, manifested in various ways, as chemical, mechanical, electrical, thermal, or radiant energy, forms which can under suitable conditions be converted one to another. (1, 6)

Enzyme: A complex protein molecule that performs as a catalyst to speed up reactions within cells. (1)

Epiphyte: A plant that grows upon another plant or comparable support but is not parasitic; e.g., bromeliads and orchids. (1, 7)

Ethylene: A plant hormone that regulates fruit ripening, flowering, and other plant growth activities. (1)

Etiolation: A condition of elongated stem growth, pale color, and weak growth usually due to lack of sufficient light. (6)

Evergreen: Retaining leaves for an indefinite period in excess of one year; not deciduous. (1, 10)

Evolution: The process by which different kinds of organisms have developed from simpler forms as a result of change and adaptation; descent with modification. (1, 3)

Exotic: Of foreign origin, not native and not fully naturalized or well established in an area. Most horticultural plants are exotic. (11)

Fern: A vascular plant that produces spores and possesses broad leaves (fronds) that are usually divided into many leaflets (pinnae). (1, 7)

Fertilization: The fusion of the sperm nucleus and the egg nucleus to produce a diploid zygote. (1, 3)

Fertilizer: A substance added to soil to provide plants with nutrient ions essential for growth. (2)

Fibrous Roots: A root system in which the roots are finely divided. Compare with tap roots. (1)

Fiddlehead: The tightly-coiled fern frond as it emerges from the ground before its expansion; a crozier; characteristic of ferns. (7)

Field Capacity: The amount of water that can be retained within a unit of soil. (2)

Flat: A shallow container, usually of wood slats or perforated plastic, which is filled with soil and planted with seeds for germination into the seedling stage. A seed tray. (3)

Flower: The reproductive structure distinctive of the angiosperms; a usually colorful plant structure consisting of sepals, petals, stamens, and carpels. (1)

Fluorescence: A visual emission produced by certain substances upon exposure to external stimulation. See fluorescent lamp for example. (6)

Fluorescent Lamp: An electric discharge lamp in tubular form in which metallic vapor or gas is stimulated to emit light. (6)

Frond: The leaf of a fern plant, usually compound. (7)

Fruit: The matured or ripened ovary of a flowering plant containing seeds. (1)

Fruiting Body: Various reproductive structures in the plant kingdom, e.g., mushrooms. (2)

Fungi: A group of non-vascular plants characterized by a thread-like vegetative mycelium and a variety of **spore-producing structures;** saprotrophic or parasitic and not photosynthetic. (1, 2)

Gamete: A haploid reproductive cell; a sperm or an egg. (1, 3)

Gametophyte: A haploid, gamete-producing plant derived from a spore. Alternates with a sporophyte plant in an alternation of generations sequence. (1, 7)

Gene: The unit of heredity, located in the chromosomes, composed wholly or in part of DNA. (1, 3)

Geotropism: The response of a plant to gravity, regulated by auxins. (1)

Germination: The beginning of growth and development of a seed or spore, sprouting. (3)

Gibberellins (GA): A group of plant hormones responsible for stem elongation, seed germination, and regulation of other plant growth. (1, 3)

Girdling: The removal or constriction of the outer tissues around a stem so that the flow of nutrients through the phloem is disrupted and the roots are starved. (10)

Grafting: Uniting a portion of one plant (the scion) with another (the stock) so that a permanent union is produced. (10)

Granite: Very hard igneous rock, visibly crystalline and consisting chiefly of quartz and feldspar. (2)

Green Manure: A green crop, such as clover or other nitrogen-providing plant, plowed under as fertilizer. (2)

Guard Cells: The specialized paired epidermal cells that surround a stoma or opening in a leaf surface. (1)

Guttation: The exudation of droplets of water from leaves. (8)

Gymnosperm: A diverse group of seed-producing, non-flowering plants, often with cones, e.g., pine, cycad, ginkgo. (1, 3)

Habitat: The natural environment of a plant or animal. (1, 7)

Haploid: Containing a single set of chromosomes as in gametes; the gametophyte generation in plants. Compare with diploid. (1, 7)

Hardwood: For stem cuttings: perennial woody, often dormant, stems of trees and shrubs. Compare with softwood. (4)

Hardy: Tolerant of cold winter temperatures; not tender. (3, 9)

Herb: A tender-stemmed seed plant, not woody; an herbaceous plant used for seasonings, perfumes, or medicinal products. (1, 9)

Herbaceous: Relating to plants without woody tissues.

Herbicide: A substance for killing plants. (5)

Herbivore: A plant-eater. (1, 4)

Heterotroph: An individual that feeds on other plants and animals; an organism incapable of making its own food, such as all animals, saprotrophs, and parasites. Compare with autotroph. (1)

Hormone: An organic substance which, in minute quantities, is capable of regulating activities in plants and animals. It is usually produced in one place in an organism and transported to another where it takes effect. (1, 3)

Horticulture: The art or science of cultivating plants.

Horizon, Soil: One of the distinctive layers in a vertical section, or profile, of a well-developed soil. (2)

Humidity, Relative: The amount of moisture an atmosphere can hold at a given temperature before precipitation occurs.

Humus: Organic matter in a highly decayed state, rich in nutrient ions and very water-retentive when added to soils. (2)

Hybrid: Resulting from a cross between two individuals differing in one or more characteristics. (3)

Hybrid Vigor: The increased vigor frequently found in hybrid individuals. (9)

Hypha: An individual threadlike strand of a mycelium of a fungus. (2)

Igneous Rock: A type of rock produced under conditions of intense natural heat and pressure such as rocks from molten magma or volcanoes. (2)

Incandescent Lamp: A lamp whose light is derived from a filament made luminescent by an electric current through it. (6)

Indoleacetic Acid (IAA): An auxin or growth-regulating plant hormone. (1)

Indusium: Membranous covering over the sorus of a fern leaf. (7)

Inflorescence: The arrangement of flowers on the axis, whether a single flower or group; a flower cluster. (1)

Insect: An invertebrate of the phylum Arthropoda with 3 body parts, 3 pairs of legs, one pair of antennae, and usually 2 pairs of wings. The most numerous group of animals, with over 700,000 described species. (5)

Insecticide: A substance for killing insects, often effective against other arthropods as well. Compare with herbicide. (5)

Insectivorous Plant: One which can utilize proteins derived from insects trapped by the plant. A carnivorous plant. (8)

Internode: The portion of a stem between nodes. (4, 10)

Ion: An atom or group of atoms that has become charged through the loss or gain of one or more electrons. See anion and cation. (2)

Juvenile Hormone: A hormone produced by insects that inhibits normal metamorphic changes and promotes the retention of larval characters. (5)

Larva: The young and immature form of an organism which is unlike the adult. In insects that undergo metamorphosis it is the form that hatches from the egg, variously called caterpillar, maggot, grub, wiggler, etc. (5)

Lateral Meristem: Dividing tissue that gives rise to secondary xylem, phloem, and parenchyma, generally increasing stem and root diameters; the cambial tissues. Compare with apical meristems. (1, 10)

Layered Bulb: A single large, globose, underground structure consisting of a short stem bearing fleshy overlapping leaves in concentric layers; a tunicate bulb. Compare with scaly bulb. (4)

Layering: A method of vegetative propagation in which a branch is placed in contact with the soil at one or more points. Adventitious roots will develop at the point or points of contact. The branch is then separated from the parent plant. See also air layering. (1, 4)

Leaching: The downward movement of minerals in water percolating through soil. (2)

Leaflet: A segment of a compound leaf. (1)

Leaf Mold: The natural accumulation of leaves and other forest litter after it is partially decayed and infiltrated with fungus mycelium. An intermediate stage in the formation of forest humus, often used as garden compost. (2)

Leaf Primordium: An outgrowth from the

apical meristem that will become a leaf. (4)

Leaf Scar: A scar left on a twig when a leaf falls. (1)

Legume: A simple, elongate, naturally splitting fruit, e.g., a bean or pea; a member of the family Leguminosae noted for the ability to harbor nitrogen-fixing bacteria in roots. (1)

Lenticels: Areas of special cells on the surface of a stem through which gases can pass. (1)

Lichen: A specialized type of plant growth produced by a symbiotic association between a fungus and an alga. (1, 2)

Life Cycle: The entire series of phases an organism accomplishes from zygote to death. (1, 7)

Lignin: A waterproof organic substance found in some plant cell walls. (1)

Loam: A friable mixture of clay, silt, sand, and humus in which plants grow well. (2)

Long-Day Plant: A plant that blooms only under conditions of extended daylength. Compare with short-day plant. (6)

Macrobiotic Seeds: Seeds of notable longevity, usually in excess of 15 years. Compare with mesobiotic and microbiotic seeds. (3)

Meiosis: A form of nuclear division in which the chromosome number is reduced by half; the division process by which spores or gametes are produced. Compare with mitosis. (1, 7)

Meristematic: Pertaining to tissue capable of dividing to give rise to new plant tissues. See apical meristem, lateral meristem. (1, 4, 10)

Mesobiotic Seeds: Seeds of somewhat extended longevity, usually of 3 to 5 years duration. Compare with macrobiotic and microbiotic seeds. (3)

Metabolism: The sum of the chemical and physical activities necessary to produce and sustain life. (1)

Metamorphic Rock: Those structurally changed by heat and pressure from their original form. (2)

Metamorphosis: In animals, a change in shape or form accomplished in development from egg to adult; notable in amphibians and certain insects. (5)

Microbiotic Seeds: Those of short longevity, usually of 1 to 3 years duration. Compare with mesobiotic and macrobiotic seeds. (3)

Microorganism: A small plant or animal not visible to the naked eye, microscopic. (1, 2)

Mildew: A white, powdery fungus growth on plants and organic matter. See downy mildew and powdery mildew. (7, 9)

Mineral: A naturally occurring element or compound, usually inorganic and crystalline, e.g., quartz, sulfur, and coal. (2)

Mite: An invertebrate of the phylum Arthropoda with 2 body parts, 8 legs, and no antennae or wings. Common plant pest often called spider mite or red spider. (5, 7)

Miticide: An agent used to destroy mites; an acaracide. (5)

Mitosis: Nuclear division usually followed by cell division resulting in two cells identical to the original cell and to each other. There is no change in chromosome number. Compare with meiosis. (1, 7)

Monocotyledon (Monocot): A flowering plant possessing one seed leaf, parallel-veined leaves, scattered vascular bundles, and flower parts in 3's. Compare with dicotyledon. (1, 3)

Monoecious: (One house) Having both staminate and pistillate flowers on the same plant, e.g., many members of the family Araceae. Compare with dioecious. (1)

Moss Peat: The organic remains of *Sphagnum*, *Hypnum*, or other mosses which have accumulated in places where decay has been retarded by excessively wet conditions as in a bog; used for soil improvement; often called peat moss. (2)

Mulch: A substance on the soil surface in a garden to control weeds and conserve moisture. Organic mulches of peat moss, bark chips, hay, etc. can be worked into the soil at the end of the season. Plastics and other decay-resistant materials should be removed for the winter. (9)

Multiple Fruit: A cluster of fruits derived from individual flowers forming a compact mass, e.g., pineapple. Compare with aggregate fruit. (1)

Mutation: A sudden inheritable change in an individual due to a change in the structure of a gene or chromosome. (3)

Mycelium: A mass of vegetative hyphae constituting the body of a fungus. (2)

Mycorrhiza: The symbiotic association between fungal hyphae and the roots of certain plants. (1, 2)

Necrosis: Death of a piece of tissue or of an organ; a diseased condition in plants resulting from the death of tissue. (2)

Nematocide: An agent used to destroy nematodes. (5)

Nematode: Any member of the phylum Nematoda, which includes elongated, cylindrical, unsegmented worms commonly called roundworms or eelworms. Common plant pests. (5)

Net Venation: The pattern of veins in a leaf in which the veins branch extensively forming a network, usually described as palmate or pinnate in pattern; primarily in dicots. (1)

Nitrogen Fixation: The conversion of atmospheric nitrogen into nitrogen compounds by special bacteria in the roots of legumes and other special plants. (1)

Node: An area on a stem where leaves are attached and buds arise. (4, 10)

Nodules: Swellings on the roots of plants inhabited by nitrogen-fixing bacteria. (1)

Nut: A hard, one-seeded, dry fruit produced from a compound ovary. (1)

Offset: A vegetatively produced offshoot, bulblet, cormel, plantlet, sucker, or other such structure; a generalized term. (4)

Optimal-Growth Curve: The curve produced on a graph using measurements taken of a plant grown under ideal or optimal conditions. (6)

Organic Gardening: Gardening with natural substances only. No inorganic or synthetic fertilizers, pesticides, or other such products are used. (5)

Organochlorine Pesticide: An agent composed of persistent, synthetic, extremely toxic, organic compounds (hydrocarbons) used against a wide range of plant pests, especially insects and mites. Examples: DDT, lindane, aldrin. (5)

Organophosphorus Pesticide: A toxic, synthetic, organic agent, less persistent in the environment than organochlorines, used to control insects and mites on plants. Examples: TEPP, malathion, diazinon. (5)

Osmosis: Diffusion through a semipermeable or selectively penetrated membrane. (1)

Ovule: A rudimentary seed that contains the egg cell; a structure in the ovary which after fertilization becomes a seed. (1)

Palmate Venation: The pattern of veins in a dicot leaf in which the several major veins extend from a common point at the base of the leaf and extend like the fingers from the palm of the hand. (1)

Parallel Venation: The pattern of veins in which all major veins extend parallel to each other for the length of the leaf; primarily in monocots. (1)

Parasite: An organism that feeds on another living organism (the host). (1, 5)

Parenchyma: Thin-walled, unspecialized plant cells that make up the soft tissues. (1, 4)

Parent Rock: The natural rock underlying an area that is usually the source of the mineral portion of the overlying soils. (2)

Peat: Partially decayed organic matter derived from marsh or bog areas. (2)

Peat Moss: Moss-derived peat used for soil improvement. Correctly called moss peat. (2)

Pedicel: The supporting stalk of a single flower. (1)

Peduncle: The main stalk of a single flower or of an inflorescence. (1)

Perennial Plant: A plant having a life span of several to many years; e.g., trees, shrubs, and flowering herbaceous plants such as *Phlox*, poppy. (1, 3)

Perianth: In flowers, the calyx and corolla taken together. (1)

Periderm: The outer layers of tissue of woody plants consisting of cork cambium, cork, and phelloderm.

Perlite: Inert, white, granular material derived from heat-expanded volcanic glass; may be used as a soil conditioner or rooting medium. (4)

Pesticide: Any substance used to destroy organisms harmful primarily to people or to desirable plants. (5)

Petiole: The stalk of a leaf that attaches it to the stem. (1, 4)

Pfr: The physiologically active form of the phytochrome molecule that absorbs far-red light in photoperiodic responses. (6)

pH: A symbol indicating the concentration of hydrogen ions in a solution. (2)

Phellem: Cork. (1)

Phelloderm: A layer of parenchyma tissue formed in woody plants inwardly by the cork cambium or phellogen. (1)

Phellogen: Cork cambium. The meristematic tissue that produces cork outwardly and phelloderm inwardly in stems and roots of vascular plants. (1)

Pheromone: A substance produced and discharged by an organism which induces a physiological response in another of the same species, such as the sexual attractants of insects. (5)

Phloem: The food-conducting tissue of a plant; part of a vascular bundle. (1, 10)

Photochemical: Chemical reactions involving light. (6)

Photon: A quantum, or unit, of radiant energy. (6)

Photoperiodism: Plant (and animal) responses to duration of periods of light and dark. (6)

Photosynthesis: The process by which plants use light energy to produce carbohydrates from carbon dioxide and water. (1)

Phototropism: A plant growth response to the direction of the source of light, regulated by auxins. (1, 6)

Phytochrome: A pigment in green plants asociated with the responses of plants to light. Refer also to Pr and Pfr. (6)

Pinnate Venation: The pattern of veins in a dicot leaf in which there is a single major vein extending the length of the leaf with smaller veins branching from it somewhat like the structure of a feather. (1)

Pioneer Community: The first community in a series of communities in a succession, e.g., bare rock inhabited by lichens and mosses, an abandoned cornfield supporting annual weeds. (1)

Pistil: The central and female portion of a flower consisting of three major parts, the stigma, style, and ovary; one single carpel or several separate or united carpels. (1)

Pistillate: Refers to unisexual flowers bearing only pistils (carpels), no stamens. (1, 12)

Pollination: The deposition of pollen on the stigma, not to be confused with fertilization. (1, 3)

Pollen: Tiny, grainlike structures containing sperm nuclei and produced by meiosis in the anther of a stamen. (1)

Porosity: In soils, the ratio of the volume of the pore space to the total volume of the soil. (2)

Potash: A potassium-containing compound in fertilizers and wood ashes; the oxide of potassium, K_2O, or potassium carbonate, K_2CO_3. (2)

Powdery Mildew: One of a number of higher fungi that form a white powdery growth on a wide assortment of plants and plant parts. The causative agent of a number of plant diseases such as those commonly seen on *Phlox*, *Zinnia*, and lilacs. (5)

Pr: The physiologically inactive form of the phytochrome molecule that absorbs red light in photoperiodic responses. (6)

Progeny: Offspring; descendants. (3)

Propagate: To cause to increase or multiply. In plants, by sexual means which involves seeds or by vegetative means using various parts of the parent plant. (1, 4)

Protozoan: A member of the phylum Protozoa comprising all unicellular animal-like organisms (a few form colonies but lack differentiated cell types or tissues). (2)

Pupa: In insects with complete metamorphosis, an inactive stage between the larva and the adult. It is usually enclosed in some form of protective structure. (5)

Pyrethroids: Synthetic compounds similar to those derived from certain plants of the family Compositae. Used as insecticides they are less toxic and persistent, and therefore safer to humans, than other synthetic pesticides. (5)

Quiescence: A state of inactivity necessitated by unfavorable environmental conditions, e.g., cold, dryness, darkness. Compare with dormancy. (3)

Receptacle: The end of the pedicel or peduncle to which parts of a flower are attached. (1)

Respiration: Cellular oxidation in which energy is released to be used for metabolic activities. (1, 3)

Rhizome: A horizontal stem at or just below ground level. (1, 4)

Root: That portion of a plant which anchors it and absorbs water and minerals. (1, 4)

Rootbound: A condition in which the roots of a plant are overly abundant for the container of soil in which it is growing, causing declining health of the plant. (7)

Root Hairs: Thin-walled extensions of an epidermal cell of a root. They increase the absorption of water by roots. (1)

Rootstock: In grafting, a rooted portion (a stock) to which a scion is applied. (1, 10). In botanical usage, a rhizome.

Runner: A thin stolon or stem that grows horizontally along the ground surface and may produce adventitious shoots and roots. (4)

Sand: Fine, crystalline grains of rock, usually of quartz. (2)

Saprotroph: An organism that derives its food from dead or non-living organic matter; a sapropohyte. (1)

Saturation: In soils, a condition in which all pore spaces are filled with water and air is absent; thoroughly soaked. (2)

Scaffold: The branches of a tree. (10)

Scaly Bulb: A globose underground structure composed of easily separated scale leaves attached to a short basal stem. Compare with layered bulb. (4)

Scion: A bud or shoot which is removed from a plant and prepared for grafting onto another plant (the stock). (1, 10)

Sedimentary Rock: Rock formed from accumulated sediments derived by erosion from other rocks, usually by wind or water. (2)

Seed: A structure formed by seed plants following fertilization and containing the embryonic plant; the fertilized ovule. (1, 3)

Seed Coat: The outer covering of a seed formed by the outer tissues of the ovule. (3)

Seedling: A newly germinated plant. (3, 7)

Selection: The perpetuation of plants that have desirable hereditary characteristics. (3)

Sepal: A segment of the calyx; the outermost parts of a flower. (1)

Short-Day Plant: A plant that blooms only under conditions of extended darkness. Compare with long-day plant. (6)

Shoot: The upper portion of a plant consisting of stems and leaves. (1)

Shrub: A woody plant of limited height usually arising from several to many stems from the ground. (10, 11)

Silt: Fine-grained soil material deposited as sediments by wind or water. (2)

Slip: A cutting for planting or grafting. (4)

Softwood: For cuttings: a generalized term for the stems of herbaceous perennials as well as the immature shoots of some shrubs and trees. Compare with hardwood. (4)

Soil Profile: The vertical exposure of soil layers or horizons. (2)

Soil Structure: The arrangement of mineral and humus particles into simple or complex soil aggregates. (2)

Soil Texture: The permeability of soil as determined by the proportions of sand, silt and clay. (2)

"Soluble Salt": A gardener's term to refer to the salts from plant fertilizers, water-softening treatments or natural rock dissolved in soil water. Accumulations of salt ions can cause root distress leading to plant death. (7)

Solution: A mixture of molecules such as salt (a solute) dissolved in a liquid such as water (a solvent). (1)

Sorus: A cluster of sporangia on a fern, sometimes called a fruit dot. (7)

Species: A group of similar organisms that are interfertile. (1)

Spectral Energy: The power derived from radiant spectrum from the sun or other light source. (6)

Sphagnum Moss: A peat or bog moss of the genus *Sphagnum*, noted for its ability when dry to absorb and hold large quantities of water. Compare with moss peat. (7)

Sporangium: A plant structure in which spores are produced; a spore case. (1, 7)

Spore: An asexual haploid cell produced by lower plants and capable of germinating into a new individual, often a gametophyte. (1, 7)

Sporophyte: The diploid, spore-producing phase in plants having alternation of generations; a diploid plant. (1, 7)

Sport: A plant or plant part that shows deviation from the normal or parent type; a mutation. (4)

Stamen: The pollen-producing male structure of a flower, consisting of an anther supported by a filament. (1)

Staminate: Refers to unisexual flowers bearing only stamens, no carpels. (1, 12)

Stem: The major axis of a plant, joined to the root at the lower end and supporting branches, leaves and reproductive structures at the upper end; the axis of the shoot. (1, 4, 10)

Stigma: The apex of the pistil or carpel, receptive to pollen. (1)

Stock: In grafting, that portion of a plant to which the scion is applied. (10) In horticulture, the mature plants from which cuttings or divisions may be taken for propagation. (4)

Stolon: A horizontal runner or specialized stem which gives rise to new shoots. (4, 8)

Stomata: Sing. **Stoma:** Openings in the epidermal tissue of a leaf to allow exchange of gases between the interior of the leaf and the outside atmosphere, controlled by a pair of guard cells. (1)

Succession: A sequence of changes through which an ecological area goes from an initial pioneer community to a final climax community. (1, 2)

Sucker: An offshoot which develops from the roots or lower part of the stem of a plant. (4)

Superphosphate: A phosphate supplement for soil derived by treating phosphate rock with sulphuric acid. (2)

Symbiosis: A close living relationship between two different organisms, which may be beneficial to only one or to both organisms. (See lichen.) (1)

Taproot: The primary root of a plant which grows directly downward, giving off lateral branches. (1)

Tender: A characteristic of plants that cannot tolerate very cold or freezing temperatures; not hardy. (3, 9)

Thinning: Of trees and shrubs, removing whole branches from the main trunk or at ground level to improve the health and appearance of the plant; a type of pruning. (10) Of seedlings, removing excess plants from the germination bed to reduce crowding and permit satisfactory growth of remaining plants. (3, 9)

Tilth: The physical condition of soil in relation to plant growth. Soils in good tilth are easily worked or tilled. (2, 9)

Tissue: An aggregation of cells of more or less similar structure and function together with their intercellular material. Principal plant tissues are meristematic, dermal, ground, and vascular. (1)

Top Dressing: A supplement, such as manure, applied lightly to the surface of a soil as a fertilizer or soil conditioner. (2)

Top Soil: The a layer of a soil profile containing abundant humus; a good-quality soil. (2)

Totipotency: The ability of a cell or small group of cells to give rise to a whole new organism. (4)

Toxic: Of or pertaining to a poison; poisonous. (5)

Transpiration: The loss of water vapor from the above-ground surfaces of plants, mainly through stomata and lenticels. (1, 7, 8)

Tropism: A growth response to an exterior stimulus which determines the direction of growth, e.g., phototropism, geotropism, hydrotropism, etc. (1, 6)

Tuber: An enlarged underground stem or root, e.g., stem, white potato; root, sweet potato. (3)

Tunicate Bulb: A large, globose underground structure consisting of a short stem bearing fleshy overlapping leaves in concentric layers; a layered bulb. Compare with scaly bulb. (4)

Turgid: Swollen, distended, or inflated by pressure from within. (9)

Turgor Pressure: Hydrostatic pressure within a cell resulting from imbibition of water by the protoplasm of the cell. (1)

Vascular Bundle: A strand of conducting tissue composed mainly of the xylem and phloem. (1)

Vascular Cambium: A cylinder of meristematic tissue in vascular plants that produces secondary xylem and phloem cells. (1, 10)

Vegetative: Growth, tissues, or functions that are concerned with normal maintenance of a plant; asexual. (1, 4)

Vegetative Propagation: Reproduction of plants by any method other than the use of true seeds, as by stem, root, or leaf tissues or sections. (4)

Vein: A vascular bundle. (1, 4)

Vermiculite: Micaceous minerals derived from heat-altered common mica. Originally developed as an insulating material, it is often used as a soil additive or as an inert rooting or seed germinating medium. (3)

Vernalization: ("Spring-ization"); the promotion of growth or flowering by exposing bulbs, seeds, or plants to low temperature and moisture as needed to accomplish the required physiological changes. (4)

Vessel: A water-conducting tube in the xylem composed of dead cells joined end to end with intervening walls perforated. (1)

Virus: An ultramicroscopic, disease-producing agent that performs as an intracellular parasite; responsible for mosaic, spot, and breaking diseases in plants. Other forms affect animals. (5)

Volva: A cup at the end of the stalk of a mushroom, at or below soil level. (2)

Weed: A plant growing where it is unwanted and generally interfering with desirable plants. (5)

Wilting Point: In soils a condition in which water is nearly absent from pore spaces and any plants present are in danger of wilting. (2)

Woody: Refers to plants with hard, fibrous, xylem-filled stems covered with thick bark; not herbaceous. (10)

Xylem: Wood or woody tissue; part of a vascular bundle. It functions principally in the conduction of water and mineral salts but also provides mechanical support and functions in the storage of water and food. Compare with phloem. (1)

Zygote: A single cell resulting from the fertilization of the egg by the sperm and capable of developing into an embryo. (1, 3, 7)

Some of these definitions are taken or modified from Steen's *Dictionary of Biology*, Barnes and Noble, Publishers, N.Y.

BIBLIOGRAPHY

ALL-PURPOSE BOOKS FOR GARDENERS

Bailey, Ralph, and Elvin McDonald, editors, **Good Housekeeping Basic Gardening Techniques**. New York: Book Division, Hearst Magazines, 1974. A good reference for gardeners. Not detailed enough on some points but has good illustrations and lots of them.

Ball, Vic, editor, **The Ball Red Book**. W. Chicago, Ill.: George J. Ball, Inc., 1975. A book of plant and equipment information especially for bedding-plant growers, but good expert information for any gardener.

Powell, Thomas and Betty, **The Avant Gardener**. Boston: Houghton Mifflin Company, 1975. A mine of useful and fascinating information. Good references and sources of supply for plant growers.

Janick, Jules, **Horticultural Science**. San Francisco: W. H. Freeman and Company, 2nd edition, 1972. A fine textbook on the technical aspects of the art and science of growing plants.

Wright, Michael, editor, **The Complete Indoor Gardener**. New York: Random House, 1974. A colorful large paperback bursting with lots of information and plant pictures for both indoor and outdoor gardens.

*Bush-Brown, James and Louise, **America's Garden Book**. New York: Charles Scribner's Sons, 1965. I couldn't garden without this one. A truly comprehensive guide. Should be updated.

Rockwell, Frederick F., and Esther C. Grayson, **The Rockwells' Complete Guide to Successful Gardening**. Garden City, New York: Doubleday & Company, Inc., 1965. A book for gardeners with the emphasis on outdoor gardens. A gardening calendar tells what to do when for all areas of the country.

***Yearbook of Agriculture, Landscape for Living**. Washington: U.S. Department of Agriculture, 1972. Good basic information about plants, gardening, landscaping, and community participation for environmental improvement.

*Riker, Tom, and Harvey Rottenberg, **The Gardener's Catalogue**. New York: William Morrow & Company, Inc., 1974. A blend of old and new information and pictures on gardening. A big paper-covered catalogue that is fun to browse through in your spare time.

Editors of Sunset Books, **Sunset New Western Garden Book**. Menlo Park, California: Lane Publishing Company, 1979. The fourth edition of a book indispensable to gardeners in the far Western States, whether novice or advanced.

BASIC BOTANY BOOKS

Delevoryas, Theodore, **Plant Diversification.** Modern Biology Series. New York: Holt, Rinehart and Winston, 1966 (paperback). An evolutionary approach to understanding plant structures, functions, and adaptations. Selected topics, very interesting.

Raven, Peter H., and Helena Curtis, **Biol-**

ogy of Plants. New York: Worth Publishers, Inc., 1970. A popular and widely used textbook.

Weier, T. Elliot, C. Ralph Stocking, and Michael G. Barbour, **Botany, An Introduction to Plant Biology**, 5th edition. New York: John Wiley and Sons, 1974. An extensively detailed textbook. Broad coverage and lots of illustrations.

Ray, Peter Martin, **The Living Plant**, 2nd edition. Modern Biology Series. New York: Holt, Rinehart and Winston, Inc., 1972. 195-page paperback of botanical essentials. Handy references.

Bidwell, R.G.S., **Plant Physiology**. New York: MacMillan Publishing Co., Inc., 1974. Textbook of detailed information on plant functioning.

Noggle, G. Ray, and George J. Fritz, **Introductory Plant Physiology**. Englewood Cliffs, New Jersey: Prentice-Hall, Inc., 1976. Clearly written up-to-date information on how plants function. Large textbook.

GROWING PLANTS UNDER LIGHTS

Bickford, Elwood D., and Stuart Dunn, **Lighting for Plant Growth**. Kent, Ohio: Kent State University Press, 1972. Technical information on lights and appropriate accessories for growing plants.

Elbert, George A., **The Indoor Light Gardening Book**. New York: Crown Publishers, Inc. 1973. Good information and ideas for the serious light-gardener.

Krans, Frederick H. and Jacqueline L., **Gardening Indoors Under Lights**. New York: The Viking Press, 1971. A good starting book for the novice light-gardener.

Cultural Guides from the Indoor Light Gardeners Society of America. List and prices from I.L.G.S.A., Inc., 297 Second St., Albany, N.Y. 12206. Inexpensive and informative little booklets on some of the aspects of growing plants under lights.

INDOOR GARDENING

Davidson, William, and T. C. Rochford, **The Complete All-Color Guide to House Plants, Cacti & Succulents**. New York: Galahad Books, 1976. An attractive catalogue of 450 houseplants with beautiful colored drawings and columns of essential information on each. An excellent reference.

Langer, Richard W., **Grow It Indoors**. New York: Saturday Review Press/E. P. Dutton and Company, Inc., 1975. Its subtitle describes it well: "A practical, personal how-to and why guide to growing successful houseplants." Good b & w drawings for identification help.

Cruso, Thalassa, **Making Things Grow: A Practical Guide for the Indoor Gardener.** New York: Alfred A. Knopf, 1969. Helpful how-to's from the first-hand experiences of one of our best-known gardeners.

Loewer, H. Peter, **Bringing the Outdoors In**. New York: Walker & Company, 1974. An unusual approach to indoor gardening using out-of-the-ordinary plants. Helpful information on plants, containers, and special constructions. Excellent drawings.

Kramer, Jack, **Gardens Under Glass**. New York: Simon and Schuster, 1969. A short book on the essentials for terrariums and dish gardens with plant suggestions.

Hoshizaki, Barbara Joe, **Fern Growers Manual.** New York: Alfred A. Knopf, Inc., 1975. Everything you ever wanted to know about ferns—and more.

Evans, Charles M., **New Plants from Old.** New York: Random House, 1976. Good instructions clearly illustrated with drawings.

Graf, A. B., **Exotica**, 9th Edition. E. Rutherford, N. J.: Roehrs Company, Inc., 1976. A "pictorial cyclopedia of exotic plants" that is the standard reference for commercial and professional plant growers. Nearly 2000 pages of photographs of and instructions for growing warm-climate ornamentals. A smaller version is available for home use.

SOILS

Gibson, J. Sullivan, and James W. Batten, **Soils, Their Nature, Classes, Distribution, Uses and Care.** Alabama: Press, 1970. This and the following two books are good reference books emphasizing the agricultural needs of plants but presenting information helpful to the home gardener too.

Bear, Firman E., **Soils in Relation of Crop Growth.** New York: Reinhold Publishing Co., 1965.

Donahue, Roy C., John C. Shickluna, and Lynn S. Robertson, **Soils, An Introduction to Soils and Plant Growth**, 3rd edition. Englewood Cliffs, New Jersey: Prentice-Hall, Inc., 1971.

Ortloff, H. Stuart, and Henry B. Raymore, **A Book About Soils for the Home Gardener.** New York: William Morrow and Co., 1972. Excellent information in an inexpensive paperback.

WEEDS AND PESTS

Anderson, Wood Powell, **Weed Science**. San Francisco: West Publishing Co., 1977. Written as a textbook, contains technical and chemical information on weeds and herbicides as well as some basic information on the related topics of botany and soils.

Westcott, Cynthia, **The Gardener's Bug Book**, 4th edition. Garden City, New York: Doubleday & Company, Inc., 1973. A classic volume no serious gardener should be without.

Johnson, Warren T., and Howard H. Lyon, **Insects that Feed on Trees and Shrubs.** Ithaca, New York: Comstock Publishing Associates of Cornell University Press, 1976. An outstanding work on insect pests and their damage. Full-color photographs of both.

"From the Plant's Point of View" Series, **How to Detect and Solve Plant Problems.** Lansing, Michigan: The John Henry Company, 1976. One of a series of little booklets available in garden stores and such. This one has excellent colored photos and drawings helpful in detecting house plant-problems.

Ware, George W., **The Pesticide Book.** San Francisco: W. H. Freeman and Company, 1978. Intended for classrooms use but an excellent compendium of information for anyone interested in pesticides.

TREES AND SHRUBS

*Brimer, John Burton, **The Home Gardener's Guide to Trees and Shrubs**. New York: Hawthorn Books, Inc., 1976. Descriptive lists and information in an easy-to-find question-and-answer format. Lots of good ideas.

*Schuler, Stanley, **The Gardener's Basic Book of Trees and Shrubs.** New York: Simon and Schuster, 1973. Descriptive lists and helpful information on buying, planting, and caring for the trees and shrubs on your property.

*Hottes, Alfred Carl, **The Book of Shrubs**. New York: A. T. De La Mare Company, Inc., 1952. Probably the best reference on shrubs, written by a master in the subject.

Baumgardt, John Philip, **How to Prune Almost Everything.** New York: William Morrow and Company, Inc., 1968. An excellent volume on the basics of pruning, plus specific instructions for

*Books with lists of plants for the home landscape.

300 different plants listed alphabetically for quick reference.

Steffek, Edwin F., **The Pruning Manual**. New York: Van Nostrand Reinhold Co., 1969. Separate chapters on various kinds of plants, e.g., hedges, small fruits, vines, etc. Excellent advice.

Perkins, Harold O., **Espaliers and Vines for the Home Gardener.** Princeton, New Jersey: D. Van Nostrand Company, Inc., 1964. Good directions for creating some garden delights.

Stowell, Jerald P., **The Beginner's Guide to American Bonsai**. Tokyo: Kodansha International Ltd., 1978. Probably the best book in English for the beginner.

Behme, Robert Lee, **Bonsai, Saikei and Bonkei: Japanese Dwarf Trees and Tray Landscapes.** New York: William Morrow and Company, Inc., 1969. For the person who wants to indulge fully in these Japanese creations.

Tukey, Harold B., **Dwarfed Fruit Trees... for Orchard, Garden and Home.** New York: The MacMillan Company, 1964. A wide range of subject matter all dealing with dwarfed fruit trees by an outstanding expert in the field.

LANDSCAPING

Ortloff, H. Stuart, and Henry B. Raymore, **The Book of Landscape Design.** New York: William Morrow and Company, Inc., 1975 (paperback). The basic principles neatly and concisely put.

Weber, Nelva M., **How to Plan Your Own Home Landscape.** New York: Bobbs-Merrill, 1976. Practical ideas and suggestions for turning plans into reality. Excellent drawings and photographs.

Eckbo, Garrett, **The Art of Home Landscaping.** New York: McGraw-Hill Book Company, 1956. A standard classic how-to book by an experienced landscaper. Well illustrated.

Church, Thomas D., **Gardens are for People.** New York: Reinhold Publishing Company, 1955. An exciting book of garden designs for a variety of gardens and gardeners, although some ideas need updating.

McDonald, Elvin and Lawrence, **The Low-Upkeep Book of Lawns and Landscape.** New York: Hawthorn Books, Inc., 1971. No frills, common-sense ideas for a well-kept yard.

Robinette, Gary O., **Plants/People/and Environmental Quality**. U.S. Dept. of the Interior, National Park Service. U. S. Gov't. Printing Office, Washington, D. C. 20402. Stock No. 2405-0479. $4.00. 1972. An environmental approach to landscaping. Attractive artwork emphasizes roles of plants and how to make them functionally effective in our lives.

Wilson, Helen VanPelt, **Own Garden and Landscape Book.** Garden City, N.Y.: Double-Day and Company, Inc., 1973. A first-hand account of the development of Stony Brook Cottage, her own legendary property.

Fairbrother, Nan, **The Nature of Landscape Design**. New York: Alfred A. Knopf, 1974. As nice an approach to a fine "philosophy" of landscaping as has ever been written. Inspiring reading.

Berrall, Julia S., **The Garden, An Illustrated History**. New York: The Viking Press, 1966. A large handsomely illustrated volume covering the development of landscape gardening from the ancient past to the present. Interesting reading. Now available in paperback.

OUTDOOR GARDENING

Cruso, Thalassa, **To Everything There is a Season.** New York: Alfred A. Knopf, 1973. Gardening essays that make delightful winter-time reading.

Cruso, Thalassa, **Making Things Grow Outdoors.** New York: Alfred A.

Knopf, 1971. Another gem from Boston's TV "plant lady".

Davids, Richard C., **Garden Wizardry.** New York: Crown Publishers, Inc., 1976. An experienced gardener shares his secrets of success.

Wallach, Carla, **Gardening in the City.** New York: Harcourt Brace Jovanovich, 1976. Good ideas for the "limited space" gardener.

Yearbook of Agriculture, Living on a Few Acres. Washington: U. S. Department of Agriculture, 1978. Good ideas and modern approaches for a successful small farm.

Schuler, Stanley, **The Gardener's Basic Book of Flowers.** New York: Simon and Schuster, 1974. How to know and grow some of the best of the flowers. Companion volume to **The Gardener's Basic Book of Trees and Shrubs.**

Price, Robert, **Johnny Appleseed, Man and Myth.** Bloomington, Indiana: Indiana University Press, 1954. The true story of how John Chapman became a legend.

Sperka, Marie, **Growing Wildflowers, A Gardener's Guide.** New York: Harper and Row Publishers, 1973. Descriptions, habitat, time of bloom, and growing conditions for over 300 wild plants that you might like to grow in your garden or terrarium. Contains list of wild-plant dealers.

GARDENING PUBLICATIONS IN SERIES

Garden and Horticulture Handbooks published as special issues of **Plants and Gardens** by Brooklyn Botanic Garden, 1000 Washington Ave., Brooklyn, N.Y. 11225. Expert advice and information on nearly 100 different topics in inexpensive 6 x 9 booklets. Photos and drawings in black and white. Highly respected by both professionals and novices. A list of topics is on the back cover of each booklet.

Countryside Books, A. B. Morse Company, 200 James Street, Barrington, Ill. 60010. Colorful and attractive booklets on common garden topics. 8 x 10 1/2 paperbacks. Written by experienced authorities in clear, no-nonsense language. Fun to read as well as informative.

Sunset Books, Lane Publishing Company, Menlo Park, Calif. 94025. These are the glamour books of garden publishing. Magazine-sized with slick paper covers and gorgeous photographs, mostly in color. Topics range from vegetable gardens to Japanese bonsai, all very well done.

"From the Plant's Point of View" Series, The John Henry Company, P.O. Box 17099, Lansing, Mich. 48901. Concise, colorful little booklets with the emphasis on houseplants. Lots of expert advice primarily from consultants at Michigan State University. Attractive format, very inexpensive, limited number of topics.

***Time-Life Encyclopedia of Gardening,** Time-Life Books, Alexandria, Va. James Underwood Crockett and other consultants and authors have contribute to this multi-volume set of hardcover books on all aspects of plants and gardening. Volumes may be purchased separately. Some are better than others.

Yearbooks of Agriculture, U. S. Department of Agriculture, Washington, D.C. Order from U. S. Government Printing Office, Washington, D.C. 20402. A new yearbook appears each year, each on a different subject. Large, hardcover books with lots of facts and figures. Current yearbooks are often available for the asking from your congressman.

PLANT SOCIETIES AND SUPPLIES

There are national societies, many with local or regional chapters, for almost every major group of plants or gardening activity. There are societies for begonias, camellias, daylilies, ferns, African violets, hollies, etc. as well as for indoor light gardening, rock gardens, organic gardening, etc. The membership fees are usually moderate and often include a subscription to their specialized publication. To find the name and current membership chairman's address of the plant society you might wish to join, check the want-ads of recent issues of gardening magazines in your library or newstand. Belonging to a society is a fine way to increase your knowledge and enjoyment of plants.

Local suppliers of plants and equipment can be located by checking the yellow pages of your telephone directory under Garden Centers, Nurserymen, or Plants—Horticulture. You may also order live plants, seeds, and supplies from mail-order nurseries, most of whom will send a catalogue upon request. Gardening magazines contain advertisements for many mail-order houses, and the appendices of many gardening books list their names and addresses. Once you make a purchase you will usually continue to receive their catalogues. In addition, if you join a society or subscribe to gardening magazines your name will sooner or later find its way onto mail-order lists. A certain amount of "junk mail" is inevitable but much of it will entice you to participate more fully in the wonderful world of plants!

INDEX

Use this index in conjunction with the Table of Contents, List of Figures, and Glossary, since not all words listed in those places are also in the index.

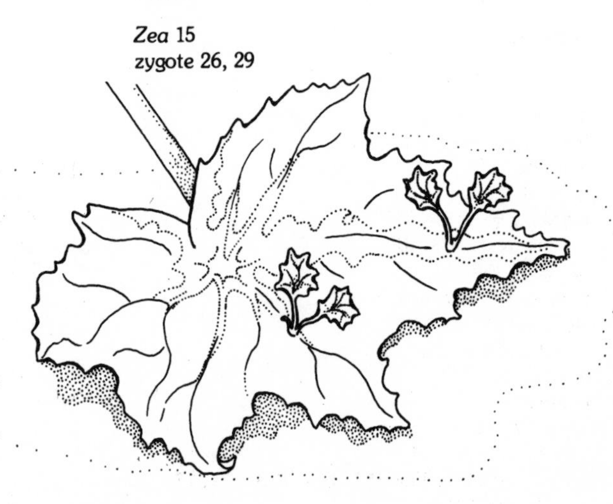

Donna N. Schumann is a professor of biology at Western Michigan University in Kalamazoo. She has developed **Living with Plants** over the last seven years for use in her popular non-major course in gardening. Her professional training includes an M.S. in paleobotany from the University of Michigan.

An enthusiastic and well-recognized gardener, Prof. Schumann belongs to numerous professional societies, including the American Horticultural Society, Botanical Society of America, Michigan Nature Society, and National Science Teachers' Association. She is a certified member of the Landscape Critics Council of the Federated Garden Clubs of America and a Member of the Board of the Michigan Botanical Club. She recently served for two years as President of the Kalamazoo Garden Club, to which she has belonged for fifteen years. She has published articles in **Plant Life** on growing and breeding amaryllids and numerous articles on the teaching of botany, and she designed exhibits and taught classes for the Cincinnati Museum of Natural History.

She is married and has three children and a grandchild. She and her husband enjoy sailing and travel.